T0310525

Construction Innovation
and Process Improvement

Construction Innovation and Process Improvement

Edited by

Akintola Akintoye
Professor and Head of School, Construction
Economics & Management
University of Central Lancashire

Jack S. Goulding
Professor, Construction Project Management and
Director, Centre for Sustainable Development
University of Central Lancashire

and

Girma Zawdie
Senior Lecturer, Science, Technology and Sustainability
University of Strathclyde

A John Wiley & Sons, Ltd., Publication

This edition first published 2012
© 2012 by Blackwell Publishing Ltd.

Blackwell Publishing was acquired by John Wiley & Sons in February 2007. Blackwell's publishing programme has been merged with Wiley's global Scientific, Technical, and Medical business to form Wiley-Blackwell.

Registered Office
John Wiley & Sons Ltd, The Atrium, Southern Gate, Chichester, West Sussex, PO19 8SQ, United Kingdom

Editorial Offices
9600 Garsington Road, Oxford, OX4 2DQ, United Kingdom
2121 State Avenue, Ames, Iowa 50014-8300, USA

For details of our global editorial offices, for customer services and for information about how to apply for permission to reuse the copyright material in this book please see our website at www.wiley.com/wiley-blackwell.

The right of the author to be identified as the author of this work has been asserted in accordance with the UK Copyright, Designs and Patents Act 1988.

Library of Congress Cataloging-in-Publication Data

Akintoye, Akintola.
Construction innovation and process improvement / Akintola Akintoye, Jack S. Goulding, Girma Zawdie.
    p.   cm.
  Includes bibliographical references and index.
  ISBN 978-1-4051-5648-6 (hardback)
1. Construction industry–Technological innovations.   2. Civil engineering–Technological innovations.   3. Competition.   I. Goulding, Jack S.   II. Zawdie, Girma, 1946–   III. Title.
  HD9715.A2A435 2012
  624.068–dc23
                                                                 2011045230

A catalogue record for this book is available from the British Library.

Wiley also publishes its books in a variety of electronic formats. Some content that appears in print may not be available in electronic books.

Set in 10/12pt Sabon by SPi Publisher Services, Pondicherry, India
Printed and bound in Malaysia by Vivar Printing Sdn Bhd

1 2012

Contents

Contributors

Akintola Akintoye
School of Built and Natural Environment
University of Central Lancashire
UK

Mustafa Alshawi
University of Salford
UK

Chimay J. Anumba
The Pennsylvania State University
USA

Oluwaseyi Awodele
Heriot Watt University
Edinburgh
UK

Graeme Bowles
Heriot Watt University
Edinburgh
UK

Charles Egbu
University of Salford
UK

Richard Fellows
Loughborough University
Loughborough
UK

Jack S. Goulding
University of Central Lancashire
UK

Umit Isikdag
Beykent University
Turkey

John M. Kamara
Newcastle University
Newcastle
UK

Murat Kuruoglu
Istanbul Technical University
Istanbul
Turkey

Anita Liu
University of Hong Kong
Hong Kong

Eric Lou
University of Salford
UK

Shu-Ling Lu
University of Reading
Reading
UK

Jamie Main
Capita Symonds Limited
UK

Wafaa Nadim
British University in Egypt
Egypt

Stephen Ogunlana
Heriot Watt University
Edinburgh
UK

Farzad Pour Rahimian
University of Central Lancashire
UK

Martin G. Sexton
University of Reading
Reading
UK

Mark D. Sharp
The Association of International Property
Professionals
London
UK

Joseph H.M. Tah
Oxford Brookes University
Oxford
UK

Jason Underwood
University of Salford
UK

Emilia Van Egmond
Eindhoven University of Technology
The Netherlands

Derek H.T. Walker
RMIT
Melbourne
Australia

Girma Zawdie
Strathclyde University
Glasgow
UK

Part I

Theory and Practice

Construction Innovation and Process Improvement

Akintola Akintoye, Jack S. Goulding and Girma Zawdie

1.1 Introduction

In order to promote and retain competitiveness, industry needs to focus on innovation and the improvement of their processes. Panuwatwanich *et al.* (2008) noted that innovation is necessary as a source of competitive advantage for firms operating in the construction industry, and consequently, that many firms are expending a significant amount of resources in an effort to acquire various forms of innovation in order to maintain and/or increase their competitiveness. Similarly, Aouad *et al.* (2010) highlighted that the competitiveness of firms inevitably depends on national and regional systems of innovation, which in turn depends on government policy. Therefore, given the constant changes and dynamism of the business environment, securing competitiveness is therefore high on the agenda of most firms. In this respect, securing innovation and process improvement is an influential lever for delivering this. On this theme, the increased complexity and sophistication of the Architectural Engineering and Construction (AEC) sector is now placing unparalleled demands on stakeholders to keep projects on time and within budget, with a new emergent theme of developing and maintaining robust and defendable innovation policies and procedures. Although it is not easy to sustain radical improvement in an industry that has historically been categorised as diverse and fragmented (Banwell, 1964; Latham, 1994; Fairclough, 2002), it has been recognised that there is a need for continuous and sustained improvement, using focused efforts to deliver the value needed by customers, along with addressing the industry challenge concerning waste and poor quality arising from existing structures and working practices (Egan, 1998). In this respect, Professor Watson (CSaP, 2011) reviewed the key concepts of innovation, noting that innovation was more than

Construction Innovation and Process Improvement, First Edition.
Edited by Akintola Akintoye, Jack S. Goulding and Girma Zawdie.
© 2012 Blackwell Publishing Ltd. Published 2012 by Blackwell Publishing Ltd.

invention or creativity, as it enveloped commercialisation, implementation and entrepreneurship as part of the innovation process – which required a change in culture to proactively promote and support innovation.

This book raises a number of wide ranging issues relating to construction innovation and process improvement, especially in the light of experience derived from construction practice in different countries with different contexts. The chapters therefore provide a rich collection of literature embracing theoretical and practice-empirical appeal, which gives credence to the pervasive and transformative effect that innovation can bring. Moreover, even in mature markets such as the AEC sector, where business behaviour is generally considered as risk averse, it highlights the increased importance and significance of embedding innovation initiatives into mainstream business practices. In this respect, construction practice is still evolving, with complex aspects underpinned by organisational and management responsibilities that seek to draw alignment across a wide range of players, not least contractors, subcontractors, suppliers and clients. This network of players in the industry is important, as they have a significant bearing on the manner in which innovative activities occur in the industry. It is also important to note that the nature of construction innovation is closely examined in terms of its impact on technological progress of the industry to date, and emergent technological trajectories.

The chapters in this book are divided into three broad themes of construction innovation relating to: Theory and Practice; Process Drivers; and Future Technologies. These three categories tease out the main salient issues on construction innovation and process improvement, and highlight the implications for future competitiveness and sustainability of the industry. These themes pose several questions for reflection, including 'What is particularly unique about construction innovation in theory and practice?', 'What are the major drivers of construction innovation?' and 'What factors are needed to support and deliver future construction technologies?'

In attempting to respond to these questions, this book sheds new light on these challenges, and provides readers with a number of ways forward, especially cognisant of the increased role of globalisation, the enhanced impact of knowledge and importance of innovation, as all these can have a significant impact on strategic decision making, competitive advantage, and sustainable policies and practices.

1.2   Innovation in Construction

Several definitions have been proffered for innovation. For example, Van de Ven (1986) regards innovation as any ideas, practices and technologies perceived to be new by the organisation involved. Slaughter (1998) defined innovation as the actual use of a nontrivial change and improvement in a process, product or system that is novel to the institution developing the change; whereas, Stewart and Fenn (2006) described innovation as the profitable exploitation of ideas, which have an important role to play in seeking competitive advantage. Innovation in construction can therefore be considered as the successful development and/or implementation of new ideas, products, process or practices, in order to increase organisational efficiency and performance (Egbu *et al.*, 1998; DTI, 2003; Ling, 2003; Sexton and Barrett, 2005; Panuwatwanich *et al.*, 2008). On the other hand, some proponents advocate that because the construction industry is largely project-based and fragmented, the patterns of innovation differ in many ways from those of other industries, and therefore, industry innovation remains hidden when co-developed at the project level (Aouad *et al.*, 2010). However, Stewart and Fenn (2006) noted that innovation in the construction industry has been recognised in three domains: product, process and organisation. Process innovation is oriented towards production methods, and organisational innovation to approaches to managing the firm and implementation of new corporate strategic orientations. They contest that in construction, innovation is mostly seen in terms of physical process and product, particularly improvements in materials. However, Blayse and Manley (2004) identified six main factors that could influence innovation in construction as:

1. clients and manufacturers;
2. the structure of production;
3. relationships between individuals and firms within the industry;
4. relations between the industry and external parties;
5. procurement systems regulations/standards; and
6. the nature and quality of organisational resources.

They noted that these influences are the key factors that drive (or in fact hinder) business innovation.

From a process perspective, the term 'processes' are 'the fundamental building blocks' of all organisations, and both process understanding and process improvement form the lifeblood of total quality organisations. Processes transform inputs, which can include actions, methods and operations, into outputs. They are the steps by which we add value, and it should be the aim of customer focused, total quality organisations, for these outputs to satisfy or exceed the needs and expectations of their customers' (DTI, 2011). On this theme, Sarshar *et al.* (2004) developed a Structure Process Improvement for Construction Enterprises maturity framework to assess organisational performance. More fundamentally, the industry acknowledged that there was a need for new process configurations and innovation through life-cycle decision analysis (which had been championed in many countries and supported by industry stakeholder groups and government bodies). Moreover, both the Latham (1994) and Egan (1998) reports identified that the construction industry needed to embrace innovation and process improvement. Similarly, in Hong Kong, the government established a Construction Industry Review Committee (CIRC) that published its report, 'Construct for Excellence', in 2001. This report, amongst other recommendations, charged the industry to collectively develop a culture of innovation which deliberately concentrated and fostered innovation, both from a technology and fostered perspective.

In Australia, Sidwell *et al.* (2004) reported on the importance of reengineering the construction delivery process, noting that the fragmented and differentiated structure of the construction industry was a major characteristic that militates against improvement. The core challenge for the construction industry was therefore to develop radical project delivery processes that concentrated on front-end issues of procurement strategies, interfaces in the process, information flows, and the elimination of non-value-adding activities.

The need for construction innovation and process improvement was further emphasised by the Egan (1998) report through the Construction Task Force, which identified as one of its terms of reference to 'examine current practice and the scope for improving the industry by innovation in products and processes'; some notable areas of which included:

- lack of research and development (R&D) investment (damaging the industry's ability to keep abreast of innovation in processes and technology);
- client dissatisfaction with consultants' performance in coordinating teams, in design and innovation (to provide a speedy and reliable service and deliver value for money);
- wasted talent (failure to recognise the significant contribution that suppliers can make to innovation);
- repeated selection of new teams (inhibits learning, innovation and the development of skilled and experienced teams, preventing the industry from developing products and an identity/brand that can be understood by its clients);
- product development requires continuity from a dedicated product team (needing product design skills, with close links to the supply chain through which the skills of suppliers and their innovations can be assessed, and with access to relevant market research);
- supply chain is critical for driving innovation (and sustaining incremental and sustained performance improvement);
- project implementation requires 'organisation and management of the supply chain to maximise innovation, learning and efficiency';
- project implementation requires 'capturing suppliers' innovations in components and systems';
- component production also includes the sustained commitment to innovation in the design of components (including the development of a range of standardised components);
- continuous learning ('upgrading, retraining and continuous learning are not part of construction's current vocabulary. There is already frustration amongst component suppliers that their innovations are blocked because construction workers cannot cope with the new technologies that they are making available. This has to change');
- improvements in innovation (more can be achieved by co-operation between clients, constructors and suppliers than through competition);
- need to encourage long-term partnering arrangements between clients and providers (to secure consistency, continuity, innovation and value for money);

- need to develop knowledge centres (through which the whole industry and its clients can gain access to knowledge about good practices, innovations and the performance of companies and projects;
- need for training (new technical and managerial skills required in order to get full value from new techniques and technologies);
- learning from other industries ('...in both manufacturing and service industries there have been increases in efficiency and transformations of companies, which a decade or more ago nobody would have believed possible);
- change in culture (changing this is fundamental to increasing efficiency and quality in construction);
- improving project processes ('...construction has two choices: ignore all this in the belief that construction is so unique that there are no lessons to be learned; or seek improvement through reengineering construction, learning as much as possible from those who have done it elsewhere');
- product development ('innovating with suppliers to improve the product without loss of reliability');
- Enabling improvement ('Substantial changes in the culture and structure of UK construction are required to enable the improvements in the project process that will deliver our ambition of a modem construction industry. These include changes in working conditions, skills and training, approaches to design, use of technology and relationships between companies);
- technology as a tool ('One area in which we know new technology to be a useful tool is in the design of buildings and their components, and in the exchange of design information throughout the construction team. There are enormous benefits to be gained, in terms of eliminating waste and rework for example, from using modern CAD technology to prototype buildings and by rapidly exchanging information on design changes. Redesign should take place on computer, not on the construction site).

The above points, whilst not exhaustive, offer a number of critical areas for reflection, all of which are covered in this book through the three core themes of: Theory and Practice, Process Drivers, and Future Technologies.

1.3　Construction Innovation: Theory and Practice

Panuwatwanich *et al.* (2008) noted that innovation diffusion in design firms could be enhanced by creating a culture for innovation using innovative leaders. Moreover, Barlow and Jashpara (1998) highlighted the importance of collaborative links between firms for stimulating organisational learning. Similarly, Gann and Salter (2010) opined how clients can act as a catalyst in the construction value chain to help foster innovation by exerting pressure on supply chain partners to improve overall performance, and also by helping them to devise strategies to cope with unforeseen changes. This is particularly important, as Vennstrom and Eriksson (2010) identified that client perceived barriers to change could be divided into three types: attitudinal, industrial and institutional; noting that 'Clients wishing to act as change agents need to be aware that their use of internal versus external project management affects their chances to influence the other construction actors and implement change and innovation'. Furthermore, Aouad *et al.* (2010) noted 'Our understanding of innovation and how it occurs in the sector is far from complete but can be enriched further by detailed work that brings together different theoretical perspectives on innovation that will enable the development of context sensitive ways of recognising and measuring innovation at different levels of resolution.'

Part I of this book therefore explores the above issues, dealing with matters relating to change management, technology, sustainable construction (SuCo), and supply chain management (SCM).

Chapter 2 highlights the occurrence of innovation through organisational learning; knowledge accumulation and knowledge sharing; conflict management and coalition building, with the aim to minimise resistance to change. The overarching factor impinging upon the process of construction innovation is the cultural context within which construction activities are pursued, so that the occurrence of innovation would be expected to be highly likely where there is 'cultural readiness' arising from exposure to new ideas and practices. Cultural readiness exists where proactive management and flexible and change-responsive organisational arrangement are evident, and where management supports coalition building through the process of providing the conditions for relationship balancing and conflict resolution. Emphasis is therefore placed on removing fragmentation, embracing change, and leveraging innovation drivers to meet and deliver business goals and new market opportunities.

Chapter 3 explores the theory and practice of construction innovation in terms of the synthesis of resource-based and market-based perspectives of innovation. The market-based view of innovation is a variation of 'demand pull' innovation, which utilises the role of institutional and market factors to stimulate innovation at the firm level. Market conditions influencing innovation possibilities include both the general business environment, and the interaction or industry-specific environment. Market conditions shape the resources that firms exploit to respond to opportunities and threats. The resource-based view of innovation is based on the understanding of firms identifying and developing resources that enable them to shape market conditions. The knowledge base of the firm, which is bolstered by the firm's relationship capital, structure capital and human capital, is crucial for interfacing organisational resources with external agencies in market relations to produce the dynamic capabilities that provide the basis for innovation and sustainable competitive advantage. It concludes by highlighting the importance of exploiting existing capabilities to produce innovation in order to enhance performance; but equally, the need to invest in explorative innovation for the future.

Chapter 4 highlights the importance and impact of culture on innovation. Innovation has a cultural context that determines the nature of the prevailing demand and supply conditions, as well as the organisation and management bases of innovation. This is especially important, as it impinges and influences the ways in which knowledge infrastructures evolve. Cultural change is therefore crucial for technological progress, as any model of innovation would be incomplete without the inclusion of cultural parameters. The role of culture in construction is discussed, noting the importance and development of co-operation structures (or social systems), and the need to include learning as one of the dimensions of innovation culture. Learning culture is therefore explored, noting the importance of the prevailing cultural climate in order to determine what changes may be needed to facilitate learning and the transfer of creative problem solving. The importance of learning (and learning from other sectors) is acknowledged as being crucial for developing an innovation culture, the mandate of which requires leaders with vision and foresight to 'plan ahead' in order to fully benefit from this.

Chapter 5 examines how innovation and international technology and knowledge transfer (TKT) relates to SuCo performance. The systemic interactions between these are investigated, particularly concerning new design concepts and their integration with building elements and processes. The efficiency and effectiveness in TKT practices are evaluated within the construction firm and across different projects, at regional, national and international levels. In this respect, knowledge accumulation and learning can be seen as a pivotal driver for innovation and TKT processes, as successful production and trade performance strongly depend on capability and motivation to innovate and quickly adopt new technologies. This can help form the building blocks for either subsequent incremental innovation or new knowledge sets for future innovation in other related areas. The construction innovation system is explored, with emphasis on collaboration and technical regime. A case study from The Netherlands is used to discuss the management of innovation and SuCo using a network-based approach. Findings highlight the importance of integration, where innovative solutions in construction go beyond the traditional ways in order to match environmental gains with economic gains.

Chapter 6 raises the importance of securing construction innovation from the vantage point of 'organisational sustainability' and the organisation's role of delivering value to its core stakeholders. It therefore provides readers with an insight into the concepts of SCM, and how this can be applied innovatively to derive value-added processes and organisational sustainability. Organisational value is discussed, highlighting the importance this can have on securing competitive advantage, especially using knowledge chains, value chains and human capital chains. Thus, an important factor here is understanding the pivotal links and dependencies that SCM can have on leveraging construction innovation from several settings, not least efficiency, value delivery, social support aspects and through sustainability issues. A case study is used to present the transformational impact of SCM, and how this led to sustainable competitive advantage. Research findings highlight the importance of supply chains, and how the industry as a whole could learn from other industries, in order to garner additional value and sustainability benefits.

1.4   Construction Innovation: Process Drivers

The drivers of construction innovation leading to process improvement were categorised by Bossink (2004) as being environmental pressure, technological capability, knowledge exchange, and boundary spanning. Whilst these innovation drivers in these categories can be considered to be active at trans-firm, intra-firm and inter-firm levels, it is equally important to acknowledge the importance that external pressures place on firms to achieve innovation. Technological capability can enable organisations to experiment with and use innovative applications and methods in their construction projects; whereas, knowledge exchange represents the development and sharing of knowledge and expertise in and between organisations, and boundary spanning represents the capability of institutions and organisations to co-innovate with other institutions and organisations. These are all essential ingredients of innovation. Given this, Widén (2010) identified two approaches for driving innovation: client-driven innovation through required innovative solutions, processes, etc, and innovation secured through procurement forms applied. On a similar theme, Stewart and Fenn (2006) noted that construction innovation tended to focus on product innovations, which did not take into account strategic innovation, nor the kind of innovation required for value-adding innovations that could procure competitive advantage. They concluded that a strategic perspective on innovation and strategic thinking was needed to motivate the organisation to look beyond the product and process to the entire system for delivering value to the customer.

Part II of this book, therefore addresses innovation and process improvement driver topics, including strategic management, concurrent engineering, risk management, innovative procurement and knowledge management.

Chapter 7 explores the importance of strategic management on process improvement, especially how strategy can be purposefully aligned to deliver innovation opportunities. It explores the fluidity of the construction sector, and the need for organisations to continually re-visit their business models, especially cognisant of market forces, and the need to align corporate competence and capability to strategic trajectories in order to secure competitive advantage. In this respect, the concepts and application of strategy is reviewed from a business performance perspective, highlighting the importance of identifying appropriate performance metrics for subsequent review. A case study is presented for discussion, which identified the strategic challenges faced by one construction company, and the solution put in place to manage competing drivers and external forces. Research findings highlight the importance of developing strategies that have a clear focus and readily identifiable objectives and deliverables, and the need to understand how strategy (strategic positioning), skills (intellectual capital), ICT (alignment to strategy) and process (understanding) all interrelate.

Chapter 8 introduces the importance of integrating risk management at the planning stage for process improvement. A major goal of most modern organisations is to meet or exceed the demands of clients, and many attempts have been made to achieve this with various degrees of success. In this respect, it is advocated that the absence of standardised process improvement methodologies are partly attributable. This chapter critically reviews the extant literature on process improvement and risk management, and identifies failures associated with project and process improvement initiatives. It advocates the need to have clearly articulated mechanisms in place to help organisational performance, as it is posited that risk predominantly emanates from uncertainties associated with pursuing certain causes of action. The need for risk management integration when planning for process improvement is therefore seen as an important lever, as organisations need to be able to understand and differentiate their core/value-creation business processes and supportive processes. Research findings also note the importance of understanding the decisions made; particularly why improvement interventions are required, as this can help create acceptance, and also help secure other desired results such as competitiveness, improved customer satisfaction levels, enhanced profitability, etc.

Chapter 9 explores the need for the construction industry to change. It introduces Modern Method of Construction and Offsite Production approaches as exemplars for discussion that, from an innovation point of view, is seen as means for both improving and changing the construction industry's thinking and practices. The manufacturing concept and offsite production approaches are explored from a strategic as well as from an implementation perspective, in order to identify the requirements needed to help overcome the industry's inherent problems. This is also advocated as a means of promoting the construction industry and overcoming skill shortages. The need for change is presented, supported by a number of initiatives taken from other industries. A conceptual model is presented for discussion based on four core dimensions: people, product, technology and process. This is promoted as a way of evaluating 'process', 'people', 'technology', 'product', 'market' and 'risk'. Research findings identify the need to secure a common 'language' across the different stakeholder groups, along with a need to share a collective view and understanding of offsite production practices.

Chapter 10 introduces the role of knowledge management in organisations, especially how this can be used to deliver innovation and secure competitive advantage. It identifies how knowledge management impacts upon innovations in project-based environments, and the challenges faced in managing this using effective knowledge management practices. These challenges are discussed, and solutions offered, particularly the need to recognise the constraints imposed on knowledge management processes by the project environments; and to determine new means of creating, transferring, sharing, implementing and exploiting individual and project knowledge. Effective knowledge management practices are advocated to include a number of core areas, including networking, Communities of Practice, storytelling, coaching, mentoring and quality circles (to share and transfer tacit knowledge through organisational/project environments). Many factors currently confront construction organisations in the management of knowledge assets, including organisational culture and maturity, strategic decision making, existing and future capabilities, financial and technological capabilities, and the effect of internal and environment stimuli. These are all seen as being inextricably linked.

Chapter 11 presents the role of procurement to innovation, particularly in relation to the criteria needed and choice of innovative procurement methods available. Traditional approaches to procurement usually involve a chain of separate firms who add value to items purchased from other organisations through 'arms-length' one-off contracts. The use of collaborative procurement methods to deliver innovation is often employed because of the shortcomings of conventional procurement methods, and collaborative relationships have brought advantages to several companies in many industries. Consequently, there is an increasing use of collaborative working relationships, including partnering, joint ventures, strategic alliances and public private partnerships. These arrangements are evaluated and discussed, using exemplars from literature to highlight success areas. The use of integrated teams is a common feature of collaborative arrangements; and by involving the team at the earliest stage in a project, improvements are advocated in quality, productivity, health and safety, cash flow, reduced project durations, and more clearly identifiable risks. It concludes by noting that different collaborative relationships can deliver real (tangible) innovation and process improvements, but that such collaboration methods need to be carefully considered to ensure that they fit into the business plans of all contributory organisations.

Chapter 12 presents the key issues and technologies needed for the adoption of concurrent engineering in construction. It explores the concepts of concurrent engineering, and explains how this can be used to secure construction innovation and process improvement in the industry. Implementation issues are identified, emphasising the need to adopt new ways of working, as this requires a change in culture and practice with respect to integration (tools, processes, teams, etc). This also requires up-front consideration of life-cycle issues in the project development process, in order to bring about change using new methods and techniques. The critical enablers of concurrent engineering adoption within the industry are seen as being predominantly 'organisational', and 'technical', albeit acknowledged as also being influenced by many other issues, including procurement. Research findings present the importance of adopting a multi-facetted approach to concurrent engineering on order to leverage key benefits, but also mitigate potential barriers to adoption.

Chapter 13 explores complexity theory, and its relationship with the various aspects of the built environment, particularly how this can be used to leverage innovation. It overviews the development of complexity and identifies where entangled complexity interacts with construction processes. Complexity on organisations is then examined, along with toolkits and models for explication. This presents the transition of complexity to innovation and performance improvement, using worked exemplars as paradigms for exploitation. A case study covering five building phases is presented for discussion. This identifies how the different types of complexity affect building processes using an aggregate scoring matrix. It concludes by noting that whilst complexity in the built environment is still maturing, there is a greater need to understand the systems, procedures, processes, and interactions of variables, in order to learn from the past and generate solutions for the future.

1.5   Construction Innovation: Future Technologies

Han (2005) noted that the future of the industry is poised to take full advantage of the ever-increasing information and automation technologies, to advance the level of quality, efficiency, technical performance and safety. Moreover, that design, construction, as well as manufacturing and onsite application techniques, are becoming more intelligent, integrated and automated. On this theme, Boddy and Abbott (2010) identified a series of drivers, challenges and solutions including:

- ICT for construction;
- industrial research leadership and sponsorship;
- offsite manufacturing;
- material developments;
- flexibility and configurability of facilities;
- professional clients;
- virtual prototyping; and
- inspiration from nature.

Similarly, Aouad *et al.* (2010) noted that the capacity of the construction industry to innovate in response to the number of external drivers and consequent challenges could be measured by the effectiveness with which appropriate solutions were developed. Some of the solutions offered included the use of ICT; asserting that innovation platforms are needed to integrate a range of technologies to better coordinate policy and procurement, in line with a construction vision that would respond to the emerging needs of the world and construction industry in the future. Given the importance of innovation in construction, it is important to acknowledge that this can be supported by a number of collaborative networks, leveraged through methods such as open innovation, micro-innovation, open source solutions, road mapping and value mapping (CSaP, 2011).

Part III of this book explores future technologies in construction – especially, how these can be harnessed and leveraged to help procure innovation and process improvement.

Chapter 14 presents the use of tangible, immersive and interactive virtual reality interfaces as a viable solution for promoting the integration of data simulation and communication through the whole design and construction processes (to improve designers' cognition and collaboration). These interfaces are acknowledged as being able to provide intuitive interaction that supports 'free' artistic expression, and bridge the gap between artistic experimentation and accurate manufacturing-oriented modelling. This is introduced as a means of proactively fostering multidisciplinary teamwork in order to enhance outcomes in collaborative arrangements. Moreover, it is asserted that this could also contribute towards transforming the conceptual architectural design phase, thereby enabling project teams to generate and test new ideas and introduce innovation into the design and construction process. A number of design visualisation and simulation tools are evaluated, along with cognitive approaches to design. A case study is introduced that investigates designers' spatial cognition, collaboration and creativity. Core findings highlight the importance of enabling stakeholders to proactively engage in these developments as part of the project development lifecycle, especially in collaborative environments that are geographically dispersed.

Chapter 15 posits that the next generation of virtual prototyping systems will need to draw heavily upon multiple configurable knowledge sources in order to cope with the complexity that traditionally characterises construction projects (as existing construction planning methods do not adequately represent or communicate the spatial and temporal aspects of construction schedules effectively). It explores the complex nature of construction projects, and evaluates various construction planning and virtual prototyping tools available, including the impact of building information models and other emergent knowledge models. A prototype system is presented for discussion, highlighting how this can be used to support innovation in modern construction management processes. Research findings advocate the need for managers to take a wider and more strategic role, especially where their capacity to visualise and understand the implications of alternative decision choices is made clearer through the convergence of emerging digital tools. This will also help support envisioning in order to improve innovation processes. To support this, a framework is advocated that can integrate multiple applications through a project model database with knowledge-based support to handle cross-application business processes.

Chapter 16 presents the importance of e-readiness to construction firms. It highlights the current challenges facing the industry, and explores the relationship between business drivers and technological solutions. Given this, it is posited that there is an exigent need to build capability and capacity in information and communication technology (ICT) with organisational structures. This requires a thorough understanding of the link between 'business process and ICT', 'people and ICT', 'business process and implementation' and 'e-readiness'. These issues are all explored, highlighting the need to define the term 'organisational readiness' in construction. Three core themes (People, Process and Technology) are presented as a viable way forward, supported by five core enablers (Leadership & Empowerment, Change Management, Business & Information Process, Policy/Strategy/Vision, and ICT Sharability/Inter-operability). Research findings highlight the importance of securing sustained e-business initiatives. However, this is posited as needing effective measures and systems in place to design, develop, deliver and evaluate these.

Chapter 17 introduces Building Information Modelling (BIM) as a formal conduit for storing and managing building information through all stages of a project's lifecycle (from conception to demolition/disassembly). It is advocated that the implementation of the BIM paradigm can enable information exchange of several applications through agreed models (i.e. schema standards), in order to optimise these phases. The development and use of BIM is evaluated, particularly from an inter-operability, integration and data sharing perspective. Two case studies using BIM are presented for discussion. These present a number of significant findings, not least the need to establish data veracity and protocols through a formal BIM strategy. It is also advocated to create more efficient team working, enhance collaboration, secure better coordination (especially in complicated projects), and add value by automating most of the information management processes. It is noted that innovation can be secured through BIM, either through direct implementation, or through newly emerging technologies (i.e. cloud computing, sensor network, web services, etc).

Chapter 18 discusses the need for the industry to evaluate the way in which it manages projects (as these have changed significantly over the past few years), especially with advances in technology, and the increased prevalence of web-based project collaboration technologies and project extranets. This has created a fundamental need to re-visit the way in which skills and competence are designed and delivered to meet business and learners' needs. Given this, learning and training developments are explored, with particular emphasis placed on the way through which learning pedagogy has evolved; especially the current trends in learning styles and models, including the game theory approach. Developments in virtual reality systems are then evaluated, and a virtual construction site simulator is presented as a case study for discussion. This presents a development framework, along with supportive rubrics for aligning learning outcomes to actual work-based scenarios. Research findings note the importance of being able to train professionals in 'safe' and 'controlled' environments. Moreover, it is posited that future developments in simulated virtual reality environments may be able to diagnose learner styles, in order to deliver bespoke training material to individual learner types.

1.6   Conclusion

Innovation in construction is essential. The industry has continued to strive to remain competitive using a variety of approaches. It is therefore asserted that a paradigm shift in thinking is now required, which proactively engages structured initiatives that purposefully align to proven innovation concepts, techniques and applications. The higher the levels of innovation garnered, the greater the likelihood that this will in turn generate economic growth (Balyse and Manley, 2004; Gambatese and Hallowell, 2011). Continuous process improvement is also likely to play a major role in this (to achieve significant productivity improvement), along with the role of people, process and technology. These need to be linked and integrated in order to ensure that innovation enablers are cogently aligned to measurable process improvement initiatives. Finally, it is important to acknowledge that the incumbent industry stakeholders and governmental bodies are likely to influence and drive construction innovation, the remit of which will require new thinking to traditional approaches and the adoption of new workable relationships.

This book therefore emphasises the role of innovation and process improvement in construction through 18 chapters. These chapters explore the theory and practice of innovation and process improvement initiatives needed, and highlight the future technologies and formal mechanisms needed to support and deliver them.

References

Aouad, G., Ozorhon, B. & Abbott, C. (2010) Facilitating innovation in construction: Directions and implications for research and policy. *Construction Innovation: Information, Process, Management*, **10**(4), 374–394.

Banwell, H. (1964) *Report of the Committee on the Placing and Management of Contracts for Building and Civil Engineering Works*. HMSO, London.

Barlow, J. & Jashpara, A. (1998) Organisational Learning and Inter-Firm Partnering in the UK Construction Industry. *The Learning Organisation*, **5**(2), 86–98.

Blayse, A.M. & Manley, K. (2004) Key influences on construction innovation. *Construction Innovation: Information, Process, Management*, **4**(3),143–154.

Boddy, S. & Abbott, C. (2010) *Review of CIB Collected Outlook Reports, CIB Publication 334*. CIB, Rotterdam. Available from http://cibworld.xs4all.nl/dl/publications/pub334.pdf (assessed August 2011).

Bossink, B.A.G. (2004) Managing drivers of innovation in construction networks. *Journal of Construction Engineering and Management*, **130**(3), 337–345.

Construction Excellence (2009) *Never Waste a Good Crisis: A Review of Progress since 'Rethinking Construction' and Thoughts for Our Future*. Construction Excellence in Built Environment, London.

CSaP (2011) *Drivers, Challenges and Approaches to Innovation in the Construction Sector*, Professor Jeremy Watson, CONNECTIONS Lecture Series, 11 March, Centre for Science and Policy, University of Cambridge, Cambridge.

DTI (2011) *Process Understanding and Improvement*. Available from http://www.businessballs.com/dtiresources/TQM_process_management.pdf (accessed July 2011).

DTI (2003) *Innovation Report – Competing in the Global Economy: The Innovation Challenge*. HMSO, London.

Egan, J. (1998) *Rethinking Construction*. UK Department of the Environment, Transport and the Regions, HMSO, London.

Egbu, C.O., Henry, J., Kaye, G.R, Quintas, P., Schumacher, T.R. & Young, B.A. (1998) Managing organisational innovations in construction. In: *Proceedings of the 14th Annual Conference of the Association of Researchers in Construction Management*, ed. Hughes W. ARCOM, Reading, 605–614.

Fairclough, J. (2002) Rethinking construction innovation and research: a review of government R&D policies and practices. Department of Trade and Industry. Available from http://www.bis.gov.uk/files/file14364.pdf (accessed August 2011).

Gambatese, J.A. & Hallowell, M. (2011) Enabling and measuring innovation in the construction industry. *Construction Management and Economics*, **29**(6), 553–567.

Gann, D.M. & Salter, A. (2010) Innovation in project-based, service-enhanced firms: the construction of complex products and systems. *Research Policy*, **29**(7/8), 955–972.

Han, H. (2005) *Automated Construction Technologies: Analyses and Future Development Strategies*. Unpublished PhD Thesis, Massachusetts Institute of Technology, Department of Architecture, USA.

Latham, M. (1994) *Constructing the Team, Joint Review of Procurement and Contractual Arrangements in the United Kingdom, Construction Industry.* HMSO, London.

Ling, F.Y. (2003) Managing the implementation of construction innovations. *Construction Management and Economics.* **21**(6), 635–649.

Panuwatwanich, K., Stewart, R.A. & Mohamed, S. (2008) Validation of an empirical model for innovation diffusion in Australian design firms. *Construction Innovation: Information, Process, Management,* 9(4), 449–467.

Sarshar, M., Haigh, R. & Amaratunga, D. (2004) Improving project process: Best practice case study. *Construction Innovation,* 4, 69–92.

Sexton, M. & Barrett, P. (2003) Performance-based building and innovation: balancing client and industry Needs. *Building Research and Information,* 33(2), 142–148.

Sidwell, A.C., Kennedy, R.J. & Chan, A.P.C. (2004) *Reengineering construction delivery process.* Report, January, Construction Industry Institute, Queensland University of Technology, Brisbane.

Slaugher, E.S. (1998) Models of construction innovation. *Journal of Construction Engineering and Management,* **124**(3), 226–232.

Stewart, I. & Fenn, P. (2006) Strategy: The motivation for innovation. *Construction Innovation,* 6, 173–185

Van de Ven, A.H. (1986) Central problems in the management of innovation. *Management Science.* **32**(5), 590–607.

Vennstrom, A. & Eriksson, P.E. (2010) Client perceived barriers to change of the construction process. *Construction Innovation,* 10(2), 126–137.

Widén, K. (2010) *Construction Innovation – Statsbygg, the Driving Client.* Report, Lund University, Division of Construction Management, Lund.

Construction Innovation through Change Management

Girma Zawdie

2.1 Introduction

Van de Ven (1986) refers to innovation as the transformation of 'new ideas into good currency'. Innovation is also viewed as a systemic phenomenon resulting from processes in which a whole range of players interact, influencing the direction of change, and also the speed with which this is effected (Freeman, 1987; Lundvall, 1992). At the heart of these interactions underlying the innovation process are evolving systems of organisation and management at both the micro and macro levels of decision making. At the micro level, the ways in which enterprises are organised and managed have significant bearing on the extent to which the enterprises engage in knowledge exchange and its use for them to be able to realise their innovative potential and establish themselves as pace setters in emerging industrial and market trends. At the macro level, innovation possibilities are contingent on the way the complex social functions of wealth creation, knowledge production and systems of regulation and governance are organised and managed to reinforce each other to generate a culture of innovation (Leydesdorff and Zawdie, 2010). Thus, for innovation to occur on a sustainable basis, change of organisation and management cultures is essential at all levels, so that social interactions and functions that underpin creativity and innovativeness are least constrained. In other words, when systems evolve as a managed and organised process, they create the condition for innovation (Jones *et al.*, 2004; Tourish and Hargie, 2004).

Innovation is sought not as an end in itself, but to provide a sustainable basis for the achievement of industrial competitiveness that is reflected in terms of improvements in the quality, cost, delivery time, as well as the social and environmental impact of production. However, innovation does not occur in a vacuum. There is *inter-alia*, a cultural and institutional context

Construction Innovation and Process Improvement, First Edition.
Edited by Akintola Akintoye, Jack S. Goulding and Girma Zawdie.
© 2012 Blackwell Publishing Ltd. Published 2012 by Blackwell Publishing Ltd.

to it, which defines the characteristics of the institutions and organisational systems within which industrial activities take place. The occurrence of innovation would therefore presume a shift in cultural and institutional modes, which introduce the flexibility required for the generation of new ideas. This would make innovation necessarily contingent upon changes in industrial culture. In this respect, industrial cultures vary across sectors with respect to the ways in which activities are organised and managed. This difference may largely explain why some categories of activity (i.e. firms in the manufacturing industry) are more innovation-active than others (i.e. firms in the construction industry) and why some are pioneers and others followers. However, openness to change through management initiatives allows organisational learning to take place, resulting in the accretion of organisational knowledge and the development of organisational core competence (Stonehouse and Pemberton, 1999). The process of learning itself evolves over time, first reactively by emulating others – a process referred to as 'single loop' learning (Argyris and Schon, 1978) – and later proactively by exercising creativity through more complex learning processes – i.e. the 'double loop' mode (Argyris et al., 1985; Argyris, 1996) and the 'triple loop' mode (Flood and Romm, 1996; Snell and Man-Kuen Chak, 1998). Where critical and creative learning of the double loop and triple loop type takes root, those who were once followers in some areas of activity can later become leaders in other areas of activity. Indeed, with the ever increasing scope for knowledge generation, knowledge transfer and exchange, and organisational learning in contemporary society and economy, industries like construction that were traditionally considered to be 'innovation-inactive' in relation to manufacturing (Marosszeky, 1999), can no longer be seen to be so, given the nature of the dynamics in learning and the innovation process.

The case for innovation is heightened because of the growing need to enhance competitiveness of activities across the industrial spectrum, not least the construction sector. In the UK, concern about innovation in the construction industry has grown, particularly over the last couple of decades, in the wake of initiatives, including the Latham Report (1994), the Egan Report (1998), the Strategic Forum for Construction (2002) and, more recently, Construction Excellence (2009). These initiatives have been at one in their drive to stimulate innovative activities in the industry through changes in organisation and management systems, and also through their consideration of the implications of these changes for the competitiveness of industry and for the design of institutional mechanisms and policy frameworks to underpin innovation and organisational learning. However, compared with firms in the manufacturing sector, construction firms, in general, have traditionally been slow in terms of innovation performance (Marosszeky, 1999). This raises questions about the structural and cultural characteristics of the construction industry – namely, the fragmentation of the industry along the supply chain, and the adversarial relations underlying activities within the industry. It also at the same time highlights the challenge for innovation and change management in construction.

In view of these questions, this chapter explores a number of issues, including cultural context, differences in firms' with respect to the propensity to innovate, supply chain dynamics, knowledge infrastructure evolution and technological trajectories (Dosi, 1982; Leydesdorff and Zawdie, 2010). The concept of innovation is explored as a system, showing how the perspectives on innovation have evolved, followed by a discussion of the cultural context of innovation and the challenge it poses for innovation management. The chapter then looks into the experience of the construction industry in the UK in the light of recent initiatives aimed at promoting a culture of innovation, and enhancing management of change at enterprise level, and discusses management strategies that could facilitate the innovation process in construction firms. The chapter concludes by highlighting issues that are critical for change management strategies in the construction business.

2.2   The Innovation Process: Evolution as a Systemic Phenomenon

Innovation in construction, as in other areas of industrial activity, is sought as a strategy for enhancing competitiveness for firms, 'value for money' for clients, and sustainability for the economy and the environment. A first step in an attempt to influence the speed and direction of innovation would be to understand the concept in its broader context. The significance of innovation is not only in the innovated product or service, but also in the innovation process itself. The latter involves a complex set of interactions between a wide range of factors that relate to the prevailing social, economic, political and technological cultures. This view, which broadens the area of focus from the output to the process aspect of innovation, aims to capture the full extent of the dynamics in the innovation process. Management of innovation is, strictly speaking, as good as the understanding of the concept itself, which has changed significantly over the years with changes in conceptual perspectives.

2.2.1   Linear Approach to Innovation

Innovation, both as a product and a process, was first formally conceptualised in the context of the socio-economic system by Joseph Schumpeter (1934), essentially as a 'supply push' phenomenon. For Schumpeter, the innovation process starts with the entrepreneur who is endowed with the capability to generate new ideas about 'new combinations' or new ways of doing things. Supply, it is implicitly assumed, will create its own demand, much in line with the Emersonian tenet that if one made a better mousetrap, the whole world would beat a path to one's doorstep. As such, Schumpeter's conception of the innovation process would evolve linearly, starting with invention (new ideas), and leading to the translation of new ideas into innovation through the process of design, engineering and pilot production. It was then assumed that innovated products and services would enter the market in fully blown form through the diffusion mechanism, thus leaving no scope for the occurrence of incremental innovations via market-driven feedback loops. Innovation in the Schumpeterian sense is thus conceived to be a radical departure from existing 'combinations' or ways of doing things. It is an 'epoch making' event driven by the entrepreneur, occurring discontinuously through 'fits and starts' – a process Schumpeter (1934) refers to as the 'gale of creative destruction'.

The linear approach based on Schumpeter's (1934) trilogy of invention, innovation and diffusion – which Gibbons *et al.* (1994) relate to their Mode 1 paradigm of innovation – is not, however, without conceptual limitations. First, while Schumpeter (1934) acknowledges the dynamism inherent in the innovation process, the conceptualisation of the process in a trilogy of events effectively isolates innovation from the wider socio-economic and cultural milieu that is crucial for the comprehensive realisation of innovation. The linear approach conceptually fragments the innovation process and strips it of the systemic framework within which all players in the innovation process are interactively linked, as in the Mode 2 paradigm of innovation (Gibbons *et al.* 1994). This is a significant omission in as much as innovation is a product of 'a series of simultaneous activities undertaken by a number of firms, often in close co-operation with the users of the innovation and with the collaboration of other institutions, such as universities and public research bodies' (Kozul-Wright, 1995). Second, by precluding 'incremental innovation' from the definition of innovation, Schumpeter's linear approach discounts the significance of the learning process and the accumulation of knowledge, and hence the dynamism at the heart of the innovation process. Most incremental innovations are a product of 'market pull' or 'demand pull', which start from the articulation of market need in the form of ideas and, once developed, are manufactured for diffusion. Incremental innovations, which may or may not be patentable, account for the bulk of innovation observed across firms, sectors and countries. Whereas, Schumpeter fell short of recognising the significance of this category of innovation, neo-Schumpeterians such as Schmookler (1966), Freeman (1987) and Lundvall (1985), among others, have given it pride of place in contemporary innovation literature. Third, the presumption that the market remains at the receiving end, and that all that matters in the innovation process is 'supply push', is empirically untenable in view of the burden of historical evidence that necessity has almost invariably been the mother of invention (Schmookler, 1966).

The linear approach to innovation nonetheless produced the first generation of innovation studies in the form of the 'supply push' or 'technology/science push' model, and the second-generation thinking in the form of the 'demand pull' model (Figure 2.1). These early generation models dominated the innovation literature in the 1950s and 1960s. However, the 'supply push' and 'demand pull' systems, which respectively constituted the first- and second-generation innovation studies, were conceived as sequentially separable stages with no feedback loops, which made them appear as competitive approaches to innovation rather than complementary aspects of

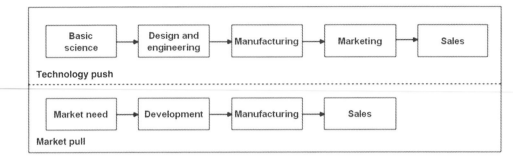

Figure 2.1 First & Second generation studies: Supply/technology push and market pull (Rothwell, 1994).

an integrated system of innovation. Lundvall (1985) viewed the two linear approaches to be no better than a 'black box', as they failed to give a full understanding of the innovation system underlying industrial activities. Moreover, the linear approach failed to capture the full extent of the complexity of the institutional context of innovation, and thus made no allowance for change management and organisational learning as the basis for innovation at the level of the firm. It is, at best, a naïve abstraction of the complex innovation process.

2.2.2   Non-Linear Approach to Innovation

A comprehensive approach to innovation is provided by what has come to be known as the innovation systems framework. This is akin to the Mode 2 paradigm of innovation proposed by Gibbons *et al.* (1994). Whereas the linear approach to innovation is based on a division of labour between 'knowledge search' and 'knowledge use', the non-linear approach is based on the understanding that knowledge cannot be independently produced in specialised research organisations and then transferred to passive users. The production and use of knowledge is a complex process, often requiring technical, social and institutional changes involving the interaction of actors across the knowledge producer-user network (Douthwaite, 2002). This approach places innovation and its producers and users in a socio-economic and cultural context and accounts for the stakeholders and partnerships, which, through their interactions, influence the innovation process. Thus, in the non-linear system of innovation, institutional and organisational factors underlying the innovation process are not assumed to be fixed or given as optimal, as in the linear system. Rather, these are expected to evolve to suit prevailing local circumstances, including stakeholders' perspectives and policy imperatives (Hall, 2002). Galbraith (1982) argues that whether it is driven by 'technology/supply push' or 'market pull', the innovation process involves knowledge of all the major elements simultaneously coupled with interactive communication paths. At an organisational level, Galbraith asserted that innovation usually occurred in groups of knowledge specialists

(a) Simultaneous coupling of knowledge by Galbraith

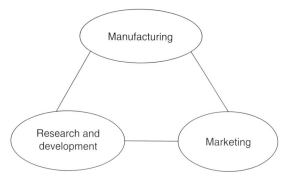

(b) The 'coupling model' of innovation by Rothwell

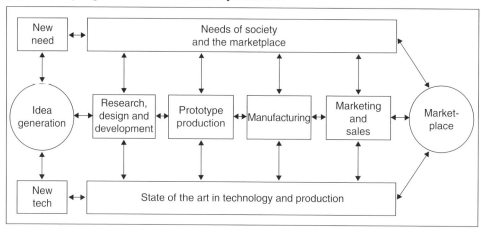

Figure 2.2   Third-generation of innovation system – the coupling model (Galbraith, 1982; Rothwell, 1994).

combined together. While the first- and second-generation innovation studies limited inputs to the innovation process from multiple knowledge sources, the third-generation studies (Galbraith, 1982) include feedback loops as communication paths generating internal inputs (e.g. idea generation, new technology, etc) and external inputs (market needs, social needs, etc) for the innovation process. This is called the 'coupling model' (Figure 2.2), in which all components connect as a complex, interactive network of technological capability, and market and social needs.

However, the coupling model does not provide information of external resources of knowledge derived from the business network and supply chain, which is particularly important, in view of the experience of large as well as small innovative firms engaging in intensive external networking activity (Rothwell, 1994). Firms would engage in external networking activities to enhance their organisational learning and knowledge sharing capability, as these are crucial factors underpinning innovation at the level of the firm. In other words, the effectiveness of external networking in terms of innovation

New product development process in Nissan

Marketing

Research and development

Product development

Production engineering

Parts manufacture (suppliers)

Manufacture

Joint group meetings (engineers/managers)

Marketing   Launch

Figure 2.3   Integrated (fourth-generation) innovation process (Rothwell, 1994).

is contingent on the integration of activities across the organisational structure of the firm. Therefore, a fourth generation of innovation studies emerged, focusing on the parallel and integrated nature of the internal features of the innovation process at the level of the firm. This parallel and integrated system of activities within the firm (Figure 2.3) provides the basis for external interactions and strategic business alliances that are featured in the third generation of innovation system. The fourth-generation model thus adds to the third-generation model, integrating suppliers collaborating in new product design and development processes at an early stage; while at the same time integrates the activities of the different in-house departments that work on the project in parallel rather than sequentially, as in the automotive and electronic industries in Japan (Graves, 1987; Dodgson *et al.*, 2002). The integration of parallel or simultaneous activities within the firm allows intensive information and knowledge exchange among the various departments, which is crucial for organisational learning, organisational synergy and innovation.

The fifth-generation model of innovation (Rothwell, 1994) involves a 'process of system integration and networking', which, though developed based on the fourth-generation model, emphasises continuous change and improvement. A distinguishing aspect of the model is its dynamic nature. It addresses strategic elements, including time and cost management, corporate flexibility and responsibility, customer focus, electronic data processing, and quality management policy that reinforce the firm's technological capability (Galanakis, 2006). Rothwell's fifth generation of innovation, which portrays innovation as a systemic and dynamical process involving the interaction of players, mainly focuses on innovation at the level of the firm, thus making the point that innovation activities are essentially firm-led. Following Rothwell's typology, some scholars have tried to conceptualise a sixth-generation innovation model. For example, Chaminade and Roberts (2002) proposed a model that included intangible factors, such as social capital, to facilitate the sharing of tacit knowledge, interactive learning and social

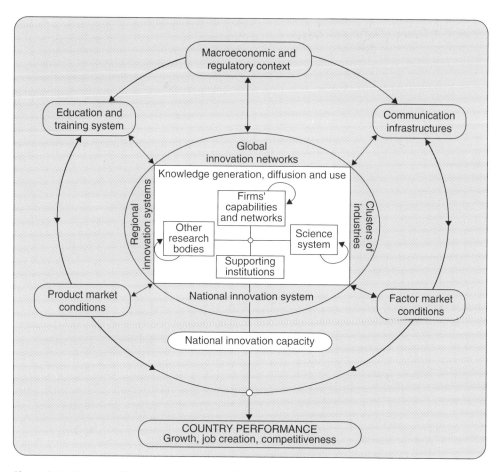

Figure 2.4 Actors and linkages in the national innovation system (OECD, 1999).

network development in an innovation system. Marinova and Phillimore (2003) also sought to extend Rothwell's typology through their contribution of the so-called 'innovative milieux' to the sixth-generation model. Their model focuses on the importance of geographical proximity and environmental conditions that facilitate knowledge generation and diffusion in certain localities. The model thus seeks to explain why innovative firms are concentrated in some locations or regions, and is similar to the concepts of innovation cluster (Porter, 1990), learning region (Florida, 1995; Morgan 1997), and regional innovation system (Cooke and Morgan, 1994; Cooke *et al.*, 1997).

The non-linear approach to innovation provided the basis for the development of the concept of 'national innovation system' (NIS) (Figure 2.4), which was first proposed by Freeman (1987) and later developed by Lundvall (1992). According to the NIS, innovation emerges from evolving systems of participants in the generation and application of knowledge. The essential elements of the system constitute interactive learning as a socially embedded process that cannot be understood without reference to its institutional and cultural context. The 'innovation system' concept thus provides a framework for exploring patterns of partnerships, revealing and managing the institutional context that governs relationships and processes, understanding research and innovation as a social process of learning, thinking about capacity development in a systems sense and linking the wide range of actors in a network with the view to promoting innovation (Hall, 2002).

The systems approach conceptualises innovation not merely in terms of output, but as a multi-dimensional, multi-directional and multi-causal process, requiring linkages and feedback mechanisms between a variety of activities both within and outside the firm (Jorde and Teece, 1990; Morroni, 1992). In the systems approach, the issue facing management of change is not whether the focus should be on the 'supply-push' or 'demand pull' aspects of innovation, but how best the complementarities between the two aspects can be exploited by matching shifts in knowledge/technology and market frontiers, since 'demand pull' and 'supply push' are, in the final analysis, part and parcel of the same dynamical innovation process (Dosi, 1982; Rothwell, 1994; Dodgson, *et al.*, 2002).

In a socio-economic system, the social functions of wealth creation, organised knowledge production and governance are fulfilled through the institution of the market integrating forces of supply and demand; the knowledge sphere, which through the provision of knowledge infrastructure in the form of R&D, skill development, etc, determine the scope for technological opportunities; and institutions of regulation and control, which, through interaction with the institution of the market, define the limits of 'selection environments' (Leydesdorff and Zawdie, 2010). These institutional actors represent the 'selection mechanisms', which, by interacting continuously and recursively, determine the scope for the management of change and technological trajectories, and act as a bridge between 'technology push' and 'demand pull' theories (Nelson and Winter, 1977, 1982; Dosi, 1982; Freeman and Perez, 1988). According to Dosi (1982), given a set of technological possibilities (paradigms) allowed by science, economic forces, together with institutional and social factors, operate as a 'selective device to choose the preferred path from a much larger set of possible ones'. Once a path (trajectory) is selected and established as a result of management of change, the trajectory evolves with 'a momentum of its own', enforcing the direction towards which "the problem solving activity" moves' (Dosi, 1982). Insofar as technological trajectories lock in selected values and systems, which could evolve into a culture of vested interests, they can pose a challenge to the management of change and, indeed, to further innovation.

2.3   Role of Culture as Challenge for Change Management and Innovation

The systems approach to innovation brings to light the importance of social processes and cultural changes as critical factors underlying the innovation process. Cultural changes affect the various aspects of social relations and modes of production. Thus, the way social and economic activities are organised and managed would be expected to reflect dominant cultures and practices that are subject to forces of change. On the other hand, innovation would occur either as a response to the needs of emergent cultures or as a reaction to a once dominant culture, which has now lost its way. Where dominant cultures prove to be resistant to change, they block opportunities for innovation. However, the focus here is on the culture of innovation. What, it may be asked, makes the culture of innovation?

2.3.1   Innovation Culture

In Schumpeter (1934), there is a clear distinction between the culture of the entrepreneurs, who lead the innovation process, and that of the imitators who follow the leaders. However, the roles of the actors in both categories are not fixed, so that those who were pioneers at one time could possibly be followers at another time in another event. Schumpeter's culture of the entrepreneur is unique and somewhat historical, insofar as it attributes the emergence of innovation to the inspiration of the occasional entrepreneurial genius, who achieves new ideas purely through the exercise of intuition and 'acts of insight' (Usher, 1954). This would make innovation the product of an entrepreneurial culture, albeit without social and historical content, so that its occurrence cannot be expected to be enhanced through policy intervention. Indeed, by this reckoning, innovation would fall outside the purview of strategic management function. Following this, Usher (1954) offered the culture of 'cumulative synthesis' as an alternative to the Schumpeterian scheme. In the Usher scheme, major 'epoch making' innovations are considered to arise from the 'cumulative synthesis' of relatively simple innovations, each of which required an individual 'act of insight' to come into being in the first place. Thus, in its entirety, the social process of innovation consists of 'acts of insight' of varying degrees of importance; and these acts converge in the course of time, culminating in a 'massive synthesis' that makes 'strategic innovation'. In comparison with the Schumpeterian view, a major advantage of the Usher scheme is that it allows conscious interventions aimed at increasing the speed or altering the direction of innovation. For example, by creating the appropriate R&D environment, the stage can be set so that only a few elements in the innovation process are left to chance. In other words, while innovation involves risk, it is not, however, essentially a purely random event. It is therefore appropriate to explain the factors that account for the making of innovation (or the absence of it) in an effort to enhance the effectiveness of management of change.

2.3.2   Organisation and Management Culture: Impact on Learning, Change and Innovation

The occurrence of change and innovation at the level of the firm is often influenced by the cultural underpinning of the organisational modes, management styles and organisational learning that define the scope or overarching environment for change within the firm. Culture influences change and innovation in proportion to the degree of resistance it poses against the counter-culture of agents of change, especially managers, project team members and consultants. Lewin (1958) explains the change process and the resistance to change envisaged thereof, as involving three steps. The first step in the process of changing behaviour associated with activities across the firm is to 'unfreeze' the existing situation. This ushers in the phase of transition, where movements take place with people having bought into the need for change. This stage examines the existing system to develop the new system through participatory approaches involving teamwork and active communication. Once the new system is in place, it would need to be embedded into the culture of the organisation in order for it to become routine. This is what Lewin (1958) refers to as the 'refreezing' stage. The task of 'refreezing' at this stage requires change agents to work actively with the people in the organisation to install the new system until the required new behaviours have replaced those that existed prior to change. Therefore, the cycle of change continues *ad infinitum* through the three steps of unfreezing, state of motion and refreezing. Following this theme, the process of change is not without incidences of conflict. It requires the organisation to be ready, not only to learn about new systems, but equally importantly not to forget some of its past (Dodgson, 1993; Prahalad and Hamel, 1994), or even obliterate all underperforming processes and structures through reengineering of business processes (Hammer, 1990; Hammer and Champy, 2003).

Figure 2.5 shows how the conflict in the crosscurrent of cultures can create opportunities for (and threats to) organisational change, thus creating

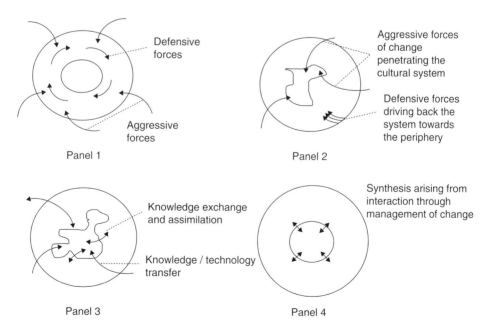

Panel 1 — Defensive forces / Aggressive forces

Panel 2 — Aggressive forces of change penetrating the cultural system / Defensive forces driving back the system towards the periphery

Panel 3 — Knowledge exchange and assimilation / Knowledge / technology transfer

Panel 4 — Synthesis arising from interaction through management of change

Figure 2.5 Culture of innovation evolution through change management (adapted from Zghal, 1996).

an environment for innovation. Panel 1 in Figure 2.5 is about the system of culture under attack by forces of change; Panel 2, about the disintegration of existing order under pressure of change; Panel 3, about a system in transition; and Panel 4, about a transformed state in which the round of innovation cycle is complete.

If change brings forth disequilibrium to the existing order of things, it is likely to be resisted by the culture that underpins the *status quo*. Hence, the conflict in the relationship between the aggressive forces of change (which often occur in the formation of new ideas) and the defensive forces of vested interest (who champion the moribund culture) as shown in Panel 1. The sustained pressure of change on the existing culture can force the breakdown of the system, as in Panels 2 and 3, thus giving way to the establishment of a transformed system, as in Panel 4. Therefore, at the firm level, culture can affect innovation through its influence on the modes of organisation and management. The organisational structure of firms can inhibit innovation when it is rigidly hierarchical, closed and restrictive, thus limiting the scope for information and knowledge exchange and for creativity to thrive. This would ensure that the dominant position of vested interests in the organisation is not compromised by the infiltration of new ideas. It is therefore in the interest of the 'insiders' (as in Panel 1) that the introduction of change by 'outsiders' is resisted.

The system of management at the level of the firm would either promote or inhibit innovation, depending on its commitment to change. Management styles, like organisational modes, are culture-conditioned. Basil and Cook (1974) identified three types of management mode, according to their implications for innovation and change. These include tradition-based, reform-oriented and change-oriented management types. Tradition-based management is characteristically closed, backward-looking and dependent on centralised and hierarchical approaches to decision making. It is therefore averse to reform, and *a fortiori* to radical changes. It is characteristically reactive and change resistant. Reform-oriented management is adaptive in character, so that it would be open to changes to the extent that these changes are capable of reforming and reinforcing existing structures and behavioural modes. On the other hand, 'change responsive' management is characteristically proactive and forward looking. It is capable of creating opportunities and encouraging the development of new ideas. This type of management is what Schumpeter (1934) would refer to as the harbinger or agency of 'creative destruction'. Change thus requires the removal of organisational and managerial barriers that would otherwise constrain creativity and the transmission of knowledge and information across individuals or groups. It therefore calls for a paradigm shift in the underlying culture and power structures at the level of the firm and in the wider cultural, institutional and policy environment that shapes the perspective of business behaviour at the micro level.

Change of organisational culture also requires organisational learning (Argyris and Schon, 1978; Fiol and Lyles, 1985; Huber, 1991; March, 1991; Dodgson, 1993). Organisational learning enhances the firm's adaptability to change, but it is also constrained by the prevalence of conflicts of interest within the organisation, as pressure is brought to bear on the organisation, not only to learn the new practices and the corresponding cultures, but also to unlearn or forget old ones (Prahalad and Hamel, 1994). Organisational learning, which results in the accumulation of organisational knowledge, occurs in three forms: single-loop learning, double-loop learning and deuteron-learning or triple-loop learning. Single-loop learning (Argyris and Schon, 1978) occurs reflexively as errors are detected and corrected without challenging the assumptions underlying the organisation's norms, procedures, policies and objectives. This particular mode of organisational learning is referred to as 'lower level learning' (Fiol and Lyles, 1985), 'adaptive learning' or learning by 'copying' (Senge, 1990a,b) and 'non-strategic learning' (Mason, 1993). Double-loop learning (Argyris and Schon, 1978; Argyris *et al.*, 1985; Argyris, 1990; 1992; 1994; 1996) occurs when, in addition to single-loop learning, the organisation's existing norms, procedures, policies and goals are challenged and modified. In this learning mode, the learning organisation is asking not merely if it is doing things the right way (which corresponds to single-loop learning), but more importantly, whether it is doing the right things. Double-loop learning changes the learning organisation's knowledge base or organisation-specific competences (Dodgson, 1993; Snell and Man-Kuen Chak, 1993; Stonehouse and Pemberton, 1999). Double-loop learning is also referred to as 'higher level learning' (Fiol and Lyles, 1985); 'generative learning' or learning to explain an organisation's capabilities (Senge 1990a,b); and 'strategic learning' (Mason, 1993). According to Mason (1993), strategic learning enables an organisation 'to make sense of its environment in ways that broaden the range of objectives it can pursue or the range of resources and actions available to it for pursuing the objectives'.

Argyris (1996) noted that most organisations have difficulties learning in the double-loop manner. Therefore, a third level of learning was proposed, namely 'duetero-learning', based on the ideas of Bateson (1973) or triple-loop learning (Flood and Romm, 1996; Snell and Man-Kuen Chak, 1998). Deutero-learning or triple-loop learning is about 'learning how to learn'. While single-loop learning is about rules (whether activities of the organisation are done right under the given set of assumptions, norms and beliefs), and double-loop learning about principles (whether the right things are being done), triple-loop learning is about strategies (how 'what is right' is decided). Triple-loop learning involves complexity thinking and as such, it is about 'increasing the fullness and deepness of learning about the diversity of issues and [the] dilemmas faced, by linking together all local units of learning in one overall learning infrastructure, as well as developing the competences and skills to use this infrastructure' (Flood and Romm, 1996). Triple-loop learning takes the form of 'collective mindfulness' (Romme and van Witteloostuijn, 1999) or 'awareness of ignorance or gaps' (Nevis *et al.*, 1995), which motivates members of the organisation to produce new structures and strategies for learning. If the appropriate environment, strategies and processes are not created, change through learning can be delayed or blocked for lack of good communication or feedback (Argyris, 1994). This problem of delays in feedback and communication, which retard organisational learning, can be mitigated by the introduction of a system of change management that is underpinned by information technology (IT) leveragability (Kittinger and Grover, 1995). Hammer (1990) would exploit the benefits arising from IT leveragability or of automation, say in the form of computerisation, not to prop up structures and processes that are outdated, but to 'reengineer' business processes through the promotion of organisational learning based on the deutero- or triple-loop learning mode. This reengineering of business processes is what constitutes change management as the 'handmaiden' of innovation.

2.4   General Framework for Change Management

Change management can be seen as 'reengineering business processes' by another name (Hammer, 1990; Hammer and Champy, 2003). Van de Ven (1986) found management of change and innovation as the 'most central concern' of chief executive officers (CEOs) of public and private firms in managing their enterprises. On this basis, a general management perspective on innovation would highlight four central problems CEOs have to contend with: the management of attention, the process problem of 'managing ideas into good currency', the structural problem of managing part-whole relationships and the strategic problem of institutional leadership.

2.4.1   Strategic Factors in Change Management

Management of attention is considered to be a crucial component of strategy for change insofar as people and organisations, if left to their own devices, would tend to be protective of existing practices rather than focusing their attention on the apparently risky task of developing new ideas. Indeed, it is difficult 'to trigger people's action thresholds to pay attention to new ideas, needs and opportunities' where organisations are well established facing little or no problems (Van de Ven, 1986). Under such circumstances, there is virtually no incentive for change; and if changes were desired, management would need to provide the conditions (incentives) that would motivate people to be proactive and attuned to making or accepting the changes. Thus, 'managing ideas into good currency' is, according to Van de Ven (1986), about implementing and institutionalising innovative ideas. This is where the 'selection mechanism' involving the coalition of interest groups and the associated social, economic and political dynamics come into play to determine the specific 'problem solving activity' through change and innovation.

The task of 'managing part-whole relationships' is based on the recognition that the fragmentation of efforts and responsibilities is an anathema to the innovation process, which involves multiple functions, resources and disciplines and calls for synergy through a process of integration. The systemic nature of innovation means that 'innovating within the parts while losing sight of the whole would be inherently dysfunctional' (Winch, 1998). The challenge for management, in this case, would be how to put the whole into the parts, and vice versa, to maximise the benefit of synergy. Management also has to provide institutional leadership that is capable of creating an infrastructure that is conducive to innovation. This is crucial, because 'innovations not only adapt to existing organisational and industrial arrangements (i.e. incremental innovations), but they also transform the structures and practices of these environments (i.e. radical innovations)' (Van de Ven, 1986). Given that innovation can be constrained by the state of organisational structure (Basil and Cook, 1974; Hankinson, 1999; Daft, 2001; Mullins, 2007) and organisational culture (Gagliardi, 1986; Silvester et al., 1997; Line, 1999; Sardi and Lees, 2001), the strategic role of management in determining the scope for innovation at the level of the firm cannot be over-emphasised. For the culture of innovation to take root at the level of the firm, it is important for management to redesign and develop or reengineer (Hammer 1990; Hammer and Champy, 2003) organisational structures and processes, so that rigid hierarchical systems of control and outdated organisational structures are superseded by systems of operation and organisational structures with built-in flexibility. This would increase the scope for interaction between members of the organisation at all levels of operation through knowledge exchange; to promote transparency and engagement in shared decision-making systems; and encourage constructive conflict with the view to replacing the culture of conformity with a culture of change. A change-responsive management system with proactive orientation should spare no effort to redefine institutional roles and missions to encourage and provide opportunities for independent action and change. This would help re-orient individuals within the organisation to accept change and be part of the change process, through the provision of institutional support systems, such as education and training schemes geared to enabling individuals to adopt a change-responsive behaviour.

In summary, therefore, the requirement for change and innovation in the construction industry is generally influenced not only by clients' requirements to secure 'value for money', but also by the need to meet public safety standards; contractors' concern with the potential risk associated with the introduction of untried and untested methods and construction materials (Harvey and Ashworth, 1993); and the concern of policy for employment expansion, economic growth effectiveness and environmental sustainability. These can all be addressed through a dynamic framework of change management. For instance, where chronic market constraints prevail mainly as a consequence of income inelasticity of demand for products, change and innovation efforts are quickly reoriented to areas of growth (income-elastic areas facing market/demand expansion).

2.4.2   Systemic Approach to Change Management

According to Kettinger and Grover (1995), implementation of change within an organisation/enterprise would be effective when issues relating to strategic initiatives, cultural readiness, learning capacity, knowledge sharing capability, IT leveragability, and network relationship balancing are addressed in a systemic framework, in order to positively impact process efficiency and quality of work-life; quality of output and customer satisfaction, and business performance gains in terms of competitiveness and growth in market share and profit margin. This would amount to a system of management that is change-responsive and capable of refreshing itself, through the process of organisational learning upon exposure to new ideas and challenges (Argyris and Schoen, 1978; March, 1991). Thus, for example, improving organisational learning capacity and adding to organisational knowledge stock, through feedback from performance gains and/or changes in the wider business environment, would be expected to have the effect of enhancing the formulation and implementation of strategic initiatives that would transform the cultural readiness of the organisation for change.

Thus for Kettinger and Grover (1995), the reengineering of business processes and structures takes place partly at the level management, where the norms, beliefs and value systems underlying the organisational culture of business set the context for change environment; and partly at the business process level, where the impacts of changes in business process management are observed. Change invariably starts with strategic initiatives manifested in the form of visions and commitments to the transformation of organisational culture. The involvement of top management is crucial to set out strategic plans and programmes and to provide the mechanisms for achieving the targets set by these plans and programmes. The uptake of strategic initiatives for change and the effectiveness of their implementation would, however, depend on the existence of a strong coalition of interests within the organisation, while effectiveness of management in coalition-building and conflict resolution would turn on its ability to create a culture of transparency, so that the change is adopted with little or no resistance. Management would enhance the effectiveness of its commitment to change by involving the services of change agents who would ensure that the change initiative is understood and effectively implemented at all levels of activity within the organisation.

For strategic changes to take root in an organisation, the organisation must be culturally ready to be able to internalise and accept change. Change and innovation can occur within an organisation subject to the limits set by the prevailing norms and belief and value systems. These underpin the pattern of organisational behaviour that bear on the decisions to take risk, make changes, innovate, share knowledge, forge links and partnerships, etc. Where behaviours like 'risk taking, openness, shared vision, respect and trust, high expectation for action, and a focus on quality', are evident, then cultural readiness would allow the promotion of change to occur. However, at this juncture, it is also important to acknowledge that culture can act as a constraint on change where 'risk avoidance, ambivalence, group think, and excessive competition' are the dominant norms (Kettinger and Grover, 1995).

Knowledge sharing within an organisation is crucial for organisational learning and, as such, provides the basis for the generation of new ideas and the occurrence of innovation in the organisation. The knowledge-sharing capability of an organisation is a function of the organisational culture, and the ways in which individuals and groups find themselves within the organisational structure. In this context, a centralised, hierarchical and fragmented organisational structure is potentially restrictive of knowledge exchange and knowledge sharing. Knowledge sharing is best promoted under a system of organisation and management that is change-responsive and proactive, which focuses on integrative approach to problem solving, and on the building of trust and transparency as a basis for collaboration between individuals and between groups within the organisation, so that the process of learning and knowledge sharing within an organisation can be expedited by the use of IT at all levels of organisational activity. Hammer (1990) and Hammer and Champy (2003) asserted that IT should be used not to entrench outdated structures and processes, but to leverage changes in organisational culture and performance improvements through the value chain, and empower individuals through change and innovation in the organisation.

Another aspect of change management is conflict resolution and coalition building within the organisation, which brings together people with diverse skills, resources and interests – removing the boundaries around them. This calls for constructive management to balance relationships and provide suitable conditions for change and innovation to occur through the synergy arising from the interactions across the various categories of people within the organisation. Balancing relationships by proactively changing the boundaries of value chains would limit the scope for conflict and would consequently enable organisations to relax the transaction cost constraint on innovation, thus determining the range of activities to be undertaken within the organisation and those tasks that are to be contracted out to agents on the supply chain. This is particularly important for the management of change and innovation in the construction industry. The resolution of conflicts that arise in the process of change can be achieved through the reengineering of business processes and structures as part and parcel of the change management practice.

Change management is also about process management aimed at improving process quality and process efficiency; optimising the quality of employees' work-life (Cranny et al, 1992); and enhancing the appeal of products and services to customers. Process improvement can be achieved using systems-oriented methods of process management, which have their origin in the Japanese 'quality movement' (Kettinger and Grover, 1995). Process improvements are often derived from incremental and radical changes, which are gauged for their performance by measures, such as cycle time, defects, productivity, cost, sustainability, environmental impact, consumer appeal and marketability, etc, where incremental changes can lead to continuous process improvements. Through the incremental approach, employees can gain learning experiences that enable them to produce better quality products with little or no variation of processes over time (Yelle, 1979; Kettinger and Grover, 1995). However, radical changes, which derive from 'double-loop' and 'triple-loop' learning, and usually involve paradigm shifts, tend to occur discontinuously in the form of totally new products and processes.

In competitive market environments, the survival and growth of business organisations depends on the extent to which they appeal to consumers or clients in terms of quality, cost, delivery time, sustainability and environmental impact performance. Change management should therefore be geared to meeting customer requirements. It is therefore important to determine the orientation of the change management strategy with respect to the significance of the roles of demand-pull and supply push, and incremental changes and radical changes. This has implications for the mode of learning and hence for the development of learning capability which, according to Schumpeter (1934), has an effect on innovation. The question that has yet to be addressed here is, 'How does the conceptual framework for the analysis of change management relate to change and innovation in the construction industry?' This is discussed in the following section.

2.5   Innovation in Construction

The construction industry has traditionally been slow in the uptake of innovation. In an attempt to explain the persistence of this feature, Winch (1998) postulated that volume production, which has historically been closely associated with innovation in manufacturing, might no longer be an appropriate method for construction. However, this view begs the question about the role of prefabrication in construction and building activities. Gann *et al.* (1992) observed innovation efforts in construction were 'disproportionately orientated towards product enhancement rather than process improvement'. Yet an important factor behind the innovation-deficit in construction is the fragmentation of the industry itself as a system, and of the actors embedded in the system who broker change and innovation. Moreover, unlike many other industries, innovation in construction, which tends to occur through problem-solving exercises, is typically project-specific, constrained by client needs and circumstances pertaining to project sites. This means that construction innovation is predominantly of the incremental type; and as Winch (1998) noted, most of the problem-solving technologies and techniques adapted and applied to meet specific client needs were based on tacit knowledge that cannot be 'learned, codified and applied to future projects'. However, what has become increasingly apparent of late is that the construction industry in many countries, not least the UK, is embracing new working practices similar to those found in the manufacturing sector. Even so, the pace of construction innovation is varied across countries. For example, the dominant culture in the UK is known to have had an effect of making contractors risk-averse and reluctant to adopt new ideas (Lorenz and Smith, 1998). Gann (1997) noted that traditional construction firms in the UK tended to compete on price rather than on quality or technical competence, and that this generally accounted for the prevalence of 'an over-developed sense of price, and an underdeveloped sense of value' in the industry. Furthermore, Winch (1998) felt the British system was the victim of the so-called 'extrapolation trap', where technologies were 'continually re-invented in a circular rather than progressive manner', thus explaining why there were plenty of new ideas, but too little by way of innovation. This view is corroborated by Lorenz and Smith (1998), who observed that the 'consultancy dominated' industrial culture in the UK was prone to the adoption of 'minimalist' solutions rather than the uptake of radical and potentially risky business options. The short-term orientation of the construction business and the preference in consequence for quick pay-off opportunities has consequently limited the scope for engagement in innovation activities.

2.5.1   Reconstructing Construction Culture

In recent years, there has been increasing awareness in the UK at both government and industry levels about the significance of innovation for the competitiveness of the industry, and the need for a radical re-orientation of policy and management practices to make innovation happen in construction. This concern has been orchestrated in a range of initiatives, including the Construction Innovation Forum at the Building Research Establishment, the Construction IT Centre for Excellence, and Government task force groups and reports (Latham, 1994; Egan 1998), which led to the establishment of the Movement for Innovation (M4I). These initiatives were principally aimed at improving value for money to clients; enhancing the innovativeness and competitiveness of construction firms; making construction activities environmentally sustainable; and minimising the risks to health and safety for users of constructed facilities. Essentially, the initiatives were out to promote innovation in construction through change management, with particular focus on change in construction culture that marginalised teamwork, partnership and co-operation. However, over the last two decades, construction innovation in the UK has occurred in two major forms – namely, SCM and value engineering. Whilst supply chain fragmentation had issues concerning innovation (Dulaimi, 1995), following Latham and Egan, among other initiatives, the trend now is to move away from the pursuit of 'competitive' tendering towards partnering arrangements, involving target cost contracts; and 'traditional' construction organisations and management systems are now losing ground in the wake of new initiatives, such as partnering, supply chain management (SCM) and value engineering.

Fellows (1997) conceptualises partnering along the supply chain as a specific type of joint venture involving long-term commitment between parties, in order to achieve specific business objectives by maximising the effectiveness of resources available to each party. However, the benefits of synergy through partnering are largely contingent on the flexibility and orientation of the organisation of firms on the supply chain – i.e. whether they are traditional, hierarchical, reactive and adaptive; or change-responsive, proactive and innovative. In the context of sustainable construction (SuCo), value management is no longer about cost cutting, but about providing clients with the best value for money. Reconstructing value management means establishing a culture that recognises value in its totality as it relates to all parties along the supply chain, and encourages the planning and monitoring of value improvements. Value management means looking into objectives before providing solutions and identifying critical areas in management that influence innovative solutions to maximise value for money and achieve competitiveness and sustainability. This has brought forth the culture of 'value engineering' aimed at maximising value for money, which takes into account the whole life of projects and covers cost effectiveness, quality performance, product safety, environmental impact, competitiveness and sustainability.

2.5.2   Changes in UK Construction Culture

The Egan agenda for 'rethinking construction', which was initiated by the Government in 1998, set the scene for construction innovation to unfold. Following Egan, the Construction Industry Task Force, which was set up by the Government, initiated the M4I to provide a network for the generation and dissemination of new ideas. The implementation of the Egan agenda called for a radical overhaul of the management ethos of construction firms. Egan challenged construction firms to adopt 'lean' production techniques and to reduce costs and time by 10% and waste (in terms of reduction of defects and accidents on site) by 20% on an annual basis, by addressing the following points:

- creating an integrated project process, based on the principle of partnering the supply chain;
- using techniques for eliminating waste and increasing value for the customer, based on the principles of value management and engineering;
- designing projects that make maximum use of standard components and processes, thus enhancing the value for money objective of construction activities; and
- replacing the conventional competitive tendering practice with target cost construction contracts, based on long-term relationships arising from partnerships along the supply chain.

The Egan agenda thus sought to transform the UK construction industry through change management – making it lean in terms of sustainability performance, and competitive in terms of growth performance. However, 10 years after its launch, the Review Team at Construction Excellence (2009), found (based on a survey of over 1,000 industry professionals) that although Egan had a big impact on some parts of the industry - and the industry as a whole was moving in the right direction - the transformation expected was still incomplete, falling short of the Egan targets. The results of the survey also revealed that while the industry's performance on safety and profitability had improved, progress in other areas was found wanting. The Review Team identified the following constraining factors: the application of business models based on short-term cycles; the persistence of the fragmented nature of the industry; the poor integration of parties and activities on the supply chain; and the lack of strategic commitment at senior management and government level (CE, 2009). These issues constrained the industry from progressing along the lines set by the Egan agenda (Zawdie *et al.*, 2003), and were partially influenced by the rigidity of the prevailing culture and lack of knowledge exchange to leverage innovation.

In a study conducted during the early phase of the implementation of the Egan agenda, 17 Scottish construction firms (including principal contractors and sub-contractors) were surveyed for the extent of cultural change impacted upon them by the Egan agenda. About 30% of these firms were aware of the main aspects of the Egan Report and had fully adopted the Egan recommendations; 35% said they were aware of the Egan agenda, having heard or read about it (but had not implemented it); and another 35% identified that they were not aware of Egan at all (Zawdie *et al.*, 2003). The results of this limited sample survey showed that the Egan agenda had a long way to go before being fully implemented in the UK. This survey highlighted some salient features about how firms respond to the challenge of innovation. Most of the firms, that were unaware of Egan, were small sub-contractors, who were generally highly risk-sensitive, and therefore reluctant to learn about the Egan agenda as the way forward. However, this did not mean that all small firms were inherently averse to innovation. Indeed, one of the small 'unaware' subcontracting firms won contracts on competitive tenders because of innovation. Furthermore, most of the firms recognised the need for cultural change in the industry, but perceived it to be an evolutionary process. They were also aware that this evolutionary and complex process could be accelerated by the development of network systems, such as links between government, universities and industry, along with supply chain partnerships including contractors and clients.

2.5.3   Change Management Initiatives for Innovation in Construction

There is growing awareness about the Egan agenda for change and innovation as the way forward for the UK construction industry. Innovation is widely understood as a factor that would, given the right circumstances, enable the industry to meet the challenges of being efficient and competitive on a global scale, as well as deliver a built environment that supports the creation of a low carbon economy. Of significance to the industry is the emergence of an environment that produces cultural readiness for change, especially by broadening the scope for organisational learning and the accumulation knowledge and experience. This would enable management to engage in strategic initiatives, resolve internal conflicts, reduce resistance to change, and provide the social capital base for innovation through the development of network partnerships that allow knowledge sharing and exchange through the supply chain.

The promotion of innovation at the firm level should therefore begin by developing strategies to make the organisation and management system forward-looking and responsive to change. A progressive approach to organisation and management would need to replace fragmented cultural systems, with flexible ones that allow firms to interact with minimum friction and minimum transaction costs. In addition, innovation-inhibiting structures, which encourage the persistence of bureaucracy and the entrenchment of vested interests, would need to make way for systems, which promote cooperation rather than conflict, and 'capacity building' rather than 'empire building'. For innovation to thrive, it is important that the *modus operandi* of firms is integrated rather than fragmented, allowing greater flexibility to respond to changes in market forces, as exemplified by the experience of car manufacturing plants in Japan (Graves, 1987). Strategies for promoting innovation would also need to provide the conditions for a progressive and proactive management system, supported by a strong management culture. The type, level and prevalence of this culture should foster open communication between management and labour, in order to envision and drive forward change. This would help establish ownership, commitment and empowerment – the rubrics of which could provide an environment that is conducive to creativity and where new ideas can be championed through an *esprit de corps*. If construction firms lacked these managerial attributes, they would not be expected to be innovative or dynamic. Nor would they be expected to be aware or capable of implementing the Egan agenda for change. However, construction innovation is more than the mere awareness of Egan. In order for it to thrive as an industrial culture, progressive management aimed at integrating activities within the firm should be complemented with the development of a wider network system, which, by exposing firms to interactive relationships of the innovation process, would enable them to realise their full innovative potential.

2.6   Conclusion

This chapter discussed the importance of innovation in the construction industry, with a particular focus on process, the role of culture, and the management of change. Innovation was identified as not being evenly distributed, with some (e.g. manufacturing) used to more frequent changes than others (e.g. construction). Notwithstanding this, there is growing evidence to suggest that this is starting to change. For example, the management of change has opened up new opportunities to re-visit skills and learning applied to business needs. This has provided the basis for the generation of new ideas and new innovation opportunities to exploit, especially through such conduits as organisational learning and knowledge management. The knowledge-sharing capability of an organisation is paramount, as a centralised, hierarchical and fragmented organisational structure can restrict knowledge exchange and knowledge sharing. Knowledge sharing is best promoted under a system of organisation and management that is change-responsive, proactive and focused upon problem solving through collaborative integration. However, short-termism can stifle entrepreneurialism and influence the prospects for change; and minimalist strategies rarely go far enough to champion major changes (Lorenz and Smith, 1998). Whilst Egan (1998) challenged construction by promoting the case for partnership and collaboration to transform the environment for decision-making and knowledge exchange, this has not fully materialised to date (CE, 2009).

For the management of change to be effective, the actors in the construction system would need to be fully integrated, so that there is a free flow of knowledge and information exchange with a top-down adoption and implementation procedure synchronised to a bottom-up learning and problem-solving approach. This is all the more important in view of the fact that construction is a complex systems industry with an 'innovation superstructure' (clients, regulators and professional institutions) and an 'innovation infrastructure' (trade contractors, specialist consultants and component suppliers), and with 'systems integrators' (including the principal architect and the principal contractor) (Winch, 1998). The industry therefore needs to remove fragmentation, embrace change, and leverage innovation drivers through new business models and strategic trajectories that streamline and align corporate energy and competence to business goals and new market opportunities.

References

Argyris, C. (1990) *Overcoming Organisational Defenses: Facilitating Organisational Learning*. Allyn and Bacon, Boston.

Argyris, C. (1992) *On Organisational Learning*. Blackwell, Oxford.

Argyris, C. (1994) Good communication that blocks learning. *Harvard Business Review*, **July–August**, 77–85.

Argyris, C. (1996) Unrecognised defenses of scholars: Impact on theory and research. *Organisation Science*, **7**, 79–87.

Argyris, C. & Schon, D.A. (1978) *Organisational Learning: A Theory of Action Research*. Addison Wesley, Reading, MA.

Argyris, C., Putnam, R. & McLain Smith, D. (1985) *Action Science: Concepts, Methods and Skills for Research and Intervention*. Jossey-Bass, San Francisco, CA.

Basil, D.C. & Cook, C.W. (1974) *The Management of Change*. McGraw Hill, London.

Bates, G. (1973) *Steps to an Ecology of Mind*. Palladin, London.

Chaminade, C. & Roberts, H. (2002) Social capital as a mechanism connecting knowledge within and across firms. Paper presented at the Third European Conference on Organisational Knowledge, Learning and Capabilities, April, Athens,.

Construction Excellence (CE) (2009) *Never Waste a Good Crisis: A Review of Progress Since 'Rethinking Construction' and Thoughts for Our Future*. Construction Excellence in Built Environment, London.

Cooke, P., Uranga, M.G. & Etxebarria, G. (1997) Regional innovation systems: institutional and organisational dimensions. *Research Policy*, **26**(4 & 5), 475–491.

Cooke, P. & Morgan, K. (1994) The creative milieu: a regional perspective on innovation. In: ed. Dodgson, M. & Rothwell, R., *The Handbook of Industrial Innovation*. Edward Elgar, Aldershot. 25–32.

Cranny, C.J., Smith, P.C. & Styone, E.F. (1992) *Job Satisfaction: How People Feel About Their Jobs and How It Affects Their Performance*. Lexington Books. New York.

Daft, R. (2001) *Organisation Theory and Design*, 7th edn. South Western College Publishing, Cincinnati, OH.

Dodgson, M. (1993) Organisational learning: A review of some literatures. *Organisational Studies*, **14**(3), 375–394.

Dodgson, M., Gann, D.M. & Salter, A.J. (2002) The intensification of innovation. *International Journal of Innovation Management*, **6**(1), 53–83.

Dosi, G. (1982) Technological paradigms and technological trajectories. *Research Policy*, **11**(3), 147–162.

Douthwaite, B. (2002) *Establishing Innovation: A Practical Guide to Understanding and Fostering Technological Change*. Zed Books, London.

Dulaimi, M. (1995) The challenge of innovation in construction. *Building Research & Information*, **23**(2), 106–109.

Egan, J. (1998) *Rethinking Construction*. Report of the Construction Task Force to the Deputy Prime Minister. DETR, London.

Fellows, R.F. (1997) The culture of partnering. In ed. Davidson, C.H., *Proceedings of the CIB W-92 International Conference on Procurement*. Montreal: CIB.

Fiol, C.M. & Lyles, M. (1985) Organisational learning. *Academy of Management Review*, **10**(4), 803–813.

Flood, R.M. & Romm, N.R.A. (1996) *Diversity Management: Triple Loop learning*. John Wiley & Sons, Ltd, Chichester, UK.

Florida, R. (1995) Toward the learning region. *Futures*, **27**(5), 527–536.

Freeman, C. (1987) *Technology Policy and Economic Performance: Lessons from Japan*. Pinter, London.

Freeman, C. & Perez, C. (1988) Structural crises of adjustment, business cycles and investment behaviour. In: ed. Dosi, G., Freeman, C., Nelson, R.R., Silverberg, G. & Soete, L., *Technical Change and Economic Theory*. London: Pinter. 38–91.

Gagliardi, P. (1986) The creation and change of organisational cultures: a conceptual framework. *Organisation Studies*, **7**(2), 17–134.

Gann, D. (1997) *Building Innovation: Complex Contracts in a Changing World*. Thomas Telford, London.

Gann, D., Matthews, M., Patel, P. & Simmonds, P. (1992) *Construction R&D: Analysis of Private and Public Sector Funding of Research and Development in the UK Construction Sector*. Department of Environment, London.

Galanakis, K. (2006) Innovation process – make sense using systems thinking. *Technovation*, **26**(11), 1222–1232.

Galbraith, G. (1982) Designing the innovating organisation. *Organisational Dynamics*, **10**(3), 5–25.

Gibbons, M., Limoges, C., Nowotny, H., Schwartzman, S., Scott, P. & Trow, M. (eds) (1994) *The New Production of Knowledge*. Sage, London.

Graves, A. (1987) Comparative trends in automotive research and development. DRC Discussion Paper No 54. Science Policy Research Unit (SPRU), Sussex University, Brighton, UK.

Hall, A. (2002) Innovation systems and capacity development: an agenda for North-South research collaboration. *International Journal of Technology Management and Sustainable Development*, **1**(3), 146–152.

Hammer, M. (1990) Reengineering work: don't automate, obliterate. *Harvard Business Review*, **July–August**, 104–112.

Hammer, M. & Champy, J. (2003) *Reengineering the Corporation: A Manifesto for Business Revolution*. Harper Collins Publishers Inc, New York.

Hankinson, P. (1999) An empirical study which compares the organisational structures of companies managing the world's top 100 brands with those managing outsider brands. *Journal of Product & Brand Management*, **8**(5), 402–415.

Harvey, R.C. & Ashworth, A.A. (1993) *The Construction Industry of Great Britain*. Butterworth-Heinemann Ltd, London.

Huber, G.P. (1991) Organisational learning: The contributing process and the literatures. *Organisational Science*, **2**(1), 88–115.

Jones, E., Watson, B., Gardner, J. & Gallois, C. (2004) Organisational communication: challenges for the new century. *Journal of Communication*, **54**(4), 722–750.

Jorde, T. & Teece, D. (1990) Innovation and cooperation: implications for competition and anti-trust. *Journal of Economic Perspectives*, **4**(3), 75–96.

Kettinger, W.J. & Grover, V. (1995) Toward a theory of business process change management. *Journal of Management Information Systems*, **12**(1), 9–30.

Kozul-Wright, Z. (1995) *The Role of the Firm in the Innovation Process*. UNCTAD Discussion Papers, No. 98, Geneva.

Latham, M. (1994) *Constructing the Team*. HMSO, London.

Lewin, K. (1958) Group decision and social change. In: ed. Maccoby, E.E., Newcomb, T.M. & Hartley, E.L., *Readings in Social Psychology*. Holt, Rinehart & Winston, New York. 197–211.

Leydesdorff, L. & Zawdie, G. (2010) The Triple Helix perspective on innovation system. *Technology Analysis and Strategic Management*, **22**(7), 789–804.

Line, M.B. (1999) Types of organisational culture. *Library Management*, **20**(2), 73–75.

Lorenz, A. & Smith, D. (1998) Britain fails to close competitiveness gaps. *Sunday Times*, London: 11 October (cited in Akintoye, *et al.*, 1999).

Lundvall, B.A. (1985) *Product Innovation and User-Producer Interaction*. Industrial Development Research Series, vol. 31. Aalborg University Press, Aalborg.

Lundvall, B.A. (1992) *National Innovation Systems: Towards a Theory of Innovation and Interactive Learning*. Pinter, London.

March, J.G. (1991) Exploration and exploitation in organisational learning. *Organisation Science*, **2**(1), 71–87.

Marinova, D. & Phillimore, J. (2003) Models of innovation. In: ed. Shavinina, L.V., *The International Handbook of Innovation*. Elsevier Science Ltd, Oxford. 44–53.

Marosszeky, M. (1999) Technology and innovation. In: ed. Best, R. & de Valence, G., *Building in Value*. Arnold, London. 342–355.

Mason, R.M. (1993) Strategic information systems: Use of information technology in learning organisation. *Proceedings of the 26th Hawaii International Conference on System Sciences*, 5–8 January. Wailea, HI.

Morgan, K. (1997) The learning region: institutions, innovation and regional renewal. *Regional Studies*, **31**(5), 491–503.

Morroni, M. (1992) *Production Process and Technical Change*. Cambridge University Press, Cambridge.

Mullins, L.J. (2007) *Management and Organisational Behaviour*, 8th edn. Prentice Hall, Essex.

National Economic Development Office (1985) *Managing Change*. NEDO, London.

Nelson, R.R. & Winter, S.G. (1977) In search of useful theory of innovation. *Research Policy*, **6**(1), 36–76.

Nelson, R.R. & Winter, S.G. (1982) *An Evolutionary Theory of Economic Change*. Harvard University Press, Cambridge, MA.

Nevis, E.C., DiBella, A.J. & Gould, J.M. (1995) Understanding organisations as learning systems. *Sloan Management Review*, **36**(2), 75–85.

OECD (1999) *Managing National Innovation Systems*. OECD publications, Paris.

Porter, M. (1990) *The Competitive Advantage of Nations*. Free Press, New York.

Prahalad, C.K. & Hamel, G. (1994) *Competing for the Future*. Harvard University Press, Cambridge, MA.

Romme, A.G.L. & van Witteloostuijn, A. (1999) Circular organising and triple loop learning. *Journal of Organisational Change Management*, **12**(5), 439–453.

Rothwell, R. (1994) Towards the fifth generation innovation process. *International Marketing Review*, **11**(1), 7–31.

Sardi, G. & Lees, B. (2001) Development of corporate culture as a competitive advantage. *Journal of Management Development*, **20**(10), 853–859.

Schmookler, J. (1966) *Invention and Economic Growth*. Harvard University Press, Cambridge, MA.

Schumpeter, J. (1934) *The Theory of Economic Development*. Harvard University Press, Cambridge, MA.

Senge, P.M. (1990a) *The Fifth Discipline: The Art and Practice of the Learning Organisation*. Currency Doubleday, New York.

Senge, P.M. (1990b) The leader's new work: Building learning organisations. *Sloan Management Review*, **32(1)**, 7–23.

Silvester, J., Anderson, N.R. & Patterson, F. (1997) Organisational culture change: an inter-group attributional analysis. *Journal of Occupational and Organisational Psychology*, **72(1)**, 1–23.

Snell, R. & Man-Kuen Chak, A. (1993) The learning organisation: Learning and empowerment for whom? *Management Learning*, **29**, 337–364.

Stonehouse, G.H. & Pemberton, J.D. (1999) Learning and knowledge management in the intelligent organisation. *Participation and Empowerment: An International Journal*, 7(5), 131–144.

Strategic Forum for Construction (2002) *Accelerating Change*. Report for the Construction Industry Council. SFC, London.

Tourish, D. & Hargie, O. (2004) The crisis in management and the role of organisational communication. In: ed. Tourish, D. & Hargie, O., *Key Issues in Organisational Communication*. Routledge, London. 1–16.

Usher, A.P. (1954) *History of Mechanical Inventions*. John Wiley & Son, Ltd, London.

Van de Ven, A.H. (1986) Central problems in the management of innovation. *Management Science*, **32(5)**, 590–607.

Winch, G. (1998) Zephyrs of creative destruction: understanding the management of innovation in construction. *Building Research & Information*, **26(5)**, 268–279.

Yelle, L. (1979) The learning curve: historical review and comprehensive survey. *Decision Sciences*, **10(2)**, 302–308.

Zawdie, G., Abukhder, J. & Langford, D. (2003) Organisational and managerial attributes of innovation among firms in the construction industry in the UK: why culture matters. In: ed. Botemps, F., *Systems Based Vision for Strategic and Creative Design*, vol. 3. Balkema, Amsterdam. 85–92.

Zghal, R. (1996) Science, technology and society: What makes the culture of innovation? In: ed. Zawdie, G. & Djeflat, A., *Technology and Transition: The Maghreb at the Crossroads*. Frank Cass, London. 105–113.

Construction Innovation: Theory and Practice

Martin G. Sexton and Shu-Ling Lu

3.1   Introduction

This chapter presents a broad overview of innovation in construction. It aims to equip readers with a set of key theories through which innovation can be understood and, in doing so, will be in a better position to steer and manage innovation in practice. The purpose therefore, is not to concentrate on current examples of say, innovative products or practices; rather, the purpose is to develop a conceptual treatment of the subject area, which can be mobilised across a broad spectrum of construction settings. To achieve this aim, the chapter is structured as follows. First, a brief discussion on the definition or, more correctly, the myriad of definitions of innovation is discussed. Second, a definitional position is developed through the umbrella concepts of market-based and resourced-based views of innovation. These two perspectives are then synthesised to produce a balanced innovation perspective. Finally, a case study is presented to illustrate a number of the issues raised.

3.2   Definitional Debate on Innovation

There are a wide range of, sometimes, conflicting definitions of innovation. In many respects, the development of a single definition of innovation is a fruitless and pointless exercise. The more meaningful challenge is to adopt a particular view of innovation appropriate for a specific context. For example, innovation within a high technology combined heat and power boiler manufacturer is likely to be significantly different to innovation within a low technology service provider such as a labour-only painting contractor. Furthermore, the diverse range of theoretical and methodological perspectives from which to investigate innovation demonstrates this contingent and

Construction Innovation and Process Improvement, First Edition.
Edited by Akintola Akintoye, Jack S. Goulding and Girma Zawdie.
© 2012 Blackwell Publishing Ltd. Published 2012 by Blackwell Publishing Ltd.

context-specific approach. Traditional economic approaches to innovation, for example, tend to concentrate on particular technologies and market signals that encourage or stifle uptake and diffusion. Thus, the innovation discipline then evolved to complement the micro- and macro-economic considerations with a managerial treatment that explored the company-level strategies and capabilities that influence the direction and nature of new product and service development. More recently, institutional analyses have been made of the role of a broader social system in influencing innovation trajectories, through functions such as regulations, norms and values. In order to find a productive path through these bodies of theory, this chapter concentrates on a key generic characteristic of innovation; namely, that innovation involves the development and implementation of a new idea in an applied setting. The 'newness' characteristic is found in the general literature (van de Ven et al., 1999; DTI, 2003) and in the construction literature (Barrett and Sexton, 2006).

The UK government defines innovation as 'the successful exploitation of new ideas' (DTI, 2003: 8; NAO, 2009). The 'new idea' dimension embraces a range of domains. Policy makers understand innovation as a broad, systemic concept that encompasses the 'successful introduction of new services, products, processes, business models and ways of working' (ESRC, 2008: 2), and in this respect, construction literature is generally consistent with the general literature. Sexton and Barrett (2003b: 626) define successful innovation as 'the effective generation and implementation of a new idea, which enhances overall organisational performance.' Similarly, the Civil Engineering Research Foundation (CERF, 2000: 3) defines innovation as 'the act of introducing and using new ideas, technologies, products and/or processes aimed at solving problems, viewing things differently, improving efficiency and effectiveness, or enhancing standards of living' (focusing specifically on construction professional practices). The newness characteristic needs to be further elaborated in order to distinguish between new to the world, or new to a given situation. Rogers (1983: 11), for example, defines innovation as 'a product or service that is perceived as new by the members of the social system', and that 'it matters little whether the idea is "objectively" new as measured by the lapse of time since its first use or discovery. The perceived newness of the idea for the individual determines his or her reaction to it. If the idea seems new to the individual, it is an innovation.'

Innovation literature carries the theme of distinguishing different types of innovation across a multitude of dimensions: incremental versus radical, technological versus administrative, and so on. Again, these distinctions are important, but these should not be considered as discrete, separate types. In practice, such distinctions become blurred and mutually constituted. Innovation, for example, is often classified into 'product innovation' and 'process innovation'. 'Product innovation' refers to the development and introduction of new or improved products and/or services, which create or meet a new demand, and which are successful in the market. In contrast, 'process innovation' involves the adoption of new or improved methods of manufacture, distribution or delivery of service that 'lower the real cost of producing outputs, although they may also give rise to changes in their nature' (Clarke, 1993: 143). The 'product' versus 'process' view of innovation has now evolved towards a more systemic view. Athey and Schmutzler (1995) assert that process innovation (cost reducing) and product innovation (demand enhancing) are complementary. Indeed, Imai (1992: 226) speculates that 'process improvement and product differentiation are now being fused'.

However, the key common theme across the definitional debate in the literature is that 'new ideas' are taken to be the foundation or the starting point for innovation. The central question that will now guide the discussion is 'what is the stimulus for these "new ideas"?'

3.3 Market-based, Resource-based and Balanced Perspectives on Innovation

There are two key perspectives on the principal drivers for innovation: the market-based view and the resource-based view. Each perspective will be discussed in turn, followed by a synthesis of the two.

3.3.1 Market-based Perspective on Innovation

The market-based view of innovation stresses the role of institutional and market factors in stimulating innovation within companies. Market conditions are viewed in a broad sense as encompassing both the general business environment and the interaction environment. The general business environment covers the full spectrum of social, legal, economic, political and technological forces and configurations. In contrast, the interaction or construction-specific business environment consists of industry structures, clients, suppliers, competitors, financiers, regulators, and so on. The broad argument offered is that market conditions provide the context or initial conditions that either facilitate or constrain the direction and quantity of firm innovation activity (Slater and Narver, 1994; Porter 1980, 1985).

In the general literature, a number of market-based innovation theorists have investigated market or environmental influences on innovation. For example, influences have been articulated as customer–supplier relations (von Hippel, 1988), network studies (Håkanson, 1989), market conditions (Ames and Hlavacek, 1988) and external knowledge infrastructures (Nelson, 1993). A current example of regulatory pressures can be seen in the UK, which is inducing innovation through the Code for Sustainable Homes (DCLG, 2008), which requires housing developers to radically rethink their business models, designs and production processes to comply with, amongst other criteria, the zero-carbon requirement by 2016. The emphasis of the market-based innovation position is that firms adapt or orientate themselves through innovation to exploit changing institutional and market conditions in an optimal fashion. The environment where client interaction occurs is defined as 'the task environment' (Kotler, 1980), whilst the environment where other firms compete with the firm, customer and scarce resources is termed as 'the competitive environment' (Kotler, 1980). Together, the task environment and competitive environments are labelled collectively as 'the interaction environment' (Sexton and Barrett, 2003: 36). In summary, the interaction environment is a significant market-based stimulus to innovation within the construction context.

3.3.2   Resource-based Perspective on Innovation

The resource-based view of innovation emphasis is that firms identify and nurture resources that enable them to generate innovation to 'shape' market conditions; rather than the market-based view which advocates that market conditions 'shape' the resources that firms develop and exploit to respond to opportunities and threats. In other words, the resource-based view of innovation stresses that resources available to the firm, rather than market conditions (market-based perspective), are the principal catalysts for innovation (Itami, 1987; Barney, 1991; Grant, 1995).

Resources (e.g. human capital, social capital and physical capital) in themselves are not seen as productive. However, dynamic environments often demand generation of new resources as the business environment changes (Chaharbaghi and Lynch, 1999). Therefore, the challenge for firms is to create sustainable competitive advantage in rapidly changing and competitive environments for resources to be integrated, coordinated and deployed as 'distinctive capabilities' (Teece *et al.*, 1997). Amit and Schoemaker (1993: 35) note that capabilities 'refer to firm's ability to deploy resources, usually in combination, using organisational processes, to affect a desired end. They are information-based, tangible or intangible processes that are firm specific, and are developed over time through complex interactions among the firm's resources.'

The development of 'distinctive capabilities' in dynamic environments is labelled 'dynamic capability' (Teece *et al.*, 1997), where the concept of 'dynamic capabilities' is defined as 'the firm's ability to integrate, build, and reconfigure internal and external competencies to address rapidly changing environments' (Teece *et al.*, 1997: 516). In a construction context, Green *et al.* (2008: 63) argue that dynamic capabilities 'relates to a firm's ability to reconfigure its resources in response to changing environments'. Therefore, it is important that the capability of organisations to adopt, adapt and transform existing technological applications and know-how from other environments into relevant and appropriate solutions, organisational processes and technological products/services to match the socio-cultural context of the construction industry sector, is crucial for bringing about innovation (Sexton and Barrett, 2003a,b, 2004; Harty *et al.*, 2007; Green *et al.*, 2008).

3.3.3   Balanced Innovation

The perception of a potential market-driven opportunity or imperative is a necessary condition for innovation, but not a sufficient one (Dosi, 1984). The adaptation and orientation to market conditions requires firms to choose appropriate strategies that are adequately resourced and implemented (Snow and Hrebiniak, 1980). In this context, an appreciation of the resource dimension has played a significant part in the cultivation of the resource-based view of innovation. A resource-based view focuses on firms' resources, specifically, to understand business strategies and provide direction for, amongst other issues, innovation (Prahalad and Hamel, 1990; Andreu and Ciborra, 1996; Grant, 1997). However, the market-driven orientation does not provide a secure foundation for formulating innovation strategies for markets that are dynamic and volatile; rather, firms' own resources provide a more stable basis on which to develop their innovation activities, and to shape their markets, to a limited extent, in their own image. These stocks of resources are defined as assets, capabilities, routines and knowledge that are tied semi-permanently or permanently to, or controlled by, a firm (Barney, 1991; Wernerfelt, 1984). They may be physical, human, technological or reputational (Hadjimanolis, 2000: 264). The argument here is that innovative firms are those that can sense and act upon internal 'precipitating events' to create and develop unique resources or configurations of resources that serve as the foundations for successful streams of innovation.

Maijoor and van Witteloostuijn (1996: 549) suggest that 'the resource-based view of the firm seeks to bridge the gap between theories of internal organisational capabilities on the one hand and external competitive strategy theories on the other hand.' Indeed, recent research seeking to synthesise these two perspectives suggest that 'while firms' resource endowments may determine strategy success, strategy choice is … restricted by market structure' (Hewitt-Dunda and Roper, 2000: 1). This argument is evident in the observation that an 'innovative firm creates two kinds of products. The primary product is the innovation itself, while the secondary product is a process created by innovations, which increases the internal capacity by building up core competencies by making the *intangible capital* quicker, more flexible, more adaptable and more capable in dealing with market pressures and unexpected external shocks' (Lööf and Heshmatic, 2000: 7). In construction literature, evidence to support this balancing was reported in the observation that 'the sifting of possible [innovation] options was rigorous, with [small construction companies] being close enough to both their markets and their capabilities to instinctively know what will work, and what will not' (Sexton *et al.*, 1999: 17).

It is therefore suggested that market-based and resource-based innovation can be gainfully linked (Figure 3.1); by extending the argument that innovating firms are those that can sense and act upon 'precipitating events', both external institutional and market conditions, and internal resource conditions in an appropriately balanced and integrated fashion. This 'precipitating event' perspective is embodied in the observation that 'innovation is fostered by information gathered from new connections; from insights gained by journeys into other disciplines or places; from active, collegial networks and fluid, open boundaries. Thus, innovation arises from ongoing circles of exchange, where information is not just accumulated or stored, but created' (Wheatley, 1992: 113). However, the optimal innovation balance of market-based and resource-based innovation is contingent upon the

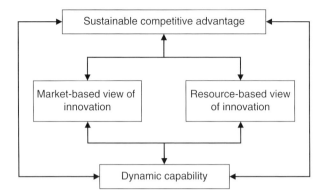

Figure 3.1   Balanced innovation.

'market pull' and/or 'resource push' implications of the prevailing precipitating events. The organisational challenge is therefore to generate the balance required to provide an appropriate innovation focus that enhances overall performance – and in so doing, dynamically links the focus and context of innovation. This view resonates with similar discussions focusing on the 'market-pull' and 'technology-push' (Coombs *et al.*, 1987) and 'research-push' and 'market-pull' (Freeman, 1982).

The market-based and resource-based views represent two stylised perspectives on how innovation can be stimulated. In practice, these two perspectives can be gainfully linked by extending the argument that there is mutual adjustment between companies 'reacting to' market opportunities and threats, and 'proactively' identifying, developing and exploiting resources and capabilities to secure a foundation for innovation in dynamic environments. The development of the optimal dynamic capabilities brings these two resources together to co-produce innovation that creates sustainable competitive advantage. This view is an extension of similar discussions focusing on the appropriate balance between market-based and resource-based views of innovation capabilities needed in small construction firms (Sexton and Barrett, 2003a).

The discussion to this point has laid out some key innovation concepts. In particular, the argument has been developed that innovation should not be seen as merely a new product or process; rather, it should be seen as a multi-layered, complex interaction that draws upon and balances market and resource-based influences through the creation of appropriate dynamic capabilities. The remainder of this chapter further develops these ideas through a case study of a small construction professional service firm.

3.4   Case Study of Innovation in a Small Construction Professional Service Firm

The aim of this case study is to demonstrate the context-specific nature of innovation in order to mobilise some of the concepts and ideas developed so far. The case study in particular, offers a definition of innovation (note, not the definition!) and identifies some key innovation management challenges for construction firms.

3.4.1   Case Study Aim and Methodology

The case study company, hereafter labelled as ArchSME (for confidentiality reasons), is a small architectural design practice based in the UK. ArchSME's principal markets are city centre and suburban residential sectors, with work varying from one-off commissions from regional clients to repeat business from national house builders.

The case study consisted of two main research phases conducted over a 22-month period. The first phase (exploratory) investigated four successful innovations and three unsuccessful innovations. Successful innovations included the development of ArchSME mission statement (innovation 1), the securing of Investors in People (innovation 2), the flow of new novel designs (innovation 3) and the company restructure (innovation 4). These were identified as being significant firm-generated innovations over the last two years. Failed innovations included the introduction and subsequent failure of in-house seminars (innovation 5), the introduction of the new materials (innovation 6) and the Learndirect project (innovation 7). These were classified as being significant innovations over the last two years that had failed. Table 3.1 identifies the summary descriptions for each of these innovations.

The second phase of this work was used to further test and develop the concept innovation model that came out of the exploratory phase. This research adopted an action research methodology, adopting the five-step process of diagnosis, action planning, action taking, action evaluation and specifying learning (Susman, 1983). The action research provided a narrative of 'real time' innovation within a company setting, and brought about a successful innovation in the case study firm. The action researcher was embedded within the case study firm full-time for a period of ten months. The action research phase began with a company workshop, the purpose of which was to discuss and evaluate the key findings from the exploratory phase and identify an action research intervention (or innovation) to be developed and implemented. The identified innovation was the design and implementation of an interim project review process innovation (innovation 8). It should be noted that an interim project review process is a procedure that consists of review activities performed by a project team to gather information on what worked well and what did not, so that current and future projects can benefit from that learning. This new process was considered as an integrated part of the ongoing preparation by the case study firm for ISO 9001 Quality Management Systems accreditation. Based on this, an interim project review process handbook (including interim project review process policy, guidelines and checklists) was jointly developed by the action researcher and the firm. This handbook was then assessed and signed off by the company senior management and integrated into the company's quality systems.

Table 3.1   Description of identified company innovations.

Types of innovation	Identified innovations	Description
Successful Innovation	Innovation 1	Mission statement is a statement that captures an organisation's purpose, customer orientation and business philosophy.
	Innovation 2	Investors in People (IiP) [1] is the national standard which sets out a level of good practice for training and development of people to achieve business goals.
	Innovation 3	New designs are novel forms of layout and structure.
	Innovation 4	Company restructure is the way in which the company of people are to coordinate work to ensure successful delivery of service to the client.
Unsuccessful Innovation	Innovation 5	Seminars are a type of meeting for an exchange ideas on a specific topic. The identified seminars included IT, project briefing, and marketing.
	Innovation 6	New materials are the building components, materials, or new products that the company has not used it before in its building designs.
	Innovation 7	Learndirect project [2] is funded by the UK government. This project aims to help the company of people to develop their IT capability in getting easy access to information about what is available.

[1] Investors in People (IiP) (http://www.investorsinpeople.co.uk/Pages/Home.aspx) is the national Standard that is widely recognised as a respected badge of quality in the UK. The standard was developed in 1990 by a partnership of leading businesses and national organisations in the UK. This standard is designed to help organisations to improve performance and realise objectives through the management and development of their people.

[2] Learndirect (http://www.learndirect.co.uk/) was launched in 2000. It has been developed by University for Industry with a remit from UK government to provide high quality learning opportunities. Its courses cover a range of subjects, including management, IT, skills for life and languages, at all levels. Most of the courses are available online, allowing people to learn wherever they have access to the internet.

3.4.2   Knowledge-based Innovation

The research findings developed a formulation of knowledge-based innovation as 'the effective generation and implementation of a new idea which enhances overall organisational performance, through appropriate exploitative and explorative knowledge capitals which develops and integrates relationship capital (RC), structure capital (SC) and human capital (HC).' This definition forms a knowledge-based innovation concept model, which is built around five variables (Figure 3.2).

1. Interaction environment is that part of the business environment that firms can interact with, and influence. The interaction environment and the firm are separated by an organisational boundary.
2. Relationship capital (RC) is the network resource of a firm. It results from interactions between individual, organisation and external supplier chain partners.

3. Human capital (HC) is defined as the capabilities and motivation of individuals within the firm and external supply chain partners to perform productive, professional work in a wide variety of situations.
4. Structure capital (SC) is made up of systems and processes (e.g. company strategies, computers, tools, work routines and administrative systems) for codifying and storing knowledge from individual, organisation and external supply chain partners.
5. Knowledge capital (KC) is the dynamic synthesis of both the 'context' and 'process' of knowledge creation and conversion within individual-organisational-individual knowledge Ba spirals, and the 'content' of RC, SC and HC.

The concept model highlights the need for senior management to strategically and systemically build, connect and energise appropriate RC, SC and HC to form KC, from which successful organisation and project innovation can flow.

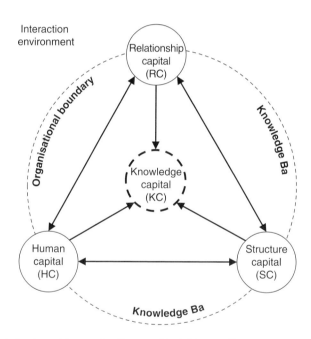

Figure 3.2   Knowledge-based Innovation Concept Model for small construction professional service firms (Lu and Sexton, *Innovation in Small Professional Practices in the Built Environment*, 1st edn, Wiley Blackwell, Oxford, copyright © 2009).

3.4.3   Two Forms of Knowledge-based Innovation

The key proposition is that the market and resource-based view of innovation can be gainfully linked, by extending the argument that there is mutual adjustment between companies 'reacting to' market opportunities and threats and 'proactively' identifying, developing and exploiting resources

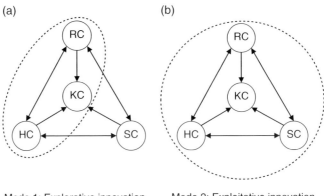

Figure 3.3   Two forms of knowledge-based innovation.

and capabilities to secure a foundation for innovation in these dynamic environments. The principal stimulus for innovation from the market-based view comes from knowledge workers' relationships with their clients, and the principal resource from the resource-based view of innovation is the knowledge worker. The development of optimal dynamic capabilities brings these two resources together to co-produce innovation to create sustainable competitive advantage.

Two types of knowledge-based innovation were identified: mode 1: explorative innovation and mode 2: exploitative innovation (Figure 3.3). New designs (innovation 3) and new materials (innovation 6) were classified as mode 1 explorative innovation. Mission statement (innovation 1), Investors in People (innovation 2), company restructure (innovation 4), seminars (innovation 5), Learndirect project (innovation 7) and interim project review process (innovation 8) were grouped into mode 2 exploitative innovation. It was found that the firm's short-term success was driven to a significant degree by 'explorative' innovation and long-term success by 'exploitative' innovation.

Mode 1 'explorative innovation' is made of three key attributes: RC, HC and KC. It focused on client facing, project-specific problem-solving, often in close collaboration with the client. Explorative innovation activity heavily relied on the capacity, ability and motivation of staff at an 'operational level' to solve client problems to generate short-term competitive advantage (i.e. project specific). Their outcomes focused on effective and efficient delivery of services to satisfy the prevailing fee-earning project needs; but were often not embedded in the organisational SC due to management attention and company resources being constantly focused on other current or near future project-specific demands. The use of new materials (innovation 6) in the case study firm, for example, was explorative in nature, being project-specific and individually driven.

Mode 2 'exploitative innovation' consisted of the four key attributes: RC, SC HC and KC. It focused predominantly on internal organisation and general client development activity (non-project-specific, fee-earning activity). The client had no direct involvement with the innovation activity. Exploitative innovation activity heavily relied on the capacity, ability and motivation of senior management at a 'social' level to improve organisational effectiveness and efficiency to generate sustainable competitive advantage. The securing of Investors in People (innovation 2) accreditation for the case study firm, for instance, initially came from senior management, who were aware that clients are interested in such accreditation. The distinctive feature of mode 2 'exploitative innovation' (compared to mode 1 'explorative innovation') was that new phenomena, systems or structures were securely embedded in the SC of the firm. The motivation for a new mission statement (innovation 1) for the case study firm, for example, came from senior management, who saw it as a way of instilling an integrating vision for its portfolio of activities. Key generic and distinctive variables for successful and unsuccessful explorative innovation and exploitative innovation are summarised in Table 3.2.

Table 3.2 shows that the key distinction between successful and unsuccessful innovation was the 'social' or 'operational' knowledge being applied to a specific innovation. 'Operational knowledge' was generated and created in 'operational level' interactions, where the focus was on solving project-specific issues/problems. These projects were either 'external' fee-earning projects, or 'internal', but specific client-driven projects. 'Social knowledge' was generated through 'social level' interactions, where the focus was on generating non-project-specific innovation that built up general organisational capability, and forged and replenished deeper client relationship over the medium to long term. Moreover, social knowledge was found to have a significant effect on feeding operational knowledge at a specific project level at a future date.

3.4.4   Definition of a Successful Knowledge-based Innovating Firm

The research findings revealed that successful explorative innovation did not necessarily need integrated SC. It was evident in the case study firm that there was too much emphasis on individual learning at the project level (explorative innovation) to the detriment of the organisational level learning (exploitative innovation). This emphasis of explorative KC over exploitative KC was not considered sustainable, as the limitation of SC will become increasingly evident as a significant restraining force for the effective integration of explorative and exploitative KCs. There was thus not an appropriate balance between explorative and exploitative innovation over time. The case study firm needed to create and continue a balance between explorative and exploitative KC barriers, which would allow the flow of KC between operational and social levels. The following definition of a successful

Table 3.2 Key generic and distinctive variables for successful and unsuccessful explorative and exploitative innovations (Lu and Sexton, innovation in small professional practices in the built environment, 1st Edition, Wiley Blackwell, Oxford, UK, copyright © 2009).

Type of Innovation	Variables	Generic Variables	Distinctive Variables	
			Successful Innovation	Unsuccessful Innovation
Explorative Innovation	**Human capital (HC)**	• The capacity, ability and motivation of staff	• Social or operational knowledge being applied to meet the project needs	• Social knowledge not being applied to meet the project needs
	Structure capital (SC)	• Team structure • Teamwork	• Team-based ideas • Teamwork • Senior management involvement through teamwork	• Individual-based ideas • Individual based work • Senior management not involved in teamwork • Limitation of relevant and updated information within the structure
	Relationship capital (RC)	• Operational RC: within internal, client and supplier interactions • Social RC: within internal, client and supplier interactions	• A combination of operational RC and social RC being applied to meet project needs	• Social RC not being applied to meet project needs
	Knowledge capital (KC)	• Social context[1]: company environments (office, meeting room) • Technical context[2]: e-mails, the internet	• A combination of social context and technical context	• Technical context
	Outcome	• Effective and efficient delivery of services to satisfy current and/or future project needs	• Project performance improvement	• Individual performance improvement

Exploitative Innovation	**Human capital (HC)**	• The capacity, ability and motivation of senior management
		• Employee participation
			• Top management support
			• Senior management implementation
			• Some employees buy in
			• Training
				• Top management not supportive
				• Senior management not driving the implementation
				• Lack of time
				• Employees not bought in
				  ○ Inappropriate encouragement
				  ○ Not related to an individual job

	Structure capital (SC)	• The administrative system • Team structure • Computer systems	• Formalised structures and documentation systems • Senior management implementation through the team structure	• No formalised structures and documentation systems • Senior management not driving the implementation through the team structure
	Relationship capital (RC)	• Operational RC: within business adviser, internal, client and supplier interactions • Social RC: within internal interactions	• A combination of operational RC and social RC being applied to meet project needs	• Social RC not being applied to meet project needs
	Knowledge capital (KC)	• Social context: company environments (office and open family culture) • Technical context: e-mails and the internet	• A combination of social context and technical context being applied to meet the project needs	• A combination of social context and technical context not being applied to meet the project needs
	Outcome	• Organisational effectiveness and efficiency	• Organisational performance improvement	• Individual performance improvement

[1] A 'social' context was used to stimulate interaction and collective 'process orientated' knowledge creation and conversion.

[2] A 'technical' context was used to support the search for external knowledge and sharing of 'asset orientated' knowledge.

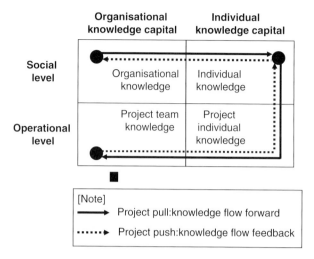

Figure 3.4   Successful innovation driven by operational focus.

knowledge-based innovating firm is therefore offered to accommodate the time dimension: 'the effective generation and implementation of a flow of new ideas, which enhance overall organisational performance over time, through appropriate exploitative and explorative knowledge capital, which develops and integrates relationship capital, structure capital and human capital.'

The time variable brings into focus the development phases of firms as they move from start-up to mature organisations. The focus and process of innovation activity will correspondingly change during the transition. It can be speculated that at the early stages of a firm's development, the emphasis is on explorative innovation, but, as the firm matures, there is an increasing need to explicitly invest in exploitative innovation to produce balanced innovation. This need was certainly evident in this case study firm.

3.4.5   Key Innovation Management Challenges

The discussion above provides insight into the nature and process of innovation for small construction professional service firms. Successful innovation in small construction professional service firms is principally characterised by 'project pull' and 'project push' individual-organisational-individual knowledge spirals, which create dynamic specific-project and/or client-driven KC. Figure 3.4 depicts specific project requirements (either external fee-producing or internal client-driven projects) 'pulling,' combining and converting, 'organisational knowledge' and 'individual knowledge' to form specific 'project individual knowledge'. Project individual knowledge is integrated and leveraged to create 'project team knowledge', which is appropriately applied to create successful innovation. The feedback individual-organisational-individual knowledge spiral is complemented by a feedback or 'project push' knowledge spiral, where new specific 'project

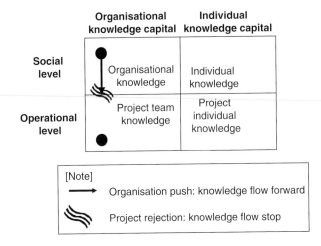

Figure 3.5   Unsuccessful innovation driven by social focus.

team knowledge' feeds back to develop 'project individual knowledge', which, in turn, further enhances 'individual knowledge' and 'organisational knowledge'. The tacit, experiential knowledge accumulation and learning is the basis for subsequent cycles of project-based innovation.

In contrast, unsuccessful innovation in small construction professional service firms is principally characterised by 'organisation push' of disjointed, unfocused 'social' non-project-specific and/or non-client-driven KC being 'rejected' by day-to-day project priorities and activities. Without a specific-project focus, innovation often fails because the individual-organisational-individual knowledge spiral does not happen. Figure 3.5 depicts that there is no specific project needs 'pulling' 'individual knowledge' and 'organisational knowledge' together. Rather, generic 'organisational knowledge' is 'pushed' into a project team setting without appropriate filtering and adaptation to meet specific project needs. Furthermore, the 'organisational knowledge' does not benefit from individual knowledge workers championing their tacit understanding. In combination, the 'organisational knowledge' is 'rejected' by day-to-day projects. Consequently, the feedback loop through individual knowledge, project knowledge and organisational knowledge does not happen.

3.5   Conclusion

This chapter presented a number of key concepts to better understand and, therefore, strategically manage innovation. A number of these ideas were mobilised and illustrated through a case study of innovation in a small construction professional service firm. From the case study, common themes can be identified that are relevant for a broad range of construction company and project settings. First, successful innovation has to balance exploitation and explorative activity. Companies have to exploit existing capabilities to produce innovation aimed at enhancing current performance. At the same time, companies have to invest in explorative innovation for the future. Second, for companies in a project-based industry such as construction, there is always the challenge to develop the necessary organisational SC to capture and diffuse innovation within and between projects. Finally, innovation is not a thing companies should engage with on a periodic basis. The pace and pervasiveness of regulatory, market and technological change is such that innovation must be embedded in all aspects of company thinking and action, all of the time.

References

Ames, B.C. & Hlavacek, J.D. (1988) *Market Driven Management: Prescription for Survival in a Turbulent World.* Irwin, Homewood, IL.

Amit, R. & Schoemaker, P.J.H. (1993) Strategic assets and organizational rent. *Strategic Management Journal,* **14**, 33–46.

Andreu, R. & Ciborra, C. (1996) Organisational learning and core capabilities development: The role of IT. *Journal of Strategic Information Systems,* **5**, 111–127.

Athey, S.E. & Schmutzler, A. (1995) Product and process flexibility in an innovative environment. *RAND Journal of Economics,* **26**(4), 557–574.

Barney, J.B. (1991) Firm resources and sustained competitive advantage. *Journal of Management,* **17**(1), 99–120.

Barrett, P.S. & Sexton, M.G. (2006) Innovation in small, project-based construction firms. *British Journal of Management,* **17**, 331–346.

Chaharbaghi, K. & Lynch, R. (1999) Sustainable competitive advantage: towards a dynamic resource-based strategy. *Management Decision,* **37**(1), 45–50.

Civil Engineering Research Foundation (CERF) (2000) *Guidelines for Moving Innovations into Practice.* Working Draft Guidelines for the CERF International Symposium and Innovative Technology Tradeshow, 14–17 August, CERF, Washington, DC.

Clarke, R. (1993) *Industrial Economics.* Blackwell Publishing, Oxford.

Coombs, R., Saviotti, P. & Walsh, V. (1987) *Economics and Technological Change.* MacMillan, New York.

Department for Communities and Local Government (DCLG) (2008) The Code for Sustainable Homes: setting new standards for sustainability in new homes, HMSO, London. Available from http://www.communities.gov.uk/publications/planning andbuilding/codesustainabilitystandards (accesses 03/11/2010).

Department of Trade and Industry (DTI) (2003) *Innovation Report – Competing in the Global Economy: The Innovation Challenge,* December, HMSO, London. Available from http://www.berr.gov.uk/files/file12093.pdf (accessed 1 October 2010).

Dosi, G. (1984) *Technical Change and Industrial Transformation.* MacMillan, London.

Economic and Social Research Council (ESRC) (2008) *Innovation Research Centre: Call Specification.* ESCR, Swindon.

Freeman, C. (1982) *Economics of Industrial Innovation.* Pinter, London.

Grant, R.M. (1995) *Contemporary Strategy Analysis: Concepts, Techniques, Applications,* 2nd edn. Blackwell Publishing, Oxford.

Grant, R.M. (1997) *Contemporary Strategic Analysis: Concepts, Techniques, Application,* 3rd edn. Blackwell Publishing, Oxford.

Green, S.D., Larsen, G.D. & Kao, C. (2008) Competitive strategy revisited: contested concepts and dynamic capabilities. *Construction Management and Economics*, **26**(1), 63–78.

Hadjimanolis, A. (2000) A resource-based view of innovativeness in small firms. *Technology Analysis and Strategic Management*, **12**(2), 263–281.

Håkanson, H. (1989) *Corporate Technological Behaviour: Corporation and Networks*. Printer, London.

Harty, C., Goodier, C.I., Soetanto, R., Austin, S., Dainty, A.R.J. & Price, A.D.F. (2007) The futures of construction: A critical review of construction future studies. *Construction Management and Economics*, **25**(5), 477–493.

Hewitt-Dundas, N. & Roper, S. (2000) *Strategic Reengineering – Small Firms' Tactics in a Mature Industry*. Northern Ireland Economic Research Centre Working paper Series No. 49, Northern Ireland Economic Research Centre, Belfast.

Imai, K. (1992) The Japanese Pattern of Innovation and Its Evaluation. In: ed. Rosenberg, N., Handau, R. & Mowery, D., *Technology and the Wealth of Nations*. Stanford Press, Stanford.

Itami, H. (1987) *Mobilizing Invisible Assets*. Harvard University Press, Cambridge, MA.

Kotler, P. (1980) *Marketing Management: Analysis Planning and Control*. Prentice-Hall, Englewood Cliffs, NJ.

Lööf, H. & Heshmatic, A. (2000) *Knowledge Capital and Performance Heterogeneity: A Firm Level Innovation Study*. SSE/EFI Working Paper Series in Economics and Finance No. 387, SSE/EFI: Stockholm, Sweden.

Maijoor, S. & van Witteloostuijn, A. (1996) An empirical test of the resource-based theory: Strategic regulation in the Dutch audit industry. *Strategic Management Journal*, **17**, 549–569.

National Audit Office (NAO) (2009) *Innovation Across Central Government*, March, TSO, London. Available from http://www.nao.org.uk/publications/0809/innovation_across_government.aspx (accessed 1 November 2010).

Nelson, P.R. (ed.) (1993) *National Innovation Systems: A Comparative Analysis*. Oxford University Press, Oxford.

Porter, M.E. (1980) *Competitive Strategy: Techniques for Analyzing Industries and Competitors*. The Free Press, New York.

Porter, M.E. (1985) *Competitive Advantage: Creating and Sustaining Superior Performance*. The Free Press, New York.

Rogers, E.M. (1983) *Diffusion of Innovations*, 3rd edn. The Free Press, New York.

Prahalad, C.K. & Hamel, G. (1990) The core competence of the corporation. *Harvard Business Review*, **May–June**, 71–91.

Sexton, M.G., Barrett, P. & Aouad, G. (1999) *Diffusion Mechanisms for Construction Research and Innovation into Small to Medium Sized Construction Firms*. CRISP–99/7, London.

Sexton, M.G. & Barrett, P.S. (2003a) A literature synthesis of innovation in small construction firms: Insights, ambiguities and questions. construction management and economics, **21**, **September**, 613–622.

Sexton, M.G. & Barrett, P.S. (2003b) Appropriate innovation in small construction firms. *Construction Management and Economics*, 21 September, 623–633.

Sexton, M.G. & Barrett, P.S. (2004) The role of technology transfer in innovation within small construction firms. *Engineering, Construction & Architecture Management*, **11**(5), 342–348.

Slater, S. & Narver, J.C. (1994) Does competitive environment moderate the market orientation-performance relationship? *Journal of Marketing*, **58**, 46–55.

Snow, C. & Hrebiniak, L.G. (1980) Strategy, distinct competence, and organizational performance. *Administrative Science Quarterly*, **25**, 317–336.

Susman, G.I. (1983) Action research: A sociotechnical systems perspective. In: ed. Morgan, G., *Beyond Method: Strategies for Social Research*. Sage, London, 95–113.

Teece, D.J., Pisano, G. & Shuen, A. (1997) Dynamic Capabilities and Strategic Management. In: ed. Foss, N.J., *Resources, Firms and Strategies*. Oxford University Press, New York, 268–287.

Van de Ven A.H., Polley, D., Garud, R. & Venkataraman, S. (1999) *The Innovation Journey*. Oxford University Press, New York.

von Hippel, E. (1988) *The Sources of Innovation*. Oxford University Press, New York.

Wernerfelt, B. (1984) A resource-based view of the firm. *Strategic Management Journal*, **5**, 171–180.

Wheatley, M.J. (1992) *Leadership and the New Science*. Berrett-Koehler, San Francisco.

Culture and Innovation

Anita Liu and Richard Fellows

Why are there construction products and processes? The essential answer is that people have needs, wants and desires for buildings and infrastructure, which they strive to satisfy. In most of today's world, that satisfaction is achieved through exercising demand in a market economic system – the ability and preparedness to pay the (financial) price determines 'who gets what'. What is indisputable for construction is the centrality of 'the user' as the primary function of 'the client' stakeholder category. Complimentarily, the supply process depends on people as the 'active factor' – by making decisions and affecting the supply processes through managing and producing, both directly (e.g. an architect producing a design drawing; or a bricklayer laying bricks) and indirectly (e.g. an engineer using structural calculation software; or hiring a carpenter). However, cultural labelling has become widely used to caricature the construction industry – usually to emphasise undesirable aspects. A *macho* culture (Fielden *et al.*, 2001) conjures associations with physical strength, danger and risk taking, rather uncaring and brutal, etc; a *claims* culture (Rook *et al.*, 2003), which suggests strong pursuit of perceived rights, heightened opportunistic behaviour and self-centred individualism. Alongside such overt cultural labelling is the common criticism of fragmentation (Construction Task Force, 1998) – the separation of functions and activities, which otherwise may be described as 'division of labour' or specialisation. Of course, the key element rendering fragmentation an important detriment is the lack of the requisite integration and coordination of the components (Lawrence and Lorsch, 1967).

In construction, it is only over recent years that culture has become acknowledged as an important factor that affects performance of both products and processes; indeed, culture is a primary determinant of the

Construction Innovation and Process Improvement, First Edition.
Edited by Akintola Akintoye, Jack S. Goulding and Girma Zawdie.
© 2012 Blackwell Publishing Ltd. Published 2012 by Blackwell Publishing Ltd.

world's built environment, and how this is perceived and used. Moreover, culture underpins peoples' actions towards sustainability and so today's culture has major implications for innovation in the future.

This chapter gives an overview of culture, its relevance to construction, and examines the relationship of culture and innovation.

4.2   Culture and Construction

As construction projects and processes become ever more complicated, so the essentials of specialisation and integration grow in importance. The stakeholder perspective and agency theory have evolved to aid understanding of human behaviour in such circumstances, as multitudes of internal and external forces act on individuals. The temporary multi-organisation (TMO) (Cherns and Bryant, 1984) constitutes a useful analytic lens and incorporates the shifting, multi-goal coalition, which operates through fluid power structuring for realisation of construction projects. In this respect, construction projects, as for any good which is produced (components are produced and then assembled), these are subject to evaluative judgements of process (project management performance) and product (project performance) – not just by participants in the processes, but by a wider array of stakeholders, all of whom are likely to have differing criteria and criteria weightings. Furthermore, the realisation processes (depicted as a generic schematic in Figure 4.1) are of long duration, in which many inter-related decisions are made, which can impact on the project. In addition, performance forecasts are often required, and it is these that form the basis for many evaluations – notably by consultants advising commissioning clients. Commonly, such forecasts are 'best-guess', conglomerate estimates of input variables but treated as certain estimates with results presented in single-figure, deterministic terms (Reugg and Marshall, 1990). Given the nature of forecasts, it is only to be expected that they are unlikely to be close to the realised outcomes (Flyvbjerg, 2007) and so give rise to dissonance and dissatisfaction.

Most participants/stakeholders are business organisations for whom performance metrics can be generally categorised as financial, technical and relational. In the current world, the technical and relational criteria are usually significantly subservient to financial performance as 'Essentially, business is about appropriating value for oneself ... only by having the ability to appropriate value from relationships with others can business be sustained... [there must be conflicts of interest between vertical participants in supply chains, just as there are between those competing horizontally]... In Western culture most suppliers are basically opportunistic' (Cox, 1999). Thus, a 'business case' usually relates to a financial evaluation – the case being 'made' if the outcome indicates sufficient profit(ability).

The scenario emerges that construction projects are realised and then operate in a competitive context in which actors tend to be self-oriented and

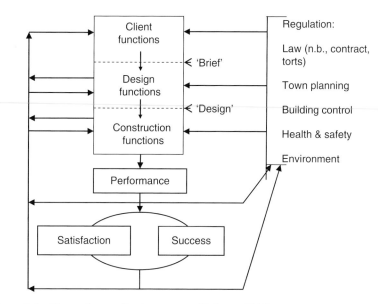

Figure 4.1   The project realisation process (Fellows, 2009).

procurement arrangements result in projects being zero-sum 'games' in which one participant gains only at the expense of other(s). A systems approach, complimented by complexity theory, notes that a system (project, firm, industry) must respond appropriately to its environment to ensure survival, leading to the conclusion that the culture of construction is shaped by the environment (society, etc) within which it operates. The traditional analytic approach of Newtonian reductionism to understand the whole by examining the parts individually and then aggregating the results of the component analyses fails to allow for the synergistic effects of inter-relationships of the parts, and so a more holistic analytic approach is called for. In this respect, cultural investigations tend to adopt a high degree of aggregation and, essentially, give express acknowledgement to embeddedness – of behaviour in project atmosphere, in organisational climate, in organisational culture, and in national culture.

It should be noted that:

- Performance leads to satisfaction of participants and hence (perspectives of) project success;
- Performance-Satisfaction-Success also produces feed-forward in the 'cycling' of project data and information, to aid realisations of future projects through participants' perception-memory-recall filtering ('experiences'); and
- A similar model applies to projects in use (beneficial occupation), but with 'Facilities Management' and 'Maintenance and Adaptation' replacing 'Design' and 'Construction' as major functionary groups.

4.2.1   Culture

Culture is a construct that concerns people in groups, notably national, organisational and social, whereas personality relates to individuals. Kroeber and Kluckhohn (1952) carried out an extensive study of the literature and found 164 definitions of culture. They define culture as '...patterns, explicit and implicit of and for human behaviour acquired and transmitted by symbols, constituting the distinctive achievements of human groups, including their embodiment in artefacts; the essential core of culture consists of traditional (i.e. historically derived and selected) ideas and, especially, their attached values; culture systems may, on the one hand, be considered as products of action, on the other as conditioning elements of future action.'

Hofstede (1994a) defines culture as '...the collective programming of the mind which distinguishes one category of people from another.' Thus, culture is learned. Much may be learned behaviourally through perceiving, copying and responding to the behaviour of others and so much culture becomes tacit knowledge as in 'intuitively' knowing how to behave, how to interpret a message. Thus, Schein (1990) identifies the essence of culture to be a pattern of basic assumptions – constituting communal values that are taken for granted by the persons comprising the cultural group. An important aspect of the development of cultures is the formation of norms and limits of acceptable behaviour relating to critical incidents – often, as lessons learned from significant mistakes – which are communicated through stories passed on between members of the community; cultures also develop through identification with leaders and what they scrutinise, measure and control (hence, what leaders regard as important).

Models often depict culture as comprising concentric layers – physiological instincts and beliefs at the core (survival imperatives, religion, morality, etc) values as the intermediate layer (the hierarchical ordering of beliefs, perhaps with trade-offs) and cultural manifestations at the outer layer (as in behaviour, language, symbols, heroes, practices, artefacts). In this respect, Schein (2004) considers levels of culture to be artefacts, espoused beliefs and values, and underlying assumptions (as the deepest level). However, an holistic representation of these inter-relationships can be seen in Figure 4.2.

Culture may also be subject to group-based analysis and categorisation, which includes the categories of national culture, organisational culture, organisational climate, project atmosphere and behaviour of persons (Figure 4.3). The boundaries of the categories are fuzzy and permeable and so the categories 'bleed' into each other. Individuals may be of a certain nationality (e.g. Brazilian), work in a particular organisation (e.g. Skanska), belong to a 'social interest group' (e.g. Greenpeace) and exhibit certain behaviour (e.g. organisational citizenship behaviour). Furthermore, environmental and situational variables can affect behaviour which therefore can be regarded as contingent. An important consequence is confusion of the categories when 'managers' wish to effect change – in particular, what has been changed and the strength and permanence of any change – culture cannot be changed by use of a '40-hour workshop' (although behaviour modifications may result).

Behaviour, Heroes, Symbols, Artifacts, Language, etc.

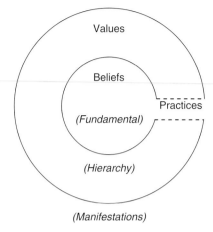

Figure 4.2   Layers of vulture (Hofstede (1980; 2001). Reproduced by permission of Geert Hofstede BV.

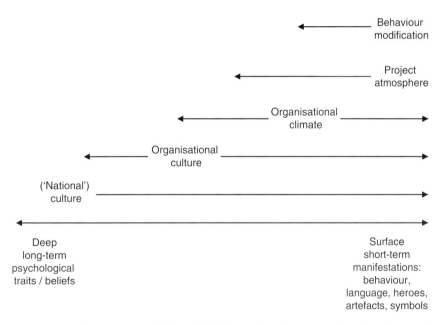

Figure 4.3   Culture spectrum (Fellows, 2006). Reproduced by permission of the Chartered Institute of Building.

Eldridge and Crombie (1974) voiced three concerns for examining culture – depth (values and commitment), breadth (coordination of the persons) and progression (coordination, development/change, over time). Often, cultures are caricatured by a small number of general perspectives, which are used to serve as simple guides to how people in those cultures behave and how visitors are expected to behave (Trompenaars and

Hampden-Turner, 1997). However, the construct, culture, contains many latent variables. To understand a culture, it is important to first determine the boundary of the culture (national, organisational, etc) and then what are the important manifest variables to examine through appropriate methods of collecting and analysing data. Such data and analyses are used to help gain insights into the deeper levels of cultural constructs – the values and beliefs (or other constructs depending on the model of culture adopted) of the members of the group(s) under study. Measurements are often comparative and so yield relative, rather than absolute, results (Hofstede, 1980).

4.2.2   National Culture

Hofstede (1980) determined the following dimensions from studying national cultures:

- *Power Distance*: 'the extent to which the less powerful members of institutions and organisations within a country expect and accept that power is distributed unequally.' (Hofstede, 1994b: 28);
- *Individualism/Collectivism*: 'individualism pertains to societies in which the ties between individuals are loose: everyone is expected to look after himself or herself and his or her immediate family. Collectivism as its opposite pertains to societies in which people from birth onwards are integrated into strong, cohesive in-groups, which throughout people's lifetimes continue to protect them in exchange for unquestioning loyalty.' (ibid: 51);
- Masculinity/Femininity: 'masculinity pertains to societies in which gender roles are clearly distinct (i.e. men are supposed to be assertive, tough, and focused on material success whereas women are supposed to be more modest, tender, and concerned with the quality of life); femininity pertains to those societies in which social gender roles overlap (i.e. both men and women are supposed to be modest, tender, and concerned with the quality of life).' (ibid: 82–83);
- *Uncertainty Avoidance*: 'the extent to which the members of a culture feel threatened by uncertain or unknown situations.' (ibid: 113); and
- *Long-Termism/Short-Termism*: 'the fostering of virtues orientated towards future rewards, in particular perseverance and thrift.' (ibid: 261)/*Short-Termism* – 'the fostering of virtues related to the past and present, in particular respect for tradition, preservation of "face", and fulfilling social obligations.' (ibid: 262–263) has been added (Hofstede, 1994b) following studies in Asia using a Chinese values survey (CVS) instrument, which detected important impacts of "Confucian Dynamism".

From this, Triandis and Gelfand (1998: 119) argue that both individualism and collectivism have horizontal (emphasising equality) and vertical (emphasising hierarchy) components – '…the most important attributes that distinguish among different kinds of individualism and collectivism are the relative emphases on horizontal and vertical social relationships.' Horizontal

Table 4.1 High context / low context (high content) cultures (Hall and Hall, 1990). Reproduced by permission of Nicholas Brealey Publishing.

Factor	High-context culture	Low-context culture
Overtness of messages	Many covert and implicit messages, with use of metaphor and reading between the lines.	Many overt and explicit messages that are simple and clear.
Locus of control and attribution for failure	Inner locus of control and personal acceptance for failure	Outer locus of control and blame of others for failure
Use of non-verbal communication	Much nonverbal communication	More focus on verbal communication than body language
Expression of reaction	Reserved, inward reactions	Visible, external, outward reaction
Cohesion and separation of groups	Strong distinction between in-group and out-group. Strong sense of family.	Flexible and open grouping patterns, changing as needed
People bonds	Strong people bonds with affiliation to family and community	Fragile bonds between people with little sense of loyalty.
Level of commitment to relationships	High commitment to long-term relationships. Relationship more important than task.	Low commitment to relationship. Task more important than relationships.
Flexibility of time	Time is open and flexible. Process is more important than product	Time is highly organized. Product is more important than process

individualists (HI) desire to be unique and distinct from groups; vertical individualists (VI) want to be distinguished, acquire status and compete with others. Horizontal collectivists (HC) emphasise common goals, interdependence and sociability; vertical collectivists (VC) emphasise the integrity of the in-group and are prepared to sacrifice their personal goals for the sake of in-group goals. Generally, collectivists belong to few in-groups, are highly loyal to in-group members, but tend to shun out-groups and their members (Gomez *et al.*, 2000); individualists can therefore be said to be loosely tied to many in-groups, and their memberships may be transitory.

Hofstede (1983) highlighted the correlation between wealth and individualism in various countries, noting '...Collectivist countries always show large Power Distances but Individualist countries do not always show small Power Distance.' Collectivist societies therefore tend to have a fairly rigid, often overt, social structure. In this respect, Hall and Hall (1990) employ high-context/low-context as the first dimension along which cultures may be analysed (Table 4.1).

In a high-context culture, many contextual elements can help people understand the meanings of messages, behaviours and other manifestations of the culture. In high context cultures, a lot is taken for granted, and so meaning

Table 4.2 Perspectives regarding time (Hall and Hall, 1990). Reproduced by permission of Nicholas Brealey Publishing.

Factor	Monochronic action	Polychronic action
Actions	Do one thing at a time	Do many things at once
Focus	Concentrate on the job at hand	Are easily distracted
Attention to time	Think about when things must be achieved	Think about what will be achieved
Priority	Put the job first	Put relationships first
Respect for property	Seldom borrow or lend things	Borrow and lend things often and easily
Timeliness	Emphasize promptness	Base promptness relationship factors

must be derived from the content of a message itself (which may appear vague, especially to a person from a high-content culture) and its interpretation in the prevailing circumstances/situation. Likewise, behaviour often tends to be indirect, and so people appear to be 'reserved'. Thus, a lot of intuition is necessary, together with a thorough understanding of both the language and the society. Conversely, in a low-context (high-content) culture, little is taken for granted. In low-context cultures, more content is needed but the resultant message is precise and explicit in its meaning, and so there is a low chance of misunderstanding; thus, people can be quite confident to interpret messages at 'face value', although such direct and obvious expression (and behaviour) can be offensive to people from high-context cultures.

Hall and Hall's second dimension of culture concerns the level of possessiveness of people over territory and objects. High territoriality includes clear and firm demarcation of (often, physical) boundaries and high needs for security to protect ownership (rights). People who are highly territorial tend to be low-context (high-content) and are likely to desire large amounts of physical space. People with low territoriality regard space, possessions and boundaries as low in importance, and so movements are easier. Low territoriality people tend to be high-context. Hall and Hall's third dimension of culture concerns how people perceive time along a continuum of monochronic/polychronic (Table 4.2). Persons who perceive time to be monochronic do one thing at a time, usually in a predetermined sequence. Such people tend to be low-context (high-content). In polychronic perceptions of time, human interaction is valued above time and material goods. That generates a low concern for 'getting things done' – things get done, 'in their own time'. Polychronic people tend to be high-context.

Trompenaars and Hampden-Turner (1997) investigated cultures' perspectives on time in sequential/synchronic terms (corresponding to mono-chronic/polychronic). They extended their study to examine how people of different cultures regard periods of time – past, present and future – to determine the relative importance of each period and how those periods relate to each other (overlaps or distances of separation). Their investigation yielded results that seem to correlate with Hofstede's fifth dimension of national cultures – long termism/short termism. In this respect, Trompenaars and Hampden-Turner (1997) suggested five value-oriented dimensions of culture which, they state, '...greatly influence our ways of doing business and managing as well as our responses in the face of moral dilemmas.' The dimensions are:

- *Universalism*: Particularism (rules-relationships);
- *Collectivism*: Individualism (group–individual);
- *Neutral*: Emotional (feelings expressed);
- *Diffuse*: Specific (degree of involvement); and
- *Achievement*: Ascription (method of giving status).

Although the dimensions relate to national cultures, they are of notable import for business activities, especially negotiations, and so not only blur the categorisation of national and organisational culture, but also reinforce the perspective of the embeddedness of organisational culture in national culture.

4.2.3   Organisational Culture

Denison (1996: 53) asserted that organisational culture is 'the deep structure of organisations, which is rooted in the values, beliefs and assumptions held by organisational members.' James *et al.* (2007: 21) describe organisational culture as 'the normative beliefs (i.e. system values) and shared behavioural expectations (i.e. system norms) in an organisation. However, Hofstede (1994b) defines organisational culture as '...the collective programming of the mind which distinguishes the members of one organisation from another;' presenting the following six dimensions for analysing organisational cultures:

1. *Process/Results Orientation*: technical and bureaucratic routines can be diverse – outcomes tend to be homogeneous;
2. *Job/Employee Orientation*: derives from the societal culture in which the organisation is embedded, as well as the influences of founders and managers;
3. *Professional/Parochial*: educated personnel identify with their profession(s) – people also identify with their employing organisation;
4. *Open/Closed System*: ease of admitting new people and of innovating; styles of internal and external communications;
5. *Tight/Loose Control*: degrees of formality, punctuality, etc, may depend on technology and rate of change; and
6. *Pragmatic/Normative*: how to relate to the environment, i.e. customers; pragmatism encourages flexibility.

Organisational cultures can therefore be said to be initiated by the founders of the organisation, given the culture of the society in which the organisation is embedded. Subsequently, the culture of an organisation is amended by others who have had major impact on the organisation's development (owners, major managers). Such people shape the values and behaviour of members of the organisation to develop the identity of the organisation through influence over the employment of staff and organisational practices. Organisational cultures (and climates) are self-perpetuating – persons who 'fit' are hired and they 'fit' because they are hired; errors of 'fit' are subject to resignation or dismissal. Organisational cultures develop to maintain effective and efficient working relationships amongst organisational members and stakeholders (both temporary and permanent). Thus, when organisations are brought into close operating contact, as with construction projects, their individual cultures are combined to produce the 'project atmosphere'. However, findings from amalgamations and takeovers are germane – that 'Usually the corporate culture of the most powerful or economically successful company dominates' (Furnham, 1997).

Both Denison (1997, 2009) and Cameron and Quinn (1999) employ competing values frameworks to model organisational cultures. In Denison's (1997, 2009) model, flexibility and stability are juxtaposed along one dimension with organisational focus – internal juxtaposed to external – on the other dimension. The resultant quadrants comprise mission, consistency, involvement and adaptability, each comprising three constituents (Figure 4.4). Denison employs a model comprising the same dimensions and quadrants but with different constituents to analyse leadership, thereby supporting a close relationship between leadership and organisational culture.

In Cameron and Quinn's (1999) model, 'flexibility and discretion' is juxtaposed to 'stability and control' on one dimension with 'internal

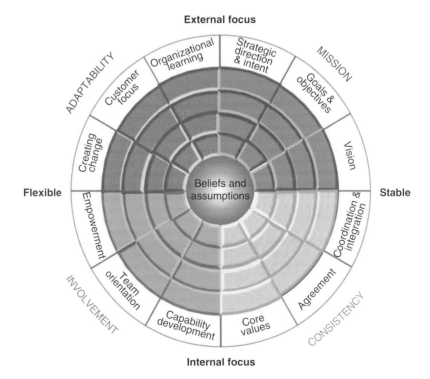

Figure 4.4   Denison's culture model. (www.denisonconsulting.com/advantage/research Model/model.aspx). Reproduced by permission of Denison Consulting.

focus and integration' and 'external focus and differentiation' juxtaposed on the other. The resultant model comprises four quadrants, each denoting a type of organisational culture – Clan, Adhocracy, Market, Hierarchy (Figure 4.5).

- *Clan:* 'Some basic assumptions in a clan culture are that the environment can be best managed through teamwork and employee development, customers are best thought of as partners, the organisation is in the business of developing a humane work environment, and the major task of management is to empower employees and facilitate their participation, commitment and loyalty' (ibid: 37).

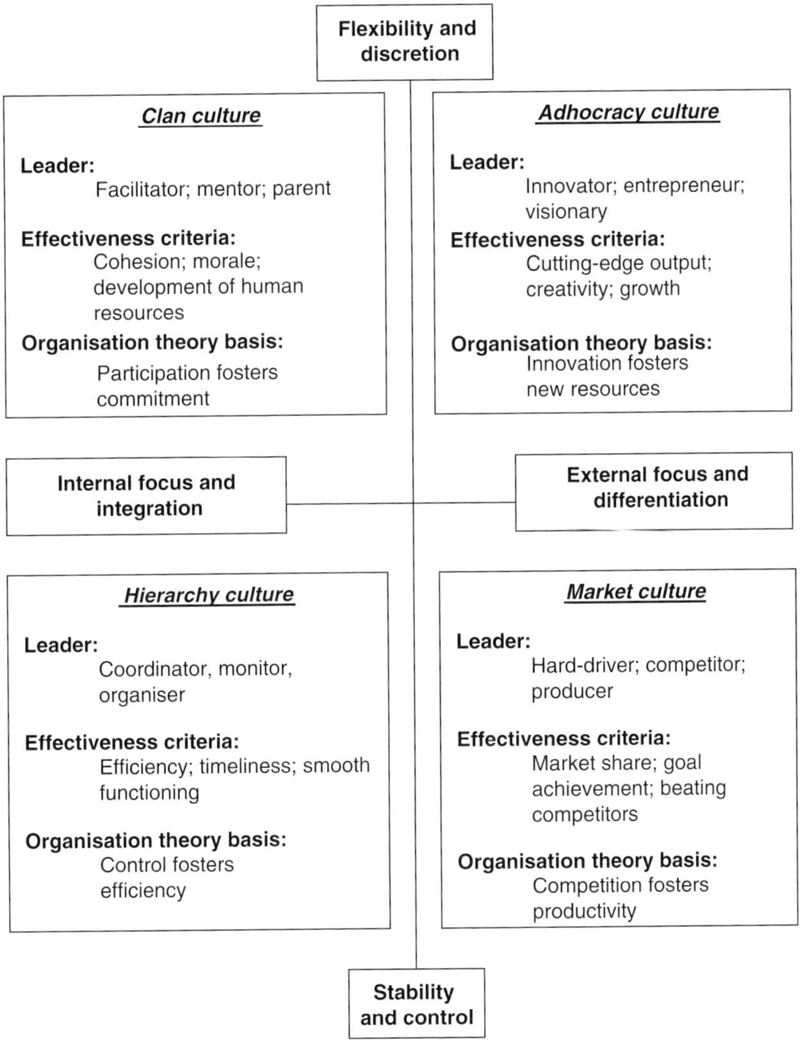

Figure 4.5 Competing values and organisational cultures model (Fellows, 2006: 57, following Cameron and Quinn, 1999). Reproduced by permission of the Chartered Institute of Building.

- *Adhocracy:* 'A major goal of an adhocracy is to foster adaptability, flexibility and creativity where uncertainty, ambiguity and/or information-overload are typical. Effective leadership is visionary, innovative and risk-orientated. The emphasis is on being at the leading edge of new knowledge, products and/or services. Readiness for change and meeting new challenges are important' (ibid: 38–39).
- *Market:* 'The major focus of markets is to conduct transactions with other constituencies to create competitive advantage. Profitability, bottom line results, strength in market niches, stretch targets and secure customer bases are primary objectives for the organisation. Not surprisingly, the core values that dominate market type organisations are competitiveness and productivity' (ibid: 35).
- *Hierarchy:* 'The organisational culture compatible with this form is characterised by a formalised and structured place to work. Procedures govern what people do. Effective leaders are good coordinators and organisers. Maintaining a smooth-running organisation is important. The long-term concerns of the organisation are stability, predictability and efficiency. Formal rules and policy hold the organisation together' (ibid: 34).

Schein (1984) determines two main types of organisational culture, which bear strong resemblance to the organic-mechanistic organisational typology of Burns and Stalker (1961): 'free flowing' – an unbounded, egalitarian organisation with minimal formal structure, to encourage debate and internal competition; and 'structured' – a bounded, rigid organisation with clear rules, procedures and requirements. Furthermore, Handy (1985) suggests a typology of organisational cultures. Power culture is depicted as a web with the major power at the centre, emphasising control over subordinates and external agents (suppliers, etc, and nature). A role culture focuses on functions/professions that provide support to top management; emphasis is therefore on rules, hierarchy and status through legality, legitimacy and responsibility. In a task culture, jobs or projects are the major foci; an organisation is regarded as a net (as in a matrix organisation), and structures, functions and activities are evaluated in respect of their contributions to achievement of the organisation's objectives. In a person culture, people interact and cluster freely and emphasise meeting the needs of members of the organisation through consensus. Handy considers that the main factors influencing which organisational culture develops are goals and objectives, history and ownership, size, technology, environment and people. Williams *et al.*, (1989) advance categories of 'Power', 'Role', 'Task' and 'People', which correspond to Handy's (1985) typology.

4.2.4   Organisational Climate

Organisational culture is an important determinant of climate (Sarros *et al.*, 2008). Moran and Volkwein (1992) argue that climate reflects the shared knowledge and meanings embodied in an organisation's culture. Hence, organisational climate is regarded by Sarros *et al.* (2008) as the expression of underlying cultural practices that arise in response to contingencies in the organisation's internal and external environment. While organisational culture focuses on the shared behavioural expectations and normative beliefs in work units, climate describes the way individuals perceive the personal impact of their work environment on themselves (Glisson and James, 2002). 'Organisational Climate is a relatively enduring quality of the internal environment of an organisation that:

a) is experienced by its members;
b) influences their behaviour; and
c) can be described in terms of the values of a particular set of characteristics (or attributes) of the organisation.' Tagiuri and Litwin (1968: 27).

Hence, climate distinguishes an organisation from other similar organisations. Organisational climate both reflects and shapes the working experiences shared by members of the organisation and so indicates their perceptions of autonomy, trust, cohesion, fairness, recognition, support and innovation through shared knowledge and meanings. Organisational climate is an important contributor to relative homogeneity amongst members through recruitment and retention.

Organisational climate is also grounded in the practices of an organisation as experienced by its members, and so is a less deep-seated construct than culture. James *et al.* (2007) differentiate organisational climate from psychological climate - i.e. the former is an aggregation of individual perceptions of the work environment, the latter refers to the perceptions individuals have of those workplaces as they reflect personal values and psychological desires. Organisational climate can therefore measure the organisation's openness to change and its provision of resources to become innovative (Sarros *et al.*, 2008).

4.3   Culture and Innovation

The understanding of innovation has changed from scientific research and development (R&D) to include changes to services, ways of working and delivery, customer insight, etc (Roper *et al.*, 2009). The National Endowment for Science, Technology and the Arts (NESTA) (2008) argues that total innovation is integrating innovation in new technologies, products and processes with innovation in new services, business models and organisational forms. Thus, to stimulate wider innovative activities, R&D must be complemented 'with a focus on stronger and broader skills for business, using taxation, regulation and procurement to boost innovation...' (Anon. 2008: 34).

In the UK, it is acknowledged in the report on *Total Innovation* (www.nesta.org.uk) by NESTA (2008) that there are six major high technology areas with world-leading strengths and they represent the majority of R&D expenditure in the UK, comprising aerospace, pharmaceuticals, automotive telecommunications, software and electronics. However, the construction sector remains a low-level innovation sector in the UK, as measured by the Innovation Index (Roper *et al.*, 2009). UK companies often focus on a linear model of innovation enforced by the traditional innovation indicators (e.g. R&D expenditure and patent production), which obscures the potentially significant forms of 'hidden innovation' – as mostly seen in high-technology sectors. Roper *et al.* (2009) also find that there are significant levels of hidden innovation in several sectors, where levels of traditional R&D investment are low. The four types of hidden innovation (NESTA, 2008) include:

1. effective use of new materials, which involve learning on the job;
2. new organisational structures and business models;
3. novel combination of existing technologies and processes; and
4. innovative solutions to small-scale problems and challenges, which happen 'under the radar'.

Hidden innovations suggest that the traditional indicators of R&D investment and the number of patents are not sufficient to provide the full picture of innovation; whereas the focus on HC (e.g. *learning*, motivation for creativity, etc) and organisational issues (e.g. culture, climate, reward systems, leadership, empowerment, etc), which are discussed in this chapter, are essential to nurture the firms towards innovation for sustaining development.

However, construction and project management research relating to culture studies are increasing in diversity. For instance, some of the covered topics include knowledge transfer, comparative culture between sectors, role of cultural boundary spanners, organisational culture and project culture (Brochner *et al.*, 2002; Chan and Tse, 2003; Liu *et al.*, 2006; Zhang and Liu, 2006; Ajmal and Koskinen, 2008; Ankrah and Proverbs, 2009; Chen, *et al.*, 2009; Ofori and Toor, 2009; Di Marco *et al.*, 2010; Cheung *et al.*, 2011). However, to appreciate the role of culture in construction and innovation, one has to examine the concepts of various 'labels' of culture - for example, innovation culture, learning culture, trip-helix culture and project atmosphere.

4.3.1   Role of Culture in Construction

Culture represents a heritage of who we are, what we know and what we believe. Innovation systems can be seen as development of social systems in achievement of their strategy, i.e. 'Processing innovation implies rules of co-operation and social behaviour of a specific cultural setting' (Pohlmann *et al.*, 2005: 3). Thus, the structure and development of innovation depends on the cultural software of a thematic and spatial concept and the political framework, so that innovation is determined by culture in the following ways (Pohlmann *et al.*, 2005):

▪ affective frames of identity and difference;
▪ cognitive frames of knowledge; and
▪ normative sets of values, norms and beliefs.

While social rules of interplay are observed and analysed as development of co-operation structures (or social systems), it is noted that the cultural software defining the innovation system varies on different levels. Palazzo (2005) argues that the organisation of transnational movements of innovation are anchored on the first level, where innovation models appear as highly visible and thematic (e.g. 'Balanced Score Card', 'business reengineering', 'CIM'. However, success depends on the second level of organisation, where expressions of new innovation models are integrated into existing frames of partly thematic cognitive maps. This leaves the underground third level of culture, where communication, attitudes and values come into play in the tacit, invisible level of stabilising innovation models, untouched.

Taking the view that the innovation process encompasses the cultural intertwining of thematic, human resources and spatial/time concepts in social systems demonstrates that no single actor can be in control of all the rules and frames that dictate the development of innovation. Hence, Gebhardt (2005) argues that innovation management is more about managerial belief systems, which are acknowledged by other actors, for example, belief systems of venture capitalists differ from the inventors, hence taking this perspective, the role of culture in innovation is important.

4.3.2   Innovation Culture

Innovation culture has become the focus of recent research on innovation and technology development. It is well acknowledged that innovation is one of the drivers of a knowledge society, and researchers (Hadjimanolis, 2010) have advocated that creation, innovation and diffusion processes are embedded in culture that vary between different countries and organisations. Culture has been studied at different levels, for example, individual, team, organisational and national, and all these levels are inter-related as national culture influences individuals, thence the organisations. Hence, innovation culture of an organisation or society, can be considered as a subset of, and embedded in, the national culture.

Innovation culture reflects the societal dynamics of self- and collective aspirations. Simply, there is an innovation culture when the cultural characteristics of an organisation encourage innovation, for example, leadership that nurtures creativity (Amabile *et al.*, 2004), communication and feedback, reward systems and innovation champions. 'Innovative culture is a way of thinking and behaving that creates, develops and establishes values and attitudes within a firm', which involves changes that 'conflict with conventional employee behaviour', but may 'in turn create highly effective cultural and managerial transformations regarding innovativeness' (Çakar and Ertürk, 2010: 345). Hadjimanolis (2010) suggests that the dimensions of innovation culture are creativity, attitude to change, attitude to risk, attitude to technology, attitude to learning, institutions, acceptance of failure and mistakes, communication and tolerance for dissenters; the continuum of the 'innovation-averse culture' and the 'ideal innovation culture' is depicted by the magnitude of each dimension. For instance, an innovation-averse culture is low on creativity, but an ideal innovation culture is high on creativity. However, innovation culture is not static and can be 'changed'; in fact, there is a constant evolutionary change in the meaning of values and norms (Inglehart and Baker, 2000) in any culture. Therefore, it is possible to change from an innovation-averse culture to a more innovation-conducive culture through encouragement of creativity, learning, communication, etc. However, changing an organisation's culture can be difficult (Johnson, 2001), if not impossible.

The tendency to focus on short-term return on investment and risk aversion can be potentially devastating to the promotion of an innovation culture and the long-term survival of the business. Little (2008, cited in Anon., 2008) advocates that during a downturn, the key to gaining value is to recognise that innovation is a significant and powerful tool for bottom line optimisation, and offers five strategies to foster innovation culture:

1. identify and understand how the organisation's unwritten rules relate to innovation behaviour;
2. proactively address risk aversion and short-termism to ensure high-return projects are maintained, even in the face of risks;
3. make sure management behaviour aligns with public declarations on the innovation agenda;
4. recognise opportunity costs and use resources creatively to pursue some riskier projects; and
5. explore open innovation partnership – past collaborators and new potential partners.

Furthermore, Hadjimanolis (2010: 97) advocated that 'innovation as an interactive process usually takes place within work groups with participation of all employees, who should receive motivation and training for creativity and innovation in a supportive work environment.' However, it should be noted that there is no single type of culture that is ideal for the development and adoption of innovation – as culture can stimulate as well as hinder innovation (van Duivenboden and Thaens, 2008). Therefore, it is crucial to 'match' the culture to the context – the organisation structure-culture fit, thereby securing continuous improvement through sustaining creativity. Finally, it is also important to acknowledge that some researchers include *learning* as one of the dimensions of innovation culture (Dobni, 2008), but others treat learning culture as a different construct. The next section examines learning culture separately.

4.3.3   Learning Culture

The role of learning is widely recognised as pivotal in innovation and the concept of organisational learning has been widely discussed (Argyris, 1992). Increasing attention has also been paid to learning organisations in human resource management and organisational development literature. Features of the learning organisation enable organisations to develop more flexible and adaptable systems that improve long-term performance (Senge, 1992; Slater and Narver, 1995). According to Kaiser and Holton (1998), the literature on learning organisations and innovation focuses on the facilitating role of the same organisational variables that enhance the adaptability and flexibility of organisations to improve long-term performance – closely related to, and influenced by, variables including culture, climate, leadership, organisational structures, systems and environment. An organisational learning culture is therefore one that values the creation, sharing and application of knowledge, and can influence specific manifestations of psychological climate in the form of a learning transfer climate (Bates and Khasawneh, 2005).

Mai (1996) suggests that learning organisations are differentiated by the degree to which they learn better or faster and they are revealed through outcomes such as creativity and innovation, which are supported by psychological climates and human resource systems that support learning. Bates and Khasawneh (2005) examined the relationship between organisational learning culture and innovation and found that the learning transfer climate is a mediating variable. More specifically, they concluded that innovation requires not only an organisational culture that allows learning and the generation of creative ideas to take place, but also a psychological climate that fosters an individual's ability to share and apply that learning. Therefore, it is suggested that organisations, which intend to pursue innovation, should analyse their culture and climate to determine what changes may be needed to facilitate learning and the transfer of creative problem solving.

4.3.4   Triple Helix Culture

The triple helix model, emphasising external linkages and collaboration, is grounded in the concept of national and regional systems of innovation, where learning (and knowledge transfer) occurs and the availability and access to knowledge is the most critical resource. It is also claimed that the triple helix model is an extension of the study of technological change advocated by Rothwell (1992), which emphasises the influence of networks, collaboration and alliances in a variety of inter-organisational relationships (Leydesdorff and Etzkwitz, 1998; Etzkowitz and Carvalho de Mello, 2004). Central to the triple helix model is the transfer and use of knowledge through networking, both intra- and inter-organisational, with the view to innovate; hence, learning and the learning transfer climate are important elements. Thus, according to Etzkowitz and Carvalho de Mello (2004), the triple helix denotes the university-industry-government in a relatively equal, yet interdependent, institutional relationship in which their roles overlap and where a triple helix culture of innovation in economic and social development is possible. Knowledge transfer is no longer considered as a linear process, but a complex set of organisational ties with overlapping boundaries (Leydesdorff and Etzkowitz, 1998), for example, universities take on entrepreneurial tasks, while firms develop an academic dimension in sharing knowledge. Saad and Zawdie (2005: 97) refer to the triple helix culture in their analysis of innovation in Algeria and assert that the 'major challenge for the development of the triple helix culture in developing countries is one of drawing a balance between the skill development objective and the knowledge accumulation objective of policy.' Learning based on actions, therefore, must be supplemented by cognition, that is, training initiatives should not be driven by short-term objectives alone. The constraints faced by developing countries concern the issues of power, bureaucracy, rigid boundaries, hierarchy, adversarial relationships and lack of trust (Saad and Zawdie, 2005). However, Pohlmann *et al.* (2005: 3) observed that the triple helix interaction was loaded with uncertainty and a lack of transparency, where organisations often 'try to carry out their own innovation strategy following their own rules and imply specific tools they believe in. As a consequence, belief systems differ to a great extent.'

A culture based on the triple helix model is closely associated with learning. According to Saad and Zawdie (2005), a learning strategy that supports a triple helix innovation system needs to challenge the traditional forms of learning to:

1. endorse the need for learning and change;
2. adopt a learning approach that involves both actions and cognition; and
3. develop a culture conducive to organisational learning.

However, according to Pohlmann *et al.* (2005), there are differences in the belief systems of the triple helix organisations; hence, if differences in their belief systems persist, it is arguable whether a triple helix culture with values and beliefs common to the participating organisations is possible.

4.3.5   Project Culture/Project Atmosphere

Organisational cultures are often initiated by the founders of the organisation, given the culture of the society in which the organisation is embedded. Consequently, the culture of an organisation is amended by others who have had major impact on the organisation's development (owners, major managers). Such people shape the values and behaviour of members of the organisation to develop the identity of the organisation through influence over the employment of staff and organisational practices.

For construction and similar project-based industries, failure to appreciate the nature of culture has spawned notions of cultural formation and change that are superficial, and so can be affected in the short term. Hence, there has arisen the concept of a 'project culture', which has been examined, even for short duration projects. Such a perspective also fosters the notion of control by management, such that a culture change can be effected rapidly and under managerial control to, usually, effect efficiency savings. While it is appropriate to consider that long-term, mega-projects are likely to develop their own cultures, the same cannot be said for smaller projects of shorter durations. On such projects, the concept of 'project culture' is essentially an amalgam of the cultures of the constituent organisations (and societies), as transmitted to the project by their more powerful agents (Liu and Fellows, 2008); and, perhaps more appropriately therefore, referred to as 'project atmosphere' (analogous to workplace atmosphere) (Hovmark and Nordqvist, 1996). In addition, acknowledging that project organisations are often of a temporary nature (Cherns and Bryant, 1984), when organisations are brought into close operating contact (as on construction projects), their individual cultures tend to combine to produce the 'project atmosphere'. Cognisant of this, Pohlmann *et al.* (2005) noted that collaborators in the triple helix innovation model of alliances could pursue their own strategy (and strategic goals), and their values and beliefs (which are fundamental elements in the construct of *culture*) may therefore differ extensively. Thus, arguably, if there are no common values or beliefs, is there still a common culture (the project culture) amongst the project organisational members?

4.3.6   Other Forms of Culture

There are other 'types' of culture associated with innovation, one of which is the cluster culture in the Kyoto Model of Innovation and Entrepreneurship (Ibata-Arens, 2008). Through this, a cluster is formed when the economic dynamism in a region has continuously produced highly profitable entrepreneurial firms leading to a self-sustaining critical mass of firms (Ibata-Arens, 2008). Hence, the term 'cluster' describes the concentration of innovative activity in certain local communities (DeBresson, 1996; OECD, 2001).

Researchers also refer to the culture of public administration (van Duivenboden and Thaens, 2008; Arnaboldi *et al.*, 2010), which has its own set of barriers and constraints to innovation. For instance, Van Duivenboden and Thaens (2008) assert that the traditional bureaucratic organisational culture within the public sector often hinders innovation, but innovations do take place within government organisations; hence, they conclude that the reciprocity iterative process of mutual influence between innovation and culture change, which co-evolve within the specific government environment. In addition, research devoted to culture change in order to achieve innovation has also gained prominence (Pitta, 2009; Hadjimanolis, 2010), along with awareness of cultural differences and characteristics attributed to Eastern societies with those in the West (note: the term 'Eastern' is used loosely in the same manner that 'Western' is), where Eastern people are often viewed as flexible/adaptive, whilst Westerners are characterised as would-be controllers. Thus, traditionally, Eastern people tend to regard themselves as subservient to natural forces and desires to be in harmony with nature (as do people from other parts of the world, e.g. Africa, Australia, etc). Change and, hence, uncertainty is accepted as inevitable, truth is determined by spiritual (religious and philosophical) principles and considerations of time are long-term oriented. Western people, in contrast, tend to regard themselves as somewhat in control of nature and so can harness many of its forces to improve human society. Following this analogy, Westwood and Low (2003) noted that it has become increasingly important to have an informed understanding of the extent to which creativity and innovation processes vary across cultures. 'The Western creativity worker is predatory: he grabs the insight for a purpose … A process-oriented, rather than a product-oriented creative person would use insight-producing states to … obtain 'enlightenment'" (Krippner and Arons, 1973: 121). In Eastern cultures, the role of creativity concerns providing personal fulfilment and enlightenment, or connection to an inner realm of reality (Kuo, 1996). However, 'there is insufficient evidence to enable definitive statements to be made about systematic differences across cultures in personality or cognitive style with respect to creativity' (Westwood and Low, 2003: 235).

4.4   Factors Affecting Innovation

There are various endogenous (internal to the organisational) and exogenous (external to the organisation) factors that can enhance innovative working in organisations. Whilst strategic management literature covers many of the exogenous factors (as well as endogenous ones), this chapter focuses on the variables internal to the organisations. For example, according to Patterson, *et al.* (2009), leadership capability, organisational culture and organisational values are amongst the most important factors and initiatives that enhance innovative working. However, not many working practices that promote innovation are being readily adopted by organisations. This section examines the factors of creativity, creative climate, leadership, empowerment and other related organisational variables.

4.4.1   Creativity and Creative Climate

Central to the idea of innovation is creativity: 'Creativity is the seed of all innovation', where creativity is defined as 'the production of novel and useful ideas in any domain' (Amabile *et al.*, 1996: 1155). 'Innovations are the practical application of creative ideas, and an organisation cannot innovate unless it has the capacity to generate creative ideas' (Westwood and Low, 2003: 236). In this respect, Miron *et al.* (2004) reviewed the literature on the culture of innovation, and determined that high autonomy, risk-taking, tolerance of mistakes and low bureaucracy were the most prevalent dimensions. However, despite a significant interaction effect of creativity and innovation culture, creative people are not always highly innovative. Most importantly, innovative performance depends on the organisation culture in which people operate (Miron *et al.*, 2004). Hence, a creative climate, supported by organisational culture, is essential.

The traditional psychological approach to creativity focuses on the characteristics of creative persons (MacKinnon, 1965), but the sociological approach takes the view that both the level and the frequency of creative behaviours are influenced by the social environment (Woodman *et al.*, 1993). Following this theme, it is found that while workplace atmosphere and workplace innovative activity are closely intertwined, individual employee creativity relates to these two dimensions to a lesser degree

(Wongtada and Rice, 2008), indicating that there is a need to convert creativity into innovation. In addition, Unsworth *et al.* (2005) found that support for innovation does not significantly predict creativity. Perhaps what constitutes a creative climate in the workplace therefore needs to be precisely defined? With this in mind, climate has be defined as a psychologically meaningful description of the work environment (James and Jones, 1979), which is affected by organisational characteristics such as culture, structure and managerial behaviour (Burke and Litwin, 1992). Campbell, *et al.* (1970: 390) defined climate as 'a set of attributes specific to a particular organisation that may be induced from the way the organisation deals with its members and its environment. For the individual member within an organisation, climate takes the form of a set of attitudes and expectancies which describe the organisation in terms of static characteristics … and behaviour-outcome and outcome-outcome contingencies.' In the context of innovation, Bates and Khasawneh (2005: 99) proposed that climate 'could be reflected in perceptions of task related support for creative learning and problem solving or in the cognitive (e.g. attitudes about change and innovation) and affective states (e.g. motivation to innovate) that ensue from these perceptions.'

While there are many classifications of organisational climates, the three types of climates that are more closely associated with innovation are learning climate, innovative climate and creativity climate. For instance, Bates and Khasawneh (2005) found that supportive learning transfer climates, in the form of individual efficacy beliefs, attitudes about change, and effort-outcome and performance-outcome expectancies, are consistent with organisational cultures that believe in and value learning as an adaptive strategy. An innovative climate is a climate wherein actors are willing to change (van Duivenboden and Thaens (2008), and is closely associated with innovation culture discussed earlier. In addition, based on Amabile *et al.* (1996), a creative climate is thus proposed as the psychological context of creativity representing the work environment perceptions that can influence the creative work carried out in organisations. In this respect, there are some instruments developed to assess the work environment for creativity, for instance, the Creative Climate Questionnaire (developed in Swedish) by Ekvall *et al.* (1983, cited in Amabile *et al.* 1996), where KEYS were developed by Amabile *et al.* (1996). Through KEYS, the *work environment scales* included work group supports, challenging work, organisational encouragement, supervisory encouragement, organisational impediments, freedom, workload pressure and sufficient resources; the *criterion scales* included two variables of creativity and productivity. Furthermore, Oldham and Cummings (1996) found that employees produced creative work when they had appropriate creativity-relevant characteristics (based on personality), and faced a favourable working environment in the organisational context of challenging job complexity, with supportive and non-controlling supervision. It was also found that work environment for creativity is closely associated with leader support (Amabile *et al.*, 2004). Other examples in creativity research can be found in the works of Farmer *et al.* (2003), Zhou and Shalley (2003) and Egan (2005).

4.4.2   Leadership

A creative climate is closely associated with leadership. Leader support is proposed to be a key element, which supports a work environment for creativity (Amabile *et al.*, 2004) and, more specifically, transformation leadership (Waldman and Bass, 1991) is related to organisational innovation. Specifically, Jung *et al.* (2003) assert that transformational leadership enhances innovation by engaging employees' personal value systems – encouraging employees to think creatively. Elenkov and Manev (2005) found that the socio-cultural context is important in the leadership-innovation relationship in 12 European countries, confirming that leaders positively influence innovation processes. The type of leadership required to change culture was transformational, because culture change needs enormous energy and commitment to achieve outcomes; thus, according to Sarros *et al.* (2008: 148), transformational leadership 'can help build a strong organisational culture and thereby contribute to a positive climate for organisational innovation and subsequently influence innovative behaviour.'

Ogbonna and Harris (2000) found a link between participative leadership and innovative culture; and Ostroff *et al.* (2003) identified leadership as an emergent process that acts on both organisational culture and climate. More recently, Sarros *et al.* (2008) established the link of transformational *leadership*, organisational *culture* and *climate* for innovation through structural equations, modelled from a sample of 1,158 managers. In effect, culture is the lens through which leader vision is manifested, which helps build the climate necessary for organisations to become innovative (James *et al.*, 2007).

The leader's vision is therefore strongly associated with organisational culture and innovation (Damanpour and Schneider, 2006) and visionary leaders are often associated with organisations having adequate resources, funding, personnel and rewards to innovate, as well as time for workers to pursue their creative ideas (Sarros *et al.*, 2008). More importantly, vision is mediated through organisational culture, although caution is drawn that the capacity of leaders to define a vision is one thing, and to have that vision accepted and acted upon (as anticipated) is quite another. Hence, it is suggested that 'articulating vision can achieve results only when its development involves those it is most intended to influence, the workers and clients of the organisation' (Sarros *et al.*, 2008: 154). In this respect, Amabile *et al.* (2004) also found that *vision* was important in leadership, and that the effect of the leader's behaviour on employees' perceptions and creativity are neither static nor uni-directional. Amabile *et al.* (2004) also proposed that certain behavioural traits should be reduced, if the work environment of creativity is to be protected, that is, 'giving assignments without sufficient regard to the capability or other responsibilities of the subordinate receiving them, micro-managing the details of high level subordinates' work, and dealing inadequately with difficult technical or inter-personal problems, whether due to technical incompetence, inter-personal incompetence, inattention, or sloth.'

4.4.3   Empowerment

Çakar and Ertürk (2010) defined empowerment as an energising process that expands the feelings of trust and control in one as well as in one's organisation, which leads to outcomes such as enhanced self efficacy and performance. Therefore, empowerment is constructed as participation in decision making and access to information is shared by the management; where managerial strategies and processes are derived from the organisational culture (Conger and Kanungo, 1988; Spreitzer, 1996). In Çakar and Ertürk's (2010) study of 93 small-sized and medium-sized enterprises (SMEs), with 743 respondents in Turkey, empowerment was positively related to innovation capability on both the individual and firm level analyses. Specifically, empowerment was seen to have a mediating effect on culture and innovation capability – being an antecedent of innovation capability and a consequence of organisational culture. Thus, it was deemed important to focus on managerial practices that enhanced empowerment in order to promote innovation.

From an empowerment perspective, high power distance cultures often expect supervisors to control information, make decisions and tell subordinates what to do (Newman and Nollan, 1996); which can also inhibit information sharing between supervisors and subordinates (Randolph and Sashkin, 2002). Hence, efficacy of empowerment is doubtful in high power distance cultures. Contrary to high individualist organisations, where employees prefer sharing information that directly relates to their jobs, employees in collectivistic cultures share resources and ideas, and are therefore prepared to participate in collective interests (Sagie and Aycan, 2003). Thus, it is also easier to achieve empowerment in cultures with high assertiveness focus (Randolph, 2000) and high uncertainty avoidance with low tolerance for ambiguity (Randolph and Sashkin, 2002). Therefore, organisational culture has a strong link with politics and the distribution of power within the organisation, which 'makes clear that introducing innovation in an organisation, through its cultural dimension, is shaped by, but also can have consequences for, concepts such as authority and organisational politics' (van Duivenboden and Thaens, 2008: 219). Dougherty and Hardy (1996) revealed that the inability to connect new products with organisational resources, processes and strategy often thwarted innovation in large mature organisations, where innovators lacked the power to make these connections, hence suggesting that organisations should reconfigure their systems of power to become capable of sustained innovation.

4.4.4   Other Organisational Variables Relating to Organisational Development

Dougherty and Hardy (1996: 1146) found that large mature organisations could not achieve sustained innovation, because innovators within them could not solve innovation-to-organisation problems - that is, 'availability of resources, processes and meaning was piecemeal [and] depended primarily on individuals,' rather than through the organisational system. Hence, focus on organisational development variables is essential for promoting innovation. Therefore, within business, organisational development emphasises employees' skills training and problem-solving capabilities, but another important factor is the development of a firm's structure and rewards (Pitta, 2009: 448) – where '*structure* is a group of systems by which individual creativity is harnessed by the organisation', to allow creativity to be channelled in the service of goals and objectives. For instance, in a case study of Portugal Telecom (Pitta, 2009), it was advocated that the organisation owed its success to a strategy of fostering innovation across the company by using clear metrics to guide each employee's contribution, thereby fostering individual innovation efforts – to create a climate and culture of innovation by using the company's internal *communication* system, and operating a *reward* system to build interest in the innovation effort.

Other organisational variables relating to innovativeness (Kitchell, 2001) include long-term corporate goals (in response to competitors and the environment), proactive information search (market intelligence gathering/ knowledge acquisition), international market extension (adaptability/ outreach) and flexibility (to reduce resistance to change). In particular, resistance to (organisational and culture) change is a significant barrier to innovation.

4.4.5   Culture Change

If culture is viewed as a sub-system of an organisation (where culture defines the success of the organisation), 'culture can be an important barrier for organisational change, while at the same time altering the culture of the organisation is a fundamental condition for realising a substantial change of the organisation' (van Duivenboden and Thaens, 2008: 219). However, if culture is viewed as an aspect system, it is a 'dimension of all the structures and processes within the organisation… Everything in an organisation can then have a cultural dimension, such as, for example, language, structure and technology' (van Duivenboden and Thaens, 2008: 219). In the case of innovation, it is particularly relevant where culture is a sub-system – innovation requires organisational change, and hence change in the culture of an organisation. In this respect, research findings point to the need for culture change in most cases where innovation is to be promoted (Pitta, 2009). Concepts of value shift and behaviour modifications can therefore be seen to be closely associated with culture change. Usually, motivation schemes are developed to modify the behaviour of employees to increase their output (productivity) by offering positive, systematic reinforcement through rewards that are valued by the employees (commonly financial). Securing the reward is contingent on both behavioural change and enhanced performance. The basis of the scheme, from the employer's perspective, is that intended changes in employees' behaviour should cause increased productivity output, which would be more profitable for the employer. The effectiveness of the reinforcements as motivators to change behaviour is however likely to vary between people; reinforcements can operate in the negative direction too – as 'punishments' for unwanted behaviour, as in employment dismissal.

Behaviour modification (BMod) can operate at the level of the organisation, for example, a change to be more ethical (e.g. purchasing timber from sustainable forests only, performing more corporate social responsibility, etc). However, it is common for organisational BMod to occur because the organisation's management believe that such changes are good holistically *per se*, which in turn will lead to enhanced financial performance of the organisation and, according to Williamsonian opportunism, consequently improve their personal position and rewards too. However, through any BMod scheme, there are likely to be issues of ethics relating to personal (usually employee) choice, and a significant aspect of this is that the causal chain tends to operate rapidly. Furthermore, a potential detriment of this is that (e.g. financial rewards as reinforcement) the effectiveness of this may be only temporary. However, the sustained use of legislation to affect BMod coupled with a programme of education, could lead to culture change in the long term through changes in people's beliefs and their practices. This is important for promoting personal and industrial safety, for example.

4.5   Conclusion

Traditional metrics for 'assessing' innovation have tended to focus on investment in R&D and its outputs. In this respect, various NESTA reports (www.nesta.org.uk) have emphasised the change in the understanding of *innovation*: 'where once it was understood to be largely the result of scientific R&D, it is now seen more widely to include changes to services, ways of working and delivery, customer insight and many other forms' (Roper *et al.*, 2009: 4) - that is, *hidden innovation*. Based on Hansen and Birkinshaw's (2007) conceptual framework of the three sequential phases of innovation (i.e. knowledge investment, innovation and value creation), two of the findings in Roper *et al.*'s (2009) report (comparing the Innovation Index of nine sectors in the UK) have implications for the construction sector, specifically:

1. there are significant levels of hidden innovation in traditionally low R&D investments sectors, where innovation capability also varies strongly between firms within those sectors; and
2. the construction sector is comparatively weak in all stages (i.e. accessing knowledge, building innovation, commercialisation) of the innovation process, where there are low levels of innovative activity, and firms are therefore less likely to access external knowledge or encourage employee team-working.

Hidden innovations include new business structures/models and innovative solutions to small-scale problems that happen 'under the radar'

(NESTA, 2008); where the focus on HC (e.g. learning, motivation for creativity, etc) and organisational issues (e.g. culture, climate, reward systems, leadership, empowerment, etc) are essential. This chapter examined these relationships, particularly how leaders can foster a creative climate with organisational support and rewards systems to empower and motivate employees to 'think out of the box'. In this respect, the construction sector has a number of hidden innovations, especially innovative solutions to small problems on site, etc, which have hitherto gone unnoticed and unrecorded. These creative solutions are fundamental in 'knowledge transfer' for capturing (experiential, etc) learning from project to project. The focus on construction research relating to knowledge transfer therefore needs to flourish outside the boundary of ICT *per se*, to capture some of the 'softer' issues or 'elements that foster hidden innovation'. Roper *et al.*'s (2009) message that low innovation sectors could have high hidden innovations is therefore encouraging in this sense. Furthermore, there is plenty of scope for 'creative problem solving', especially when projects have always been enveloped in many uncertainty variables.

Finally, learning (including learning from other sectors) and creativity is crucial for an innovation culture. Leaders with *vision* can therefore be considered as pivotal drivers here, especially visionary leaders with the foresight to 'plan ahead' in order to fully benefit from this.

4.6   Acknowledgements

Part of the contents of this Chapter are based on a research grant funded by the Research Grants Council of Hong Kong, China ("The influence of creative climate and learning in fostering innovation culture in construction", RGC ref.no. 715111). The support of the RGC is gratefully acknowledged.

References

Ajmal, M.M. & Koskinen, K.U. (2008) Knowledge transfer in project-based organizations: an organizational culture perspective. *Project Management Journal*, 39(1), 7–15.

Amabile, T.M., Conti, R., Coon, H., Lazenby, J. & Herron, M. (1996) Assessing the work environment for creativity. *Academy of Management Journal.* 39(5), 1154–1184.

Amabile, T.M., Schatzel, E.A., Moneta, G.B. & Karmer, S.J. (2004) Leader behaviours and the work environment for creativity: Perceived leader support. *The Leadership Quarterly.* 15, 5–32.

Arnaboldi, M., Azzone, G. & Palermo, T. (2010) Managerial innovations in central government: not wrong, but hard to explain. *The International Journal of Public Sector Management.* 23(1), 78–93.

Ankrah, N. & Proverbs, D. (2009) Factors influencing the culture of a construction project organisation. *Journal of Engineering, Construction and Architectural Management*, 16(1), 26–47.

Anon (2008) Reviewing innovation effort. *Strategic Direction*, 24(19), 32–34.

Argyris, C. (1992) *On organisational learning.* Blackwell Publishers, Cambridge, MA.

Bates, R. & Khasawneh, S. (2005) Organizational learning culture, learning transfer climate and perceived innovation in Jordanian organizations. *International Journal of Training and Development.* **9**(2), 96–109.

Bröchner, J., Josephson, P.E. & Kadefors, A. (2002) Swedish construction culture, quality management and collaborative practice. *Building Research & Information,* **30**(6), 392–400.

Burke, W. & Litwin, G. (1992) A causal model of organizational performance and change. *Journal of Management.* **18**(3), 523–545.

Burns, T. & Stalker, G.M. (1961) *The Management of Innovation.* Tavistock Publications, London.

Çakar, N.D. & Ertürk, A. (2010) Comparing innovation capability of small and medium sized enterprises: examining the effects of organizational culture and empowerment. *Journal of Small Business Management.* **48**(3), 325–359.

Cameron, K.S. & Quinn, R.E. (1999) *Diagnosing and Changing Organizational Culture,* Addison-Wesley Longman, Reading, MA.

Campbell, J.P., Dunnette, M.D., Lawler, E.E. & Weick, K.E. (1970) *Managerial Behaviour, Performance and Effectiveness.* McGraw Hill, New York.

Chan, E.H.W. & Tse, R.Y.C. (2003) Cultural considerations in international construction contracts.' *Journal of Construction Engineering and Management,* **129**(4), 375–381.

Chen, P., Josephson, P.E. & Kadefors, A. (2009) Cross-cultural understanding of construction project managers' conceptions of their work. *Journal of Construction Engineering and Management,* **135**(6), 477–487.

Cherns, A. & Bryant, D.T. (1984) Studying the client's role in construction management, *Construction Management and Economics,* **2**, 177–184.

Cheung, S.O., Wong, P.S.P. & Wu, A.W.Y. (2011) Towards an organizational culture framework in construction. *International Journal of Project Management,* **29**(1), 33–44.

Conger, J.A. & Kanungo, R.N. (1988) The empowerment process: integrating theory and practice. *Academy of Management Review.* **13**(3), 471–482.

Construction Task Force (1998) *Rethinking Construction: The Report of the Construction Task Force to the Deputy Prime Minister, John Prescott, on the scope for improving the quality and efficiency of UK Construction.* HMSO, London.

Cox, A. (1999) Power, value and supply chain management. *Supply Chain Management,* **4**(4), 167–175.

Damanpour, F. & Schneider, M. (2006) Phases of the adoption of innovation in organizations: effects of environment, organisation and top managers. *British Journal of Management.* **17**, 215–236.

DeBresson, C. (1996) Why innovative activities cluster. In: ed. DeBresson, C. *et al., Economic interdependence and innovative activity: an input output analysis.* Edward Elgar, Cheltenham, UK. 149–164.

Denison, D.R. (1996) What is the difference between organizational culture and organizational climate: *Academy of Management Review.* **2**(3), 619–654.

Denison, D.R. (1997) *Corporate Culture and Organizational Effectiveness.* Denison Consulting, Ann Arbor, MI.

Denison, D.R. (2009) *Organizational Culture and Leadership Surveys: The Denison Model,* Denison Consulting, Ann Arbor, MI. Available from http://www.denison-consulting.com/advantage/researchModel/model.aspx (accessed 15 June 2011).

Di Marco, M., Taylor, J. & Alin, P. (2010) The emergence and role of cultural boundary spanners in global engineering project networks. *Journal of Management in Engineering,* **26**(3), 123–132.

Dobni, C.B. (2008) Measuring innovation culture in organizations: The development of a generalized innovation culture construct using exploratory factor analysis. *European Journal of Innovation Management*, **11**(4), 539–559.

Dougherty, D. & Hardy, C. (1996) Sustained product innovation in large, mature organizations: Overcoming innovation-to-organization problems. *Academy of Management Journal.* **39**(5), 1120–1153.

Egan, T.M. (2005) Factors influencing individual creativity in the workplace: an examination of quantitative empirical research. *Advances in Developing Human Resources.* **7**(2), 160–181.

Ekvall, G., Arvonen, J. & Walkenstrom-Lindblad, I. (1983) *Creative Organisational Climate: Construction and Validation of a Measuring Instrument.* Report 2, Stockholm: Swedish Council for Management and Organisational Behaviour (cited in Amabile *et al.*, 1996).

Eldridge, J. & Crombie, A. (1974) *A Sociology of Organizations.* Allen and Unwin, London.

Elenkov, D.S. & Manev, I.M. (2005) Top management leadership and influence on innovation: the role of socio-cultural context. *Journal of Management*, **31**(3), 381–402.

Etzkowitz, H. & Carvalho, de Mello, J.M. (2004) The rise of a triple helix culture – Innovation in Brazilian economic and social development. *International Journal of Technology Management and Sustainable Development*, **2**(3), 159–171.

Farmer, S.M., Tierney, P. & Kung-McIntyre, K. (2003) Employee creativity in Taiwan: an application of role identity theory. *Academy of Management Journal*, **46**, 618–630.

Fellows, R.F. (2006) Understanding approaches to culture. *Construction Information Quarterly*, 8(4), 159–166.

Fellows, R.F. (2009) Culture in supply chains. In: ed. Pryke, S., *Construction Supply Chain Management: Concepts and Case Studies.* Wiley-Blackwell, Chichester/ 42–72.

Fielden, S.L., Davidson, M.J., Gale, A.W. & Davey, C.L. (2001) Women, equality and construction, *Journal of Management Development*, 20(4), 293–304.

Flyvbjerg, B. (2007) Truth *and Lies about Megaprojects, Inaugural Speech, Faculty of Technology, Policy, and Management.* Delft University of Technology, 26 September 2007. Available from http://flyvbjerg.plan.aau.dk/Publications2007/InauguralTUD21PRINT72dpi.pdf (accessed 9 August 2010).

Furnham, A. (1997) *The Psychology of Behaviour at Work: The Individual in the Organization.* Psychology Press, Hove.

Gebhardt, C. (2005) The impact of managerial rationality in the organisational paradigm. Role models in the management of innovation. *Technology Analysis and Strategic Management*, **17**(1), 21–34.

Glisson, C. & James, L.R. (2002) The cross level effects of culture and climate in human service teams. *Journal of Organizational Behavior*, **23**, 767–794.

Gomez, C., Kirkman, B.L. & Shapiro, D.L. (2000) The impact of collectivism and in-group / out-group membership on the generosity of team members. *Academy of Management Journal*, **43**(6), 1097–1100.

Hadjimanolis, A. (2010) Methods of political marketing in (trans) formation of innovation culture. *Journal of Political Marketing.* **9**(1), 93–110.

Hall, E.T. & Hall, M.R. (1990) *Understanding Cultural Differences.* Cultural Press, Yarmount, ME.

Handy, C.B. (1985) *Understanding Organisations*, 3rd edn. Penguin, Harmondsworth.

Hansen, M. & Birkinshaw, J.M. (2007) The innovation value chain. *Harvard Business Review.* **85**(6), 121–131.

Hofstede, G.H. (1980) *Culture's Consequences: International Differences in Work-Related Values.* Sage Publications, Beverley Hills, CA.

Hofstede, G.H. (1983) The cultural relativity of organizational practices and theories. *Journal of International Business Studies*, **14 (Fall)**, 75–89.

Hofstede, G.H. (1994a) The business of international business is culture. *International Business Review*, **3(1)**, 1–14.

Hofstede, G.H. (1994b) *Cultures and Organizations: Software of the Mind*. Harper Collins, London.

Hofstede, G.H. (2001) *Culture's Consequences: Comparing Values, Behaviors, Institutions, and Organizations Across Nations*, 2nd edn. Sage, Thousand Oaks, CA.

Hovmark, S. & Nordqvist, S. (1996) Project organization: Change in the work atmosphere for engineers. *International Journal of Industrial Ergonomics*, **17(5)**, 389–398.

Ibata-Arens, K. (2008) The Kyoto model of innovation and entrepreneurship: regional innovation systems and cluster culture. *Prometheus*. **26**, (1), 89–109.

Inglehart, R. & Baker, W. (2000) Modernization, cultural change and the persistence of traditional values. *American Sociological Review*. **65**, 19–51.

James, L.R. & Jones, A.P. (1979) Organisational structure: a review of structural dimension and their conceptual relationship with individual attitudes and behaviours. *Organisational Behavior and Human Performance*. **16**, 74–113.

James, L.R., Choi, C.C., Ko, C.M., McNeil, P.K., Minton, M.K., Wright, M.A. & Kim, K. (2007) Organisational and psychological climate: a review of theory and research. *European Journal of Work and Organisational Psychology*. **17(1)**, 5–32.

Johnson, G. (2001) Mapping and re-mapping organizational culture: a local government example. In:, ed. Johnson, G. & and Scholes. K., *Exploring public sector strategy*. Pearson Education Ltd, Harlow.

Jung, D.I., Chow, C. & Wu, A. (2003) The role of transformational leadership in enhancing organizational innovation: hypotheses and some preliminary findings. *Leadership Quarterly*, **14(4–5)**, 525–544.

Kaiser, S. & Holton, E. (1998) The learning organization as a performance improvement strategy. In: ed. Torraco, R., *Proceedings of the Academy of Human Resource Development Conference*. 75–82 (cited in Bates and Khasawneh, 2005).

Kitchell, S. (2001) Corporate culture, environmental adaptation, and innovation adoption: a qualitative/quantitative approach. *Journal of the Academy of Marketing Science*, **23(3)**, 195–205.

Krippner, S. & Arons, M. (1973) Creativity: person, product or process. *Gifted Child Quarterly*, **17(2)**, 116–123.

Kroeber, A.L. & Kluckhohn, C. (1952) Culture: a critical review of concepts and definitions. In: *Papers of the Peabody Museum of American Archaeology and Ethnology*, vol. 47. Harvard University Press, Cambridge, MA.

Kuo, Y.Y. (1996) Taoistic psychology of creativity. *Journal of Creative Behavior*, **30(3)**, 197–212.

Lawrence, P.R. & Lorsch, J.W. (1967) *Organization and environment: managing differentiation and integration*. Harvard University, Boston, MA.

Leydesdorff, L. & Etzkwitz, H. (1998) The triple helix as a model for innovation studies. *Science and Public Policy*, **25(3)**, 195–203.

Little, A. (2008) *Innovation Culture: maintaining a strong innovation culture during a downturn*. Available from www.adl.com/innovationculture (accessed 24 August 2010).

Liu, A.M.M. & Fellows, R. (2008) Organisational culture of joint venture projects: a case study of an international JV construction project in Hong Kong. *International of Human Resources Development and Management*, **8(3)**, 259–270.

Liu, A.M.M., Zhang, S. & Leung, M.Y. (2006), A framework for assessing organizational culture of Chinese construction enterprises. *Engineering, Construction and Architectural Management*, **13(4)**, 327–342.

MacKinnon, D.W. (1965) Personality and the realization of creative potential. *American Psychologist*, **20**, 273–281.

Mai, R. (1996) *Learning partnerships: how leading American companies implement organisational learning.* Irwin, Chicago, IL.

Miron, E., Erez, M. & Naveh, E. (2004) Do personal characteristics and cultural values that promote innovation, quality and efficiency compete or complement each other? *Journal of Organizational Behavior*, **25**(2), 175–199.

Moran, E.T. & Volkwein, J.R. (1992), The cultural approach to the formation of organizational climate. *Human Relations*. **45***(1)*, 19–48.

NESTA (2008) *Total innovation*. NESTA, National Endowment for Science, Technology and the Arts. Available from http://www.nesta.org.uk/publications/reports/assets/features/total_innovation (accessed 24 August 2010).

Newman, K.L. & Nollan, S.D. (1996) Culture and congruence: the fit between management practices and national culture. *Journal of International Business Studies*, **27**(4), 753–779.

OECD (2001) *Innovative clusters: drivers of national innovation systems*. Organisational of Economic Cooperation and Development, Paris.

Ofori, G. & Toor, S.U.R. (2009) Research on cross-cultural leadership and management in construction: a review and directions for future research. *Construction Management & Economics*, **27**(2), 119–133.

Ogbonna, E. & Harris, L.C. (2000) Leadership style, organizational culture and performance: empirical evidence from UK companies. *International Journal of Human Resource Management*, **11**(4), 766–788.

Oldham, G.R. & Cummings, A. (1996) Employee creativity: personal and contextual factors at work. *Academy of Management Journal*, **39**(3), 607–634.

Ostroff, C., Kinicki, A.J. & Tamkins, M.M. (2003) Organizational culture and climate. In: ed, Weiner, I.B., Borman, W.C., Ilgen, D.R., & Klimoski, R.J., *Handbook of Psychology: volume 12: Industrial and Organizational Psychology*. Wiley, Hoboken, NJ.

Palazzo, G. (2005) Postnational constellations of innovativeness: a cosmopolitan approach. *Technology Analysis and Strategic Management*. **17**(1), 55–72.

Patterson, F., Kerrin, M., Gatto-Roissard, G. & Coan, P. (2009) *Everyday innovation – how to enhance innovative working in employees and organisations*. NESTA, National Endowment for Science, Technology and the Arts. Available from http://www.nesta.org.uk/publications/reports/assets/features/everyday_innovation_how_to_enhance_innovative_working_in_employees_and_organisations (accessed 24 August 2010).

Pitta, D.A. (2009) Creating a culture of innovation at Portugal Telecom. *Journal of Product & Brand Management*, **18**(6), 448–451.

Pohlmann, M., Gebhardt, C. & Etzkowitz, H. (2005) The development of innovation systems and the art of innovation management – strategy, control and the culture of innovation. *Technology Analysis and Strategic Management*, **17**(1), 1–7.

Randolph, W.A. (2000) Re-thinking empowerment: why is it so hard to achieve? *Organizational Dynamics*, **29**(2), 94–107.

Randolph, W.A. & Sashkin, M. (2002) Can organizational empowerment work in multinational settings? *Academy of Management Executive*, **16** (1), 102–115.

Reugg, R.T. & Marshall, H.E. (1990) *Building Economics: Theory and Practice*. Van Nostrand Reinhold, New York.

Riley, M.J. & Brown, D.C. (2001) Comparison of cultures in construction and manufacturing industries. *Journal of Management in Engineering*, **17**(3), 149–158.

Rooke, J., Seymour, D.E. & Fellows, R.F. (2003) The claims culture: A taxonomy of attitudes in the industry. *Construction Management and Economics*, **21**(2), 167–174.

Roper S., Hales C., Bryson, J.R. & Love, J. (2009) *Measuring sectoral innovation capability in nine areas of the UK economy*. NESTA, National Endowment for Science, Technology and the Arts. Available from (http://www.nesta.org.uk/publications/reports/assets/features/measuring_sectoral_innovation_capability_in_nine_areas_of_the_uk_economy (accessed 24 August 2010).

Rothwell, R. (1992) Successful industrial innovation: critical success factors for the 1990s. *R&D Management*, **22**(30), 221–239.

Saad, M. & Zawdie, G. (2005) From technology transfer to the emergence of a triple helix culture: the experience of Algeria in innovation and technological capability development. *Technology Analysis and Strategic Management*, **17**(1), 89–103.

Sagie, A. & Aycan, Z. (2003) A cross cultural analysis of participative decision-making in organisations. *Human Resources*, **56**(4), 453–473.

Sarros, J.C., Cooper, B.K. & Santora, J.C. (2008) Building a climate for innovation through transformational leadership and organizational culture. *Journal of Leadership and Organizational Studies*, **15**(2), 145–158.

Schein, E.H. (1984) Coming to an awareness of organisational culture, *Sloan Management Review*, **25**(2), 3–16.

Schein, E.H. (1990) Organisational Culture. *American Psychologist*, **45**, 109–119.

Schein, E.H. (2004) *Organizational Cutlure and Leadership*. John Wiley & Sons, Inc, New Jersey.

Senge, P. (1992) Building the learning organization. *Journal for Quality and Participation*, **March**, 30–38.

Slater, S. & Narver, J. (1995) Market orientation and the learning organization. *Journal of Marketing*, **59**, 63–74.

Spreitzer, G.M. (1996) Psychological empowerment in the workplace: dimensions, measurement, and validation. *Academy of Management Journal*, **38**, 1442–1465.

Tagiuri, R. & Litwin, G.H. (eds) (1968) *Organizational Climate*. Graduate School of Business Administration, Harvard University, Boston, MA.

Triandis, H.C. & Gelfand, M.J. (1998) Converging measurement of horizontal and vertical individualism and collectivism. *Journal of Personality and Social Psychology*, **74**(1), 118–128.

Trompenaars, F. & Hampden-Turner, C. (1997) *Riding the Waves of Culture: Understanding Cultural Diversity in Business*, 2nd edn. Nicholas Brealey Publishing, London.

Unsworth, K.L., Wall, T.D. & Carter, A. (2005) Creative requirement: a neglected construct in the study of employee creativity. *Group and Organization Management*, **30**, 541–560.

van Duivenboden, H. & Thaens, M. (2008) ICT-driven innovation and the culture of public administration: a contradiction in terms? *Information Polity*, **13**, 213–232.

Van Marrewijk, A. (2007) Managing project culture: The case of the environ megaproject. *International Journal of Project Management*, **25**, 290–299.

Waldman, D. & Bass, B.M. (1991) Transformational leadership at different phases of the innovation process. *Journal of High Technology Management Research*, **2**, 169–180.

Westwood, R. & Low, D.R. (2003) The multicultural muse: Culture, creativity and innovation. *International Journal of Cross Cultural Management*, **3**(2), 235–259.

Williams, A., Dobson, P. & Walters, M. (1989) *Changing culture: New organizational approaches*. Institute of Personnel Management, London.

Wongtada, N. & Rice, G. (2008) Multidimensional latent traits of perceived organizational innovation: differences between Thai and Egyptian employees. *Asia Pacific Journal of Management*, **25**, 537–562.

Woodman, R.W., Sawyer, J.E. & Griffin, R.W. (1993) Toward a theory of organisational creativity. *Academy of Management Review*, **18**, 293–321.

Zhang, S.B. & Liu, A.M.M. (2006) Organisational culture profiles of state-owned construction enterprises in China. *Construction Management and Economics*. **24**(8), 817–828

Zhou, J. & Shalley, C.E. (2003) Research on employee creativity: a critical review and proposal for future research directions. In: ed. Marocchio, J.J. & Ferris, G.R., *Research in personnel and human resource management*. Elsevier, Oxford.

Innovation, Technology and Knowledge Transfer for Sustainable Construction

Emilia van Egmond

Emilia van Egmond

5.1 Introduction

This chapter presents a series of innovation and sustainability issues facing the construction industry. In this respect, it poses the following question to set the discussion and narrative in context: 'How is innovation and international technology and knowledge transfer (TKT) related to sustainable construction performance?' In order to address this, the theoretical views on innovation, TKT and sustainable construction are presented. The chapter concludes with lessons learnt, and how the synergies from these can be leveraged in a meaningful way.

5.1.1 Construction Performance and Challenges

Construction plays an important role in national economies, contributing significantly to the Gross Domestic Product (GDP), fixed capital formation, government revenue and employment. Indeed, there is hardly any sector in the economy where the significance of construction activities and construction technologies is not felt. This explains the close relationship between the state of construction and socio-economic situations in nearly every country, not least those in the developing world (Zawdie and Langford, 2002). In terms of production output, which is still heavily concentrated (77%) in the high income countries, construction, with its many forward and backward linkages, is one of the largest industries for providing opportunities for employment (ILO, 2001). Yet, construction everywhere faces problems and challenges that are amplified by the rapid increase of

Construction Innovation and Process Improvement, First Edition.
Edited by Akintola Akintoye, Jack S. Goulding and Girma Zawdie.
© 2012 Blackwell Publishing Ltd. Published 2012 by Blackwell Publishing Ltd.

demand for construction due to population growth and the effects of industrialisation. Thus, for example, construction has been challenged to meet the enormous need for housing, particularly in rapidly urbanising areas in developing countries, where a large amount of the population still lives in deplorable conditions. Construction is resource-intensive and notorious for its contribution to greenhouse gas emissions and waste generation (Muller, 2000; Macozoma, 2002; OECD/IEA, 2009). In addition, construction firms have to face globalisation-induced international competition and pressures to meet the social, economic and environmental imperatives of sustainability. Thus, improved sustainable performance is essential if the construction industry is to move forward. This calls for a radically new *modus operandi* in construction – namely, sustainable construction (SuCo).

5.1.2   Sustainable Construction

SuCo encompasses the simultaneous pursuit of a balanced social equity, environmental quality and economic prosperity (people, planet and profit) in the built environment in such a manner as to meet the needs of the present generation without compromising the ability of future generations to respond to their demands for a sustainable built environment (WCED, 1987; Kibert, 2005). This definition, based on the Report of the Brundtland Commission of 1987, embraces two major aspects:

1. basic needs – food, clothing, shelter – and the need for an acceptable life standard above the absolute minimum; and
2. limits to the capability of construction to fulfil the needs of present and future generations.

SuCo therefore requires construction firms to deal with the *'natural limits'* of finite natural resources (energy, materials, water) and *'man-made limits'*, which are related to the development, diffusion accessibility, adoption and implementation status of technology and knowledge (T&K) in construction processes.

Construction firms typically have to innovate in order to improve their performance to meet the increasingly complex and fast changing customer needs for a sustained quality of the built environment (Figure 5.1). Thus, SuCo requires designers, engineers, building material producers and contractors to bring about new design concepts, building elements and components as well as building processes, and integrate these into construction projects. Cognisant of this, policy makers are, for their part, expected to provide an adequate framework that facilitates improvements in construction.

Innovation theories (Nelson and Winter, 1982) suggest a shift in the technology regime, the consequence of which would profoundly and irreversibly change the pattern of behaviour or perception underlying the current mode of construction practice.

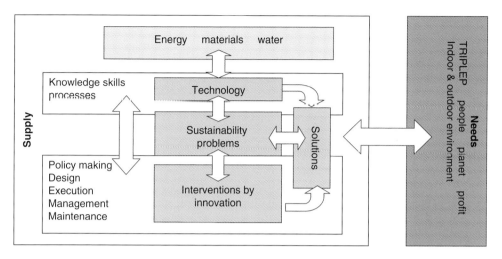

Figure 5.1   Sustainable Construction (adapted from Stofberg and Duijvesteijn, 2006).

5.2   Innovation, Technology and Knowledge Transfer Practices

Innovation and sustainability in construction needs to balance a range of important factors, from performance metrics, through to capability drivers and competence building.

5.2.1   Technology, Innovation and Production Performance

Innovation refers to the total cycle from *invention*, that is, the development of new knowledge and technologies (products and production processes) – and the *diffusion, adoption and application* of these (Rogers, 1995). The growth performance of the developed market economies over the last 150–200 years has largely been on the emergence and diffusion of fundamental technological advances, and changes of production systems in manufacturing industries linked to periods of economic expansion (Kondratiev, 1925). Innovation is therefore crucial for the production performance of industries and their competitiveness as well as for the socio-economic development of countries. Conventional mono-disciplinary economic approaches have been used to analyse and explain the role of innovation and TKT in competitiveness and economic growth. In addition, capital accumulation (in terms of advancement of technologies and innovation) has been seen as the engine for improved competitiveness of enterprises and economic growth in Western countries through traditional growth theories. The basic assumption was that a motivated *profit-maximising, cost minimising and output maximising* entrepreneur had to make choices amongst various production technologies in a perfect competition environment (Schumpeter, 1934). In this respect, product–life-cycle theories identified that the drive of firms to innovate was fostered by their need to survive and maintain their competitive position through incremental innovation (Abernathy and Utterback,

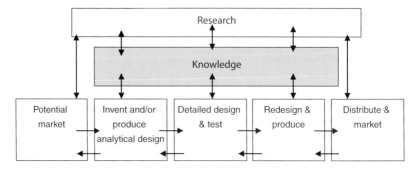

Figure 5.2 Interactions in the process of technology development and innovation (adapted from Kline and Rosenberg, 1986).

1978). Traditionally, innovation, be it incremental or radical, was conceived as a linear process. *Product innovations* were seen as successfully developed, introduced, diffused and used product technologies, and *process innovations* as successfully developed, introduced, diffused and used production process technologies (Rogers, 1995). However, Kline and Rosenberg (1986) noted innovation to be complex and not 'chain/sequential', which occurred as a process 'parallel' to or 'integrated' with upstream activities (with suppliers), and downstream linkages (with users), or in the context of alliances, with other firms with different phases overlapping (Figure 5.2).

5.2.2   Technology and Knowledge Transfer

New technologies and knowledge are not necessarily generated by in-house R&D, but can also be acquired through TKT, that is, through:

1. simple acquisition of T&K from elsewhere in the country (or from abroad); or
2. acquisition of parts of T&K components that are missing and combine these with further in-house development and production processes.

It is therefore generally more profitable and less risky for firms to acquire knowledge and technologies, for instance, through collaboration in production processes (e.g. joint-venture agreements, or licence packages) than to rely on high, often uncertain R&D investments. Moreover, due to the increasingly shorter and unpredictable life cycle of products or services, firms are increasingly expected to innovate at a faster pace. This is no longer feasible in 'traditional' production organisations. In many firms, conditions are simply not present to stimulate entrepreneurship and creativity (Chesborough, 2003), and the increased complexity of the market has brought pressure to bear on firms and organisations to be more innovate and competitive in networks (even beyond the boundaries of their own sector). In this respect, ICT can facilitate the exchange, sharing and acquisition of T&K, with increased opportunities of being able to transmit, store and manipulate large quantities of information.

The relative openness of the global economy has created opportunities for an international exchange of advanced technologies and knowledge. The essence of international TKT is that technologies, either in hardware or software forms (i.e. knowledge and information), can be transferred over spaces/borders from one source (e.g. R&D laboratories, firms, universities, state and local governments, third-party intermediaries) to other recipients in the international market (Dunning, 1981). Most technologies and knowledge flow through commercial contracts, such as:

a) *trade contracts*: the purchase and sales of, e.g. capital goods and intermediate goods, consultancy, turn-key projects;
b) *exchange of know-how contracts*: R&D collaboration, sales of patents, licences, trademarks, management and consultancy agreements; and
c) *private investments*: foreign direct investments, wholly owned subsidiaries, joint ventures, other forms of collaboration between firms (Katz and Shapiro, 1985; Lall, 1987).

Knowledge also may flow *free of charge* in codified form embodied in capital goods, technical manuals, blueprints and instruction books, either as a by-product of market transactions, or through the maintenance and reverse engineering of new products. However, the imperfection of the international technology market has been blamed on the monopolistic and oligopolistic tendencies and asymmetric power relationships between technology recipients and suppliers (Dunning, 1981). These theories emphasise that such tendencies and the resulting market imperfections are further reinforced due to government interventions (trade regulations, subsidies, and taxes) and the implementation of legal systems of protection (patents). On the other hand, where appropriately designed in the context of good governance, regulatory policy and legal mechanisms – i.e. intellectual property rights (patents, copyrights, trade secrets, etc), then this could foster TKT and innovation activities.

T&K components are often transferred in commercial transactions as part of a contractual package deal, accompanied by particular terms and conditions that are built into contractual agreements, which give both the supplying and recipient companies the opportunity to control the TKTs. Conditions attached to TKT packages vary from limited ones, through to a full and complex set of restrictions and controls, such as: the use of the technology; markets for the end-products of the used technologies; and suppliers of necessary raw materials, spare parts, etc. Conditions determine whether the components – e.g. machines, equipment, documents (blueprints, designs of products, manufacturing instructions, process flow charts, equipment specifications, etc) or personnel in charge of specific operations/training schemes are all included in the package deal, or whether missing components have to be acquired separately from another source. For example, a licence agreement could be provided without any instructions, documentation or assistance on the use of the patented technology. Acquiring a complete ready-to-use package (turn-key) means that risks can be minimal, where the quality is secured for the recipient. In this respect, conditions can be ascribed that allow the supplying firm to establish a quasi-monopoly over components of non-monopolised knowledge (Stewart, 1979; Rosenberg, 1982; Lall, 1985).

Packaging associated with discontinuities in knowledge places restraints on organisational learning and on the development of 'organisational memory' (Stiglitz, 1999). Hence the case for unpackaging (i.e. breaking down a technology into several components to purchase separately – possibly from different suppliers), as this would help to bring down costs, limit the conditions of the transactions, open opportunities for recipients to increase technological capabilities through learning by doing, maintain control over TKTs, and decrease dependence on a single supplier. However, in practice, unpacking turned out to be rather expensive, time consuming, was vulnerable to failures and involved usually 'old' technologies. Moreover, the sector did not always support unpackaged deals due to the influence of market positioning drivers.

5.2.3   Innovation System

The linear approach to conceptualising innovation does not offer a clear insight into the content and process of innovation, TKTs, or on the existence of differences in the scope of innovation across firms, sectors or countries (Rosenberg, 1976). Technology is often understood in relation to production processes as a system of interrelated know-how, skills and knowledge embodied in products and processes. Thus, a distinction is made between:

- *Product technology*: the knowledge and skills embodied in the output of a production process, which find expression in the technological attributes of products; and
- *Process technology*: the knowledge and skills embodied in the 'transformer' of inputs into the required goods and services in production processes, where the transformer is composed of four inextricable components: technoware (materials, equipment, tools); humanware (people); infoware (documented facts, blueprints, etc); and orgaware (firm, institute) (ESCAP, 1989).

Literature on innovation covers work on issues such as adoption and the diffusion of innovations, international technology transfer, and the theory of economic growth (including issues of catching-up or relative stagnation). Developments in evolutionary economics (Nelson and Winter, 1982; Lall, 1992; Metcalfe, 1995; Stiglitz, 1999; Dosi, Nelson and Winter, 2000), technological innovation studies (Nelson and Winter 1982; Teece *et al.*, 1997; Dosi, Coriat and Pavitt, 2000; Dosi, Nelson and Winter, 2000) and organisational and management studies (March and Simon 1993; March 1991; Tidd *et al.*, 2006) have coalesced to nurture the theoretical basis of empirical studies. As opposed to neo-classical thinking, evolutionary thinking assumes that innovations are not an outcome of *ex-ante* economic reasoning, but analogous to biological evolution – i.e. outcome of the evolutionary processes of social needs and competition, T&K development and government intervention, which together constitute

Table 5.1 Technology intervention and promotion agents (Egmond, 2006).

Technology Intervention & Promotion Agent	Form & Field of Technology Promotion	Resulting in
National government	Formulation and implementation of technology policies	Favourable climate by means of subsidies, incentives, tax-holidays etc. to enhance TD&I
Educational and training institutions; consultancy companies	Training and education	Human resources development (Humanware)
Documentation & information centres, libraries, statistical organizations, patent and registration offices, museums; consultancy companies	Storing and lending of documented facts Training and informing	Improved knowledge & insight (Infoware + Human ware)
Testing, certification and standardization laboratories R&D & financing organizations; design, engineering & management consultants	up-grading & standardization of technologies assistance for setting up of management and organizational structures	Improved and standardized Technologies (Technoware) Improved organisational management, structure and culture (Orgaware)

the innovation system. An innovation system can therefore be said to encompass 'interconnected institutions', which form a social network of inter-related actors (individuals, organisations and enterprises) 'who share a common field of knowledge and interest regarding innovation in a certain domain' and who 'create, store and transfer the knowledge, skills and artefacts which define new technologies' (Nelson and Winter, 1982; Metcalfe, 1995; Dosi, Coriat and Pavitt, 2000).

A number of innovation agents intervene in the processes of development, diffusion and utilisation of new technologies (Table 5.1). This is in line with the actor-centred social construction of technology theories, which suggest that an innovation process can produce different outcomes depending on the social circumstances in which it takes place (Bijker *et al.*, 1987). Similarly, Rogers (1995) postulated that innovation and TKT are accomplished through human interactions and communication between members of a community of practice, which gives rise to particular innovation trajectories that are sustained by industrial interests vested in it, along with assumptions about user needs, and the costs of making system changes (March, 1991).

5.2.4   Technological Capability, Knowledge and Learning

Capability building through knowledge accumulation and learning is central to innovation and TKT processes, since successful production and trade performance strongly depend on capability and motivation to innovate and quickly adopt new technologies. This view is supported by research that shows a high correlation between the success of countries in assimilating foreign technologies and success of the same countries in terms of the educational attainment of their population (Dosi *et al.*, 1990). Thus, *technological capabilities* encompass the total stock of resources including technologies, knowledge and skills that can be found at different levels in innovation systems – which can be categorised into three functional classes: *investment capability* (to identify, negotiate, purchase and manage suitable and feasible technologies and knowledge); *production capability*; and *innovation capability* (to create, diffuse, adopt and implement new knowledge and technological solutions) (Lall, 1992). In the absence of a strong innovation system, the development of technological capabilities is likely to be constrained, and so *a weak* endogenous *knowledge and technology base* would persist. In such circumstances, an industry (e.g. construction) may fail to use its scarce resources efficiently. This would make its performance hardly competitive in market conditions (Stewart, 1979; Rosenberg, 1982).

More recently, *knowledge* has been recognised as a key aspect of innovation capability and TKTs (Stiglitz, 1999). The argument is that individuals and organisations working in a specific innovation system create new technologies and knowledge by combining different knowledge sets, thanks to the increasing levels of knowledge and technological skills. Continuous interactive learning – i.e. the *exchange, accumulation, integration and mobilisation* of both knowledge (*know why*) and know-how (*craftsmanship, skills to carry out a job*) is expected to enhance this process (Von Hippel, 1988; Lundvall, 1992; Lundvall *et al.*, 2006). The distinction between knowledge and know-how is here emphasised by the argument that knowledge and know-how are needed in combination to effectively implement a new technology (Stiglitz, 1999). Therefore, the processes of T&K mastery – (i.e. *'learning'* or up-grading of *technological knowledge, know-how and skills formation*) – and learning processes have become the centre of analysis and debate consequent upon the recognition of the outstanding role of knowledge in the development of innovation capability (Lundvall *et al.*, 2006).

Various learning mechanisms, by which knowledge and skills flow from a source to a recipient, have been identified. These include:

a) learning by doing, using, reverse engineering, imitation, searching (formalised activities such as R&D), adoption;
b) learning by interacting with upstream or downstream sources of knowledge (suppliers or users) or in co-operation with other firms;
c) learning by absorption of new developments in science and technology; and
d) learning by querying what and how competitors and other firms in the industry are doing (inter-industry spillovers).

Therefore, it can be seen that learning mechanisms often overlap. For example, learning by searching may take place jointly with learning from advances in science and technology (Lall, 2003). Learning processes also involve personal interactions, observation and practical experience in specific contexts. The process of embodying knowledge in people (*learning*) and in products and processes (*application and integration*) is generally costly in time and resources. Various authors (e.g. Polanyi, 1967) have argued that this is because knowledge is to varying degrees *tacit*. Stiglitz (1999) considers tacit knowledge and the ways it is managed to be particularly important for a company's competitive performance; but 'translation' of tacit knowledge into fully explicit or codified statements (information) is barely possible, since it is hard to explicitly delineate or codify certain procedures and behavioural patterns. Stiglitz (1999) also highlights the importance of an extensive basis of contextual knowledge for the proper accumulation and integration of technologies and knowledge acquired through TKTs. Such knowledge might either be incomplete in the recipient enterprise to correctly apply the new materials, technologies and knowledge, or it has not yet developed sufficiently. It is not only understanding of the transferred hardware bits and pieces that might be lacking, but also the knowledge and skills to handle these, to manage the organisation around the production process with the technologies and knowledge that are acquired from elsewhere. The required pool of tacit knowledge cannot easily be transferred or just 'downloaded' from one firm to another. Indeed, it takes a certain learning period for any newly introduced technology to be correctly applied (Lall, 1992; Stiglitz, 1999).

The above indicates that the key to improved performance and competitiveness does not lie in simply the acquisition of technologies and knowledge from elsewhere. The true source of competitiveness lies in the stock of knowledge and experience, and through a deliberate investment in a complete array of knowledge regarding how, why, where and when to use technologies. These processes imply a *continuous accumulation and integration of knowledge and skills* in cyclical fashion through invention, diffusion, adoption and application of novelties. This can enable the creation of new possibilities through the combination of different sets of knowledge and skills. Currently, effective knowledge management is increasingly associated with various processes of continuous learning, which involve the acquisition and mobilisation of knowledge and technological skills of individuals and organisations to create product, process and organisational innovations (Tidd *et al.*, 2006).

5.2.5   Technological Regime and Routines

Learning, innovation and TKT practices can be considered as routine procedures or regularly followed courses of action that are mechanically and unconsciously applied (Polanyi 1967; Stiglitz 1999; Geels, 2004). These routines evolve from the *technological regime* of innovation systems. Technological regime is a social construct, a pattern made of knowledge, rules, regulations conventions, consensual expectations, collective memories, experience, assumptions or thinking shared by the stakeholders, which guide the design and further the development of innovations. It sets the achievement boundaries for innovative activities,

as well as the directions (natural trajectories) along which diffusion, adoption and application of new technologies take place (Dosi *et al.*, 2000). In this respect, Malherba (2002) noted that an innovation system with strong inter-relationships between stakeholders, who share a common field of knowledge and interest (i.e. technological regime), can facilitate innovation and TKTs. Nelson and Winter (1982) note that asymmetries between industries, industrial dynamics and innovativeness can be interpreted on the grounds of *technological regimes* in *innovation systems*. New technologies are successful when they fit well in the prevailing technology regime. Technological regimes are sector specific and (to a large extent) typical for individual firms, thus making it difficult for firms to imitate one another (Patel and Pavitt, 1997; Breschi *et al.*, 2000; Marsili and Verspagen, 2002). At national and international levels, technological regime is reflected in particular through political, legal and economic settings. These are seen as important facilitating frameworks that may create a favourable climate for innovation, TKTs, competitive production and societal development (ESCAP, 1989).
Technological regimes define:

1. the properties of the knowledge base;
2. technological opportunities;
3. appropriation of innovations; and
4. cumulativeness of technological advances in innovation systems (Malerba and Orsenigo, 1996; Breschi *et al.*, 2000).

The *properties of knowledge base* relate to the *nature* and *availability* of technologies and knowledge - that is, the know-how, skills, experience and perceptions. Knowledge can be seen as being partly institutionalised, which includes standards, norms, rules and regulations. *Technological opportunities* indicate the extent to which an industry can draw from the knowledge base and the technological advances of its suppliers, and major scientific advances in universities and R&D institutions. High opportunities can be found in an innovation system where collaboration, communication and knowledge exchange amongst the actors are least constrained. The existence of opportunities provides a powerful incentive for potential innovators to undertake innovative activities. A *high appropriation* indicates a high protection of innovations from imitation by means of patents, secrecy, lead times, costs and time required for duplication, learning curve effects, superior sales efforts, and differential technical efficiency due to scale economies through which firms can reap profits from innovative activities. *Cumulativeness of technical advances* reflects continuities in innovative activities and increasing returns as well as the existence of a T&K base that forms the building blocks for either subsequent incremental innovations or a completely new knowledge set that can be used as a basis for other innovations in related areas (Murray and O'Mahony, 2007). Innovation systems characterised by continuities in innovative activities and increasing returns are considered to have high levels of cumulativeness.

5.2.6 Strategic Niche Management

Kemp *et al.* (2001) noted that innovative technologies often need to be 'pampered' in the initial stage of their development, for fear of undersupply, or not even being supplied at all due to high uncertainty, high up-front costs, or because the wider social benefits arising from these technologies are not adequately valued by the market. Innovations would thus need to be managed by a 'niche manager', who could facilitate their diffusion, adoption and implementation. A novel technology can be seen as a technological niche, which forms together with other existing technologies and knowledge a *technological network* in a particular innovation system. Nodes of the technological network are technologies – each with specific embodied knowledge and skills, upon which another adjacent technology can be created. The position of a technology in the network is therefore a niche in the domain. The properties of the technology determine whether it fits the *technological regime* of the innovation system; and the expected rate of return on the investment determines its market potential. Different actors in the innovation system - for example, policy makers, regulatory agencies, local authorities, citizen groups, private companies, industrial organisations, special interest groups, etc. - may take responsibility for the *niche management* of the new technology, and promote and stimulate its diffusion.

The discussions so far have raised the importance of being able to successfully garner benefits from innovation, cognisant of both the internal and external microcosms, and embedded routines therein. Based on this, a conceptual framework was developed (Figure 5.3) to contextualise the study of innovation and TKTs underpinning SuCo performance. Thus, Figure 5.3, is based on the notion that innovation is determined by three building blocks of an innovation system:

1. a network of interacting actors;
2. a particular technological regime; and
3. routines.

Technological regime defines the routines, which involve learning processes as well as the knowledge base, technological opportunities, appropriation of knowledge and technologies and cumulativeness of the

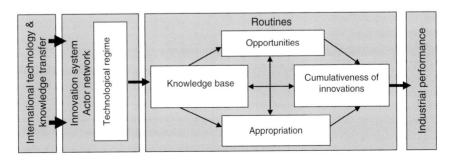

Figure 5.3   Conceptual framework for the development of innovation.

innovations. Learning occurs through collaboration, communication and knowledge exchange between the actors in the innovation system and may result in the development and supply of complementary inventions as well as in institutional, organisational and management adaptations. An innovative technology, however good technically, will only be diffused, adopted and successfully implemented if it fits well in the prevailing technological regime that characterises the professional practice (routines) of the actors in the innovation system (Douthwaite, 2002). Yet the diffusion, adoption and implementation and use of any innovative technology take a certain learning period (Stiglitz, 1999). However, this can be accelerated by a 'niche manager'.

5.3   Innovation, Technology and Knowledge Transfer in Construction

The occurrence of innovation in construction has created significant opportunities for improvements in construction performance in terms of reduction of costs, speed of construction, and construction quality and sustainability. Innovations also help to mitigate the effect of changes in weather conditions on activities at construction sites, enhance the coordination and planning of construction activities, and create new market opportunities by enhancing the competitiveness of construction firms. Innovation in construction has been stimulated largely by increasing construction resource costs and a growing lack of on-site skilled labour.

5.3.1   Construction Technologies and Innovation

Innovation in the construction sector differs from innovation in other economic sectors, as often 'old' technologies are used alongside newly developed technologies and materials, whereas, in most other industries older technologies are completely replaced by the new ones. Construction processes are nonetheless observed in many countries evolving with increasing mechanisation, rationalisation, systematisation, standardisation and automation of construction. This shows:

1. the shift of building construction activities from the site to the factory; and
2. the standardisation of industrially produced building products, components and technologies.

What seems to have happened in construction is that combinations of innovative solutions based on accumulated technological and knowledge advances were adopted in an attempt to move from craft-based construction to a systematic construction process where resources can be utilised more efficiently. In fact, what is apparent from a construction innovation perspective is the *convergence of technologies and knowledge* from different areas and disciplines. The main principles applied to increase sustainability in construction seem to have been placed on the reduction, reuse and recycling of resources, along with the elimination of toxic substances, etc (Kibert, 2003).

The majority of innovations are meant to improve construction sustainability through the introduction of new products to increase energy efficiency, by substituting traditional materials and products with innovative ones, such as eco-materials and products. Other innovations are in areas such as design and engineering, transport and equipment, ICT, robotics, and new business and procurement models. The focus to date has largely been on innovation in residential construction. In other segments, such as office building construction, only marginal improvements have been made to achieve improved SuCo performance (Rovers, 2007). Most of these construction innovations have been achieved incrementally via the traditional practice of 'innovation-by-addition' to existing technologies on component level. As such, they offer only *partial solutions* to meet the building project requirements. The total *construction process* itself has remained relatively unchanged, involving the following separate stages:

1. development and production of building materials and elements;
2. development and production of building design and engineering; and
3. construction process development and execution.

Thus, on-site construction practices take place in more or less parallel processes (Lichtenberg, 2002, 2005). On the other hand, improvement in SuCo performance requires 'innovation-by-integration' – i.e. integrated building product and process innovations. This means that new products and new forms of construction processes should be developed systemically in an integrated way. Innovation following these principles can therefore bring about:

1. improvement of the construction process in terms of reduced costs and time and improved quality;
2. flexible and lifespan-based buildings to reduce operational costs and safeguard value for the client; and
3. reduction in environmental impact (Brand *et al.*, 1999; Lichtenberg, 2005).

5.3.2   International Technology Knowledge Transfer in Construction

The dominant providers of technologies and services in the Architectural, Engineering and Construction (AEC) sector are invariably firms from high-income countries, and firms working in international markets depend on these markets for 32–50% of their incomes. Most of them offer packages composed of a wide range of services in AEC (UNCTAD, 2009). The (quasi-) monopoly position of T&K suppliers in the market often allows them to prevent local partners from becoming competitors. T&K spill-over effects in the host countries depend on the type of contract and the conditions attached to it. The bargaining position of recipients of T&K is therefore influenced by legal protection systems and capabilities to search, select and negotiate. In practice, the quality and price of T&K might be determined through crude bargaining prior to competition (Stewart, 1979; Lall, 2003; Tidd *et al.*, 2006). Local firms in host countries are often only involved in lower value-adding segments of the respective value chains. However, countries such as China, India and Brazil have of late increased their stake in the international construction market since the turn of the century (UNCTAD, 2009). On this theme, Chinese contractors are increasingly engaged in construction technology flows, working with a relatively high level of control in their overseas operations and consequently, a low level of actual TKTs in the host country. In Africa, for example, more than 50% of the Chinese construction firms work as a subsidiary of a transnational corporation or sole venture company, and only 3% collaborate in joint ventures, whilst more than 50% of their workforce (most management staff, technicians and skilled labour) and all plant and equipment originate from China (Javernick *et al.*, 2007).

With a growing number of companies from developing countries entering into *ad-hoc* co-operation agreements with companies of high-income countries, international TKTs have, through foreign direct investments, become the largest source of external financing. It is a crucial instrument and a powerful source of knowledge, capability building and trade expansion, which often is fully exploited in these countries that are characteristically weak in terms of bargaining position, purchasing power, investment, R&D and absorption capacity, and physical infrastructure provision. A majority of developing countries are often left disadvantaged in TKT negotiations. Technology choices are, for instance, usually led by successful advanced technology examples from high-income countries, which enhance the potential adoption of products and processes that may not always be appropriate to local conditions. However, construction has an extensive scope for TKTs from other projects, since construction operations are often transparent, easy to copy and capable of giving opportunities for job-site training (Tatum, 1986). At the same time, a major limitation in construction with respect to innovation is that activities are characteristically *sui generis*, which means that there is, at least in theory, hardly any incentive for innovation, insofar as innovation under such circumstances involves limited replication and extended usage of technologies from previous projects. The application of successful innovative solutions in other projects is even more limited due to the temporary nature of project alliances, with the involvement of a relatively large number of individual parties, and relatively low level of communication (Barlow, 2000; Tjandra and Tan, 2002). Therefore, firms face uncertainties, and their risk vulnerability reduces their willingness to share and exchange information and knowledge about their project experience and reinforces their preference for tried and tested technologies (Franco *et al.*, 2004). This reduces the scope for organisational learning and capability building (Pries and Janszen, 1995; Barlow, 2000; Vakola and Rezgui (2000); Kumaraswamy and Shrestha, 2002; Thomassen, 2003; Bresnen *et al.*, 2004). In this respect, small- and medium-scale enterprises generally appear to have a poor record in terms of continuous professional development, life-long learning and innovation (Davey *et al.*, 2002).

Recent changes in the structure of construction project execution have raised barriers to formal continuous technological capability building and training. Subcontractors, who are now the real employers, tend to be small firms with limited resources and limited organisational and technical capacity. However, although most construction companies regard *training* as an important form of TKT and means of improving their existing knowledge base, they are reluctant to engage in training exercises, partly because of the perceived risk that they would lose trained workers to other firms (or other countries), and partly because of the cost of training, which (at least in the short run) can increase the price of their bids (Egbu *et al.*, 2003). Thus, workers are only persuaded to undergo training if they are paid for 'lost time' (ILO, 2001). Notwithstanding this, the importance of *debriefing* and project evaluation as a *learning mechanism*, is noted in several publications highlighting the links between learning, innovation and benefits arising from these in terms of improvement of business performance (Tjandra and Tan, 2002; Borgatti and Cross, 2003). Yet, given the marginal project profits often associated with construction projects and the time pressures due to the intensity of project activities, ex-post evaluation of the projects is rarely undertaken to document project achievements (Keegan and Turner, 2002). Therefore, knowledge gained in projects is generally neither secured nor diffused across the construction sector. This reduces the scope for learning, for the creation of organisational memory, for the diffusion of the knowledge, and for capability building.

Knowledge acquisition often takes place on the job, especially through learning from the experience of others; but many construction firms still fail to capture tacit knowledge or experience (Egbu *et al.*, 2003). The nature of knowledge, and particularly the difficulty involved in codifying knowledge, is an important constraining factor in the diffusion process. For example, in a number of developing countries, even basic construction knowledge and skills are still mainly acquired through learning-by-doing and learning-by-using, through informal traditional apprenticeship systems (although vocational training schools do exist). The informal apprenticeship system is often not well developed, with deficiencies such as:

1. restricted learning opportunity (only learning-by-doing);
2. a narrow and static range of skills, often long kept within families, clans or tribes without being passed on to 'outsiders' (Debrah and Ofori, 2001); and
3. lack of training about new technologies and techniques (ILO, 2001).

The result is that projects often show delays in delivery and weakness in the quality construction output, thus making the learning-by-doing system a rather costly and time-consuming exercise with little pay back. On this theme, Bell *et al.* (1984) assert that firms cannot rely only on learning-by-doing in order to develop the technological capabilities; they must invest in training and other knowledge creation mechanisms. This requires a deliberate allocation of resources, which is seen as an unnecessary expense rather than an investment by construction firms, workers and clients (ILO, 2001). This reflects that construction firms are working in fields of tension – trying to reconcile short-term cost efficiencies with long-term benefits. Jashpara (2003) concluded that competitive environments have a positive effect on learning and organisational performance, since learning is often focused on efficiency and proficiency to improve the competitive advantage of construction firms.

In contrast, it has been asserted that better performance can be reached in construction through engagement in long-term supply relations. This arrangement would allow knowledge creation, knowledge acquisition and sharing to be closely integrated into learning processes (Egbu *et al.*, 2003). Strategic alliances are an important means for the survival of organisations, and one of the key factors that would help improve performance and satisfy clients, whilst providing opportunities to form learning alliances which encourages mutual (and reflective) learning between partners. Co-operative alliances can create a shared vision of mutuality from which a learning organisation would be expected to evolve. This is endorsed by various authors who indicate that working in a networked context – i.e. as in Communities of Practice or Joint Ventures, which offers the opportunity to combine resources and skills beyond the capability of one single organisation (Kumaraswamy 1995; Egmond *et al* 2007). This also has the additional benefits of being able to spread financial and other risks to better capture learning and use these assets to improve the partnership's competitive advantage (Walker and Johannes, 2003). In addition, Hakansson *et al.* (1999) noted that learning that takes place in a networked context can be influenced by the nature of connections between the parties. Thus, TKTs and embedding tacit and un-codified knowledge (project experience) in firms appear to be more problematic in construction than in other sectors. Kumaraswamy and Shrestha (2002) noted that this might be partly due to conservatism along with the fragmented nature of the industry. In this respect, learning in project-based organisations (as in construction) is still in its infancy. It is therefore important to disseminate lessons learnt, as this is acknowledged to be crucial for business (Birkenbach, 2006), especially where low levels of knowledge adoption and diffusion exist (Ofori 1990; Nam and Tatum, 1997).

5.4   The Construction Innovation System

Long-term project partnering relationships are scarcely taken on board in construction (Barlow and Jashpara, 1998; Bresnen and Marshall, 2000), and linkages are mostly project bound, and are therefore temporary by nature. It is therefore important that the importance of collaboration and communication is understood from the outset.

5.4.1   Actor Network

Production chains in construction tend to be established by means of integrated tendering on the basis of functional demand specifications and a selection process based on best price-quality ratio; where construction innovation (as a system) includes a large variety of actors, which can loosely be grouped as:

- *Innovation support actors:* e.g. (inter)national/regional/local authorities and communities, standards and regulating institutions, professional associations; testing services companies, educational institutions, certification bodies;
- *Innovation supply actors:* including:
  1. researchers;
  2. construction industry participants, with designers, general contractors, specialist contractors, workforce; and
  3. materials and technologies suppliers; and
- *Innovation demand actors*: public and private investors (individuals, service providers, housing corporations, clients, owners, end-users) (Egmond, 2009).

An outline of a typical actor network can be seen in Figure 5.4. This represents the actors involved in the innovation supply and support processes, along with supply/demand continuum. Given this relationship, Winch (1998) noted that large contractors could be considered as mediators between R&D institutes to develop new technologies and knowledge.

In summary, the low and fragmented project bound levels of interaction of firms with other organisations (e.g. universities, technology service providers, governmental agencies, clients, etc) can restrict knowledge sharing; which not only impedes innovation, but also has an impact on project efficiency and efficacy (Malherba, 2002). Thus, there is a need to promote the common interests of actors in order to maximise innovation opportunities.

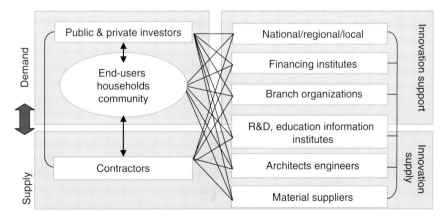

Figure 5.4   A construction innovation system actor network (Egmond, 2009).

5.4.2   Construction Collaboration and Capability

The *knowledge base* in construction is often scattered amongst distinct institutes, organisations and firms, where most of the T&K is tacit and not codified, making diffusion somewhat problematic. In contrast, the manufacturing industries relating to construction account for the majority of inventions and innovations in construction (Manley 2001; Miozzo and Dewick, 2004). This makes construction organisations rather vulnerable to market forces, which can also act as an innovation constraint – especially when under pressure to meet resource requirements in response to the social agenda for SuCo.

Construction project execution tends to follow consistent routines, with a large number of participants working side by side, with a high rate of interdependency and complexity of tasks in varying collaborations. There is also a lack of alignment between the actors in construction projects, each working with its own targets and motivations under pressure of deadlines and budget constraints. This translates into dysfunctional teams, poor levels of co-operation, miscommunication, overlaps, inefficiencies, lack of mutual respect, lost opportunities for optimum use of resources, the result of which makes it difficult for innovation and TKTs to occur (Pries and Janszen, 1995; Barlow, 2000, Lichtenberg, 2002; 2005). This is amplified by the traditional tendering system and the rather detailed specifications in tender documents, which in certain circumstances can discourage contractors from searching for innovative solutions.

The trend over the last couple of decades or so has been one in which large construction firms have moved into international markets, through mergers and acquisitions, often working mainly as service companies, concentrating on management and coordination functions, finding clients and marketing products. These changes were needed to deal with the increased complexity, size and technical sophistication of construction projects and processes as well as the increasing pressures in the international market. This stimulated firms to form consortia through such vehicles as Partnership Agreements and Strategic Alliances. However, the formation of these are often short-lived, being temporary by nature, meaning that capability building opportunities may not always be overtly evident (Wells, 2001; Bertelsen and Müller, 2003). Moreover, conditions may exist which do not support host firms' efforts to develop their capabilities (Abbott, 1985; and Carrillo, 1994). This is particularly the case in developing countries, where large projects are carried out by a handful of large foreign firms with a considerable amount of sub-contracting and other forms of collaboration between foreign and local firms, and where many of the construction technologies are imported (UNCTAD, 2000, ILO, 2001). Innovation here is therefore basically in the hands of foreign companies.

5.5   Technological Regime in Construction

Typically, the marginality of profits and the risk of unforeseen failure and damage during project execution have reinforced conservatism and the reluctance of construction firms to engage in ventures for technological change (Nam and Tatum, 1988; Ofori, 1990; Thomassen, 2003). Indeed, the experience of project implementation professionals appears to have nurtured the perception that innovations have had a negative impact on project performance, on the assumption that any alterations in a project design or implementation would involve lengthy discussions, thus involve additional time and cost (Slaughter, 1998).

The prevalence of constraints on knowledge exchange and sharing has undermined the enthusiasm for innovation. For example, clients that lack awareness and understanding of the significance of innovative solutions tend to be risk-averse and conservative in their business behaviour, and would therefore show preference for tried and tested materials and technologies. This effectively discourages manufacturers and builders from changing their practices through engagement in innovation ventures (Pries and Janszen, 1995; Miozzo and Dewick, 2004). The problem with respect to the adoption of new technologies and the willingness to share the benefits of innovation is exacerbated where construction project stakeholders are significantly dissimilar in terms of their innovation awareness, interests, objectives and motives (Ofori 1996; Malherba 2002). Building codes and regulations, and technical standards in construction can also act as barriers, which constrain innovation and international TKTs in construction along with the prevalence of a variety of contractual agreements, licensing and qualification requirements, etc (Nam and Tatum, 1997; Slaughter, 1998; Barlow, 2000; UNCTAD, 2000). Innovation and TKTs can however, be fostered through appropriate regulatory frameworks, which support predictable technology and policy regimes and good governance, with few restrictions and regulations regarding intellectual property rights (patents, copyrights, trade secrets, etc).

5.6   Opportunities, Appropriation and Cumulative Effect of Innovation

Opportunities to improve competitiveness and sustainability constitute an important driving force for building product and material manufacturers to actively develop (or look for) new ideas and technologies, given the fact that they can profit from economies of scale. Contextual factors such as the aging and shrinking construction work force and the progressively declining number of young people who enter construction have created *technological opportunities* for firms in many Western countries to innovate and apply innovatively industrialised construction processes, which require fewer specialised trades and people. Moreover, the uniqueness of each construction project and the increasingly complex and fast changing market conditions have put pressure on firms to deliver better quality goods and services, thus providing an impetus for innovation. In this respect, there are few opportunities in construction for appropriating benefits, mainly because of the interdependency in construction arising from the complementary innovations by different actors in the industry (Chesborough and Teece, 1996). Furthermore, whilst construction operations are predominantly transparent and easy to copy, they provide an extensive scope for diffusion of inventions and technological improvements from other projects and industries. However, the risk-averse nature of the industry means that innovation would rarely be adopted without the evidence of a proven track record (Hillebrandt, 1984; Tatum, 1986). This situation calls for a paradigm change, with long-term formalised collaboration between construction actors who would be willing to share innovation benefits as well as risks.

Construction technologies have developed over time by means of cumulative innovation, but this has happened at a rather slow pace compared to other economic sectors. Cumulative innovation is possible, especially where this builds on the existence of a knowledge base that is comprehensible. In this respect, tacit knowledge may be available (and comprehensible), but if it is neither accessible nor acquirable, then this will act as a barrier. This, along with the project-based characteristic of construction and the diversity of stakeholders, increases the potential for divergent interests, which can limit the scope for learning and *cumulative innovation* (Nelson, 2004).

5.7 Managing Innovation for Sustainable Construction: The Dutch Case

This section discusses the findings of a study about the factors underlying the state of SuCo in The Netherlands through the application of a network-based approach. However, it has to be noted that in The Netherlands, success in SuCo has not been as impressive as, for instance, in some of the Scandinavian countries(Hamelin, 2007).

In The Netherlands, SuCo innovation has largely focused on increasing energy efficiency and substituting traditionally used materials and building products for better ones. The main principles here are to reduce, reuse or recycle resources to protect nature in all activities, to eliminate toxic substances in construction and apply life-cycle analysis in construction decision making, using for example, innovative energy technologies. In this respect, these initiatives are seen to:

1. prevent unnecessary use of energy;
2. maximise renewables (e.g. solar boilers); and
3. deliberately use clean and high performance non-renewables (e.g. high performance boilers for central heating).

However, many of these investments are not yet completely cost-effective, and therefore not commercially appealing. Notwithstanding this, it is important to acknowledge the role that 'industrialised buildings' can play in driving up quality and value, while at the same time, saving resources and cutting down construction costs (Hendriks, 1999). The application of such building systems has been seen as a three-pronged strategy, to improve construction with the resulting benefits of:

1. flexibility for the client;
2. efficiency in industrial production with reduced material requirement, reduced costs and time, and improved output quality; and
3. demountability to decrease waste for society (www.sev.nl).

In this respect, demountability enables the separate replacement of components (with various life spans), the consequence of which not only extends the life of the building as a whole, but also decreases waste generation associated with demolition.

An elaboration of these ideas led to the development of the Slimbouwen® concept (Dutch for Smart Building and a trademark). This concept asserts that the solution for SuCo should be found in product innovations, which are integrated in the process, and through organisational innovations. Following this concept, the construction process could be transformed into a sequential process (Lichtenberg, 2005). To achieve this, the building is subdivided into four main parts, which can be prefabricated:

a)   foundation, skeleton and floors;
b)   building envelope;
c)   services (vertical through shafts, horizontally through hollow floors); and
d)   in-fill (floor finishing, partitions and ceilings).

The separation of these services package from the main structure is a basis for obtaining flexibility and adaptability, which further benefits the exploitation of this approach.

Slimbouwen® is not a building system *per se*, but can be seen as an innovation development strategy that responds to problems often encountered in traditional building construction. The Slimbouwen strategy forms a guiding framework for the development and production of innovative designs, products, building materials and construction practices in an integrated manner, whilst the functional and economic life span of the structures and recycling and deconstruction of the building components and materials are taken into account (Figure 5.5). Benefits arising out of this concept are multi-layered, from reduced construction periods, through to enhanced competitiveness. It is also possible to construct considerably lighter buildings at a much higher and predictable quality level, with increased flexibility and adaptability (to meet changing demands and market requirements). As such, the Slimbouwen innovation can reduce environmental impact, whilst achieving more sustainable and cleaner construction. For example, the materials used in a dwelling construction project in The Netherlands, weigh about 50% of comparable traditionally built houses of the same volume (Lichtenberg, 2005).

5.7.1   Technological Regime, Diffusion and Knowledge Transfer in Dutch Construction

Many innovations for SuCo in The Netherlands have been developed by organisations through relationships with universities and research centres. Some of these relationships developed through funded research projects in which innovative sustainability concepts have been applied. Knowledge and experience developed in these projects have also been used to further develop

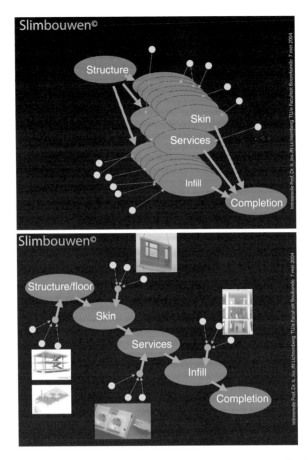

Figure 5.5 Traditional parallel process and sequential Slimbouwen building process (Lichtenberg, 2005). Reproduced by permission of J. Lichtenberg.

and improve design tools for the use of materials, energy, buildings and the built environment (Anink and Mak, 1993; Stofberg and Duijvesteijn, 2006). In addition, it has been recognised that the application and wider diffusion of an innovative construction concept such as Slimbouwen requires an early strong cross-industrial collaboration between stakeholders, along with a multidisciplinary approach during design and production, with appropriate changes being made in some of the traditional construction roles.

For this purpose, Slimbouwen also functions as a shareware platform, providing a knowledge and information infrastructure for firms as the basis for the development of their strategies. The basic concept can be further developed and translated into integrated innovative designs, building components and materials for:

1. services: energy, water supply, sanitation and ventilation;
2. floor, supporting structure and built-in assembly kits;
3. building envelope and specific façade elements; and
4. the transformation of construction project execution.

Principal stakeholders include those involved in the production chain of buildings: knowledge institutions and universities, governmental organisations and end-users. As such, it brings coherence into the fragmented development efforts of several actors in construction; with a shared and collective purpose – to accomplish SuCo through innovative designs, building components and materials. Slimbouwen is one example of this.

Many organisations and clients mention their expectations of acquiring a 'green image' through energy saving as being an important motive. However, whilst some successful pilot projects can be evidenced, the target of complete diffusion and adoption of the new T&K, as well as of an absolute reduction in resource use within construction, has only partially been achieved (Klunder, 2002). It therefore needs to be acknowledged that successful implementation and diffusion of innovative and sustainable solutions also requires process innovation and regime shift. In this respect, whilst governmental bodies are likely to play an important role in making this happen (and Building Regulations appear to be one of the most important reasons for applying energy saving innovations in The Netherlands), further work is still needed to reinforce and encourage uptake.

5.7.2 Lessons Learned

The Dutch case identified that government support was able to influence current reluctance, beliefs, expectations and standards in order to create loyalty to SuCo and encourage innovation. Both theory and empirical evidence coalesce, acknowledging that T&K forms an important component of any innovation system. However, strategies are needed to improve the knowledge, expectations and beliefs concerning innovation for SuCo, which should involve:

a)   voicing and shaping expectations about the new technologies and knowledge through, e.g. demonstration projects; and
b)   stimulation of active T&K exchange amongst the actors in the innovation systems about design and engineering specifications, user characteristics and their requirements, environmental issues, industrial development options, government policies, regulatory frameworks and governmental role concerning incentives for diffusion and implementation (Kemp *et al.* 2001).

These strategies should also create an increased awareness of the potential of novel technologies in the market, thereby increasing the market needs for innovation.

SuCo innovation is a continuous learning process influenced by several drivers, not least the political, social and economic setting of a country. This notion supports the arguments for the need to differentiate policies to formulate specific policy agendas adapted to the socio-economic setting and to the particular features of the construction innovation system and the T&K

base (Aubert, 2004). However, the T&K sets in construction are not overtly exchangeable, and would to a large extent need to be developed. Moreover, learning is hampered by traditional systems, with high degrees of fragmentation, which limits transfer opportunities (Desmyter, 2007). Whilst several technological opportunities are underutilised, it is evident that promotion of the common interests of the actors in the innovation system regarding SuCo is needed. Therefore, policies should seek to remove barriers that limit interaction, communication, and knowledge exchange through formal conduits such as research, design, engineering and marketing, etc, in order to bring about a regime shift that diffuses knowledge (learning) in the actor network.

5.8   Conclusion

Innovation in construction occurs through the accumulation and convergence of technologies and knowledge from various areas and different parties (albeit at a relatively slower rate than in other economic sectors). However innovation and TKTs have been limited, and this has had a constraining effect on SuCo performance. Construction firms are therefore challenged to proactively change their practices and policies to focus on innovations for material and energy saving in buildings and waste reduction during traditional construction processes. Improving SuCo performance requires designers, building material producers and contractors to bring about innovative design concepts, building elements and components as well as adaptations in the building processes. These changes can be integrated into construction projects to achieve the optimum application of sustainability principles during all stages of a building's life cycle. Hence, SuCo necessitates innovative solutions in construction that go beyond the traditional and generally accepted way of building, – where environmental gains would be expected to be overtly matched with economic gains.

Innovation evidence from manufacturing is a good starting point for reflection, as many sustainability and innovation issues can readily be used to improve the construction sector. However, this requires a panacea approach to current thinking, which would involve creating strong relationships and long-term collaboration with supply chain partners focused on one common interest. This requires trust, transparent systems and collective thinking regarding the existing knowledge, skills and opportunities available to meet these ideals. In this respect, the Dutch experience identified one such approach leveraged by government support. However, it is also necessary to stimulate SuCo and innovation through policies that support collaborative engagement.

References

Abbott, P.G. (1985) *Technology Transfer in the Construction Industry 'Infrastructure and Industrial development*. Special Report No. 223. The Economist Intelligence Unit, The Economist Publications Ltd.

Abernathy, W.J. & Utterrback, J.M. (1978) Patterns of Industrial Innovation. *Technology Review*, **80**, 41–47.

Anink, D. & Mak, J. (eds) (1993) *Handleiding Duurzame Woningbouw. Milieubewuste materiaalkeuze bij nieuwbouw en renovatie (Rotterdam) SEV.* Stuurgroep Experimenten Volkshuisvesting, Rotterdam.

Aubert, J. (2004) *Promoting Innovation in Developing Countries: A Conceptual Framework.* World Bank Institute, Washington, DC.

Barlow, J. & Jashpara, A. (1998) Organisational learning and inter-firm partnering in the UK construction industry. *The Learning Organisation*, **5**(2), 86–98.

Barlow, J. (2000) Innovation and learning in complex offshore construction projects. *Research Policy*, **29**, 973–989.

Bell, M. (1984) Learning and the accumulation of industrial technological capacity in developing countries. In: ed. Fransman, M. & King, K., *Technological Capability in the Third World*. Macmillan, London.

Bertelsen, P. & Müller, J. (2003) Changing the outlook: explicating the indigenous systems of innovation in Tanzania. In.Muchie M.,Gammeltoft P., and Lundvall B.-Å. (eds.) (2003), Putting Africa First: The Making of African Innovation System, Aalborg, Denmark: Aalborg University Press.Bijker, W., Hughes, T. & Pinch, T. (1987) *The Social Construction of Technological Systems: New Directions in the Sociology and History of Technology*. MIT Press, Cambridge, MA.

Birkenbach, F. (2006) Innovation management in project-based firms, PhD Thesis, Erasmus University.

Borgatti, S.P. & Cross, R. (2003) A relational view of information seeking and learning in social networks. *Management Science*, **49**(4), 432–445.

Brand, G., Van den Rutten, P. & Dekker, K. (1999) IFD Bouwen, *Principes en Uitwerkingen.TNO. Netherlands Organization for Applied Scientific Research.* Report 1999-BKR-R021. TNO Bouw, Delft.

Breschi, S., Malerba, F. & Orsenigo, L. (2000) Technological regimes and Schumpeterian patterns of innovation. *The Economic Journal*, **110**: 388–411.

Bresnen, M. & Marshall, N. (2000) Partnering in construction: a critical review of issues, problems and dilemmas. *Construction Management and Economics*, **18**, 229–237.

Bresnen, M., Goussevskaia, A. & Swan, J. (2004) Embedding new management knowledge in project-based organizations. *Organization Studies*, **25**(9), 1535–1555.

Carillo, P. (1994) Technology transfer: A survey of international construction companies. *Construction Management and Economics*, **12**, 45–51.

Chesborough, H.W. & Teece, D.J. (1996) When is virtual virtuous? Organizing for innovation. *Harvard Business Review*, **January–February**, 65–73.

Chesbrough, H. (2003) *Open Innovation*. Harvard Business School Press, Boston, MA.

Davey, C.L., Powell, J.A. & Cooper, I. (2002) Action learning in a medium-sized construction company. *Building Research & Information*, **30**(1), 5–15

Debrah, Y. & Ofori, G. (2001) The State Skills Formation and Productivity enhancement in the construction industry: the case of Singapore. *International Journal of Human Resource Management*, **12**(2), 184–202.

Desmyter, J. (2007) *Barriers to Eco-innovation in the Construction World, Belgian Building Research Institute (BBRI)*, Belgium paper presented during the European Forum on Eco-Innovation, Markets for Sustainable Construction. Brussels, 11 June. Available from http://ec.europa.eu/environment/ecoinnovation2007/pdf/ (accessed 11 June 2007).

Dosi, G., Coriat, B. & Pavitt, K. (2000) *Competences, Capabilities and Corporate Performance*. Final Report Dynacom Project.

Dosi, G., Pavitt, K. & Soete, L. (1990) *The Economics of Technical Change and International Trade.* Harvester Wheatsheaf, Hertfordshire, UK.

Dosi, G., Nelson, R.R. & Winter, S. (2000) *The Nature And Dynamics Of Organizational Capabilities.* Oxford University Press, Oxford.

Douthwaite, B. (2002) *Enabling Innovation: A Practical Guide to Understanding and Fostering Technological Innovation.* Zed Books, London.

Dunning, J.H. (1981) *International Production And The Multinational Enterprise.* Allen & Unwin, London.

Egbu, C., Kurul, E., Quintas, P., Hutchinson, V. & Anumba, C. (2003) *Knowledge production, resources and capabilities in the construction industry.* Workpackage 1- Final Report, Knowledge Management for Sustainable Construction.

Egmond-de Wilde De Ligny, E.L.C. van (2006) *Lecture Notes International Technology Transfer.* TU Eindhoven, The Netherlands.

Egmond-de Wilde de Ligny, E.L.C. van (2009) *Industrialisation Innovation And Prefabrication In Construction.* Working paper EvE/06-2008/BCC/TUe presented to OECD, Paris.

Egmond-de Wilde De Ligny, E.L.C. van Vulink, M. & Benda, S. (2007) *Capacity Building through Joint Ventures in the Construction Industry.* Proceedings 3RD International conference on multi-national joint venture for construction works, 1–2 November, Bangkok.

ESCAP (1989) *A Framework for Technology Based Development Planning: An Overview of the Framework for Technology-based Development,* vol. I. Technology Atlas Project, United Nations ESCAP.

Franco, L.A., Cushman, M. & Rosenhead, J. (2004) Project review and learning in construction industry: Embedding a problem structuring method within a partnership context. *European Journal of Operational Research,* **152,** 586–601.

Geels, F.W. (2004) From sectoral systems of innovation to socio-technical systems Insights about dynamics and change from sociology and institutional theory. *Research Policy* **33,** 897–920.

Hakansson, H., Havila, V. & Pedersen, A. (1999) Learning in Networks. *Industrial Marketing Management,* **28,** 443–452.

Hamelin, J.P. (2007) *Vision of the Industry towards 2030 and Beyond.* European Construction Technology Platform (ECTP), France paper presented during the European Forum on Eco-Innovation. Markets for Sustainable Construction Brussels, 11 June 2007.

Hendriks, N. (1999) *Industrieel, Flexibel en Demontabe.* Bouwen IFD: Ontwerpen op veranderbaarheid, Eindhoven University of Technology.

Hillebrandt, P. (1984) *Economic Theory and the Construction Industry,* 2nd edn. Macmillan, London.

ILO (2001) *The Construction Industry in the Twenty-first Century: Its Image, Employment Prospects and Skill Requirements.* ILO Geneva.

Jashapara, A. (2003) Cognition, culture and competition: an empirical test of the learning organisation. *The Learning Organisation,* **10(1),** 31—50.

Javernick, A., Levitt, R.E. & Scott, W.R.Y. (2007) *Understanding Knowledge Acquisition, Integration and Transfer by Global Development, Engineering and Construction Firms.* Collaboratory for Research on Global Projects Working Paper #0028, 1–17. Stanford University.

Katz, M. & Shapiro. C. (1985) On the licensing of innovations. *RAND Journal of Economy,* **16(4),** 504–520.

Keegan, A. & Turner, J.R. (2002) The management of innovation in project-based firms. *Long Range Planning,* **35(4),** 367–388.

Kemp, R., Schot, J. & Rip, A. (2001) Constructing transition paths through the management of niches. In: ed. Garud, R. & Karnøe, P. *Path Dependence and Creation*, Mahwah, NJ: Lawrence Earlbaum Associates pp. 269–99.

Kibert, C.J. (2003) *Forward: Sustainable Construction at the Start of the 21st Century*. Special Issue article in: The Future of Sustainable Construction. International Electronic Journal of Construction (IeJC).

Kibert, C.J. (2005) *Sustainable Construction: Green Building Design and Delivery*. John Wiley and Son, Hoboken, NJ.

Kline, S.J. & Rosenberg, N. (1986) An overview of innovation. In: ed. Landau, R. & Rosenberg, N., *The Positive Sum Strategy: Harnessing Technology for Economic Growth*. National Academy Press, Washinton, DC. 275–305.

Klunder, G. (2002) *Hoe milieuvriendelijk is duurzaam bouwen? De milieubelasting van woningen gekwantificeerd*. Delft University Press, The Netherlands.

Kondratiev, N.D. (1925) *The Major Economic Cycles* (in Russian). Moscow. Translated and published as *The Long Wave Cycle* by Richardson & Snyder, New York, 1984.

Kumaraswamy, M.M. (1995) *Technology Exchange through Joint Ventures*. International Conference on 'Technology Innovation and Industrial Development in China and Asia Pacific towards the 21st Century', Hong Kong, November, Proceedings. 254–261.

Kumaraswamy, M.M. & Shrestha, G.B. (2002) Targeting technology exchange for faster organisational and industry development. *Building Research and Information*, 30(3), 183–195.

Lall, S. (1985) Trade in technology by a slowly industrializing country: India, In: ed. Rosenberg, N. & and Frischtak, C., *International Technology Transfer: Conc, in Concepts, Measures and Comparisons*. Praeger, New York.

Lall, S. (1987) *Learning to Industrialize*. Macmillan, London.

Lall, S. (1992) Technological capabilities and industrialization. *World Development*, 20(2), 165–186.

Lall, S. (2003) Foreign direct investment, technology development and competitiveness: issues and evidence. In: ed. Lall, S. & Urata, S., *Competitiveness, FDI and Technological Activity in East Asia*, Edward Elgar, Cheltenham. 12–56.

Lichtenberg, J.J.N. (2005) Slimbouwen [in Dutch] Boxtel: Æneas, Uitgeverij van vak-informatie voor de bouw en energie wereld, The Netherlands.

Lichtenberg, J.J.N. (2002) *Ontwikkelen van Projectongebonden Bouwproducten*. PhD thesis, Kelpen-Oler: A+.

Lundvall, B, & Gu, S. (2006) China's innovation system and the move toward harmonious growth and endogenous innovation. *Innovation: Management, Policy & Practice*, 8, 1–2.

Lundvall, B-Å. (ed.) (1992) *National Systems of Innovation: Towards a Theory of Innovation and Interactive learning*. Pinter, London.

Macozoma, D.S. (2002) *Secondary Construction Materials: An Alternative Resource Pool for Future Construction Needs*. Concrete for the 21st Century Conference. Eskom Conference Centre, Midrand, Gauteng.

Malerba F. (2002) Sectoral systems of innovation and production. *Research Policy*, 31, 247–264.

Malerba, F. & Orsenigo, L. (1996) Schumpeterian patterns of innovation are technology specific. *Research Policy*, 25, 451–478.

Manley, K. (2001) Frameworks for understanding interactive innovation processes. *The International Journal of Entrepreneurship and Innovation*, 4(1), 25–36.

March, J.G. (1991) Exploration and exploitation in organizational learning. *Organization Science*, 1(2): 71–87.

March, J.G. & Simon, H. (1993) Organizations revisited. *Industrial and Corporate Change*, 2(3): 299–316.

Marsili, O. & Verspagen, B. (2002) Technology and the dynamics of industrial structure: an empirical mapping of Dutch manufacturing. *Industrial and Corporate Change*, 11, 791–815.

Metcalfe, S. (1995) Technology systems and technology policy in an evolutionary framework. *Cambridge Journal of Economics*, 17, 25–46.

Miozzo, M. & Dewick, P. (2004) *Innovation in Construction*. Edward Elgar, Cheltenham.

Muller, R. (2000) *Green Building Basics*. US Department of energy. Available from www.sustainable.doe.gov/buildings/gbarttoc.htm

Murray, F & O'Mahony, S. (2007) exploring the foundations of cumulative innovation: Implications for organization science perspective. *Organization Science*, 18(6), 1006–1021.

Nam,C.H. and Tatum, C.B.(1997) Leaders and Champions for Construction Innovation, *Construction Management and Economics*, 15 (4), 259—270.

Nam, C.H. & Tatum, C.B. (1988). Major characteristics of constructed products and resulting limitations of construction technology. *Construction Management and Economics*, 6, 133–148.

Nelson, R. (2004) The market economy, and the scientific commons. *Research Policy*, 33(3), 455–471.

Nelson, R.R. & Winter, S.G. (1982) *An Evolutionary Theory of Economic Change*. Harvard University Press, Cambridge, MA.

OECD/IEA (2009) *World Energy Outlook 2008*. OECD, Paris. Available from http://www.worldenergyoutlook.org/docs/weo2008/WEO2008.pdf

Ofori, G. (1996) International contractors and structural changes in host-country construction industries: Case of Singapore. *Engineering, Construction and Architectural Management*, 3(4), 271–288.

Ofori, G. (1990) *The Construction Industry: Aspects of its Economics and Management*. Singapore University Press, Singapore.

Patel, P. & Pavitt, K. (1997) The technological competencies of the world's largest firms: complex and path dependent, but not much variety. *Research Policy*, 26(2): 141–156.

Polanyi, M. (1967) *The Tacit Dimension*. Doubleday & Co. Reprinted 1983, Peter Smith, Gloucester, MA.

Pries, F. & Janszen, F. (1995) Innovation in the construction industry: the dominant role of environment. *Construction Management and Economics*, 13(1), 43–51.

Rogers, E.M. (1995) *Diffusion of Innovations*, 4th edn. The Free Press, New York.

Rosenberg, N. (ed.) (1976) *Perspectives on Technology*. Cambridge University Press, Cambridge.

Rosenberg, N. (1982) *Inside the Black Box: Technology and Economics*. Cambridge University Press, Cambridge, MA.

Rovers, R. (2007) What is Sustainable building? In *Report on 2nd European Forum on Eco-Innovation Markets for Sustainable Construction EU Environmental Technologies Action Plan (ETAP)*. Brussels 11 June. Available from http://ec.europa.eu/environment/ecoinnovation2007/pdf/report_lowres.pdf

Schumpeter, J.A. (1934) *The Theory of Economic Development*. Harvard University Press, Cambridge, MA.

Slaughter, E.S. (1998) Models of construction innovation. *Journal of Construction Engineering and Management*, 124(3), 226–232.

Stewart, F. (1979) *International Technology Transfer: Issues and Policy Options*, World Bank Staff Working Paper No. 344. World Bank, USA, 22.

Stiglitz, J. (1999) *Knowledge as a Global Public Good*. World Bank. Available from www.worldbank.org/knowledge/chiefecon/articles/undpk2/

Stofberg, F. & Duijvestein, K. (2006) *Basisdocument Wat is duurzaam bouwen?SenterNovem*, 2nd edn. Ministerie van VROM.maart.

Tatum, C.B. (1986) Potential mechanisms for construction innovation. *Journal of Construction Engineering and Management*, **112**(2), 178–191.

Teece, D.J., Pisano, G. & Shuen, A. (1997) Dynamic capabilities and strategic management. In: ed. Foss, N.J., *Resources, Firms and Strategies*. Oxford University Press, New York. 268–287.

Thomassen, M.A. (2003) The economic organization of building processes., PhD thesis, BYG-DTU, Technical University of Denmark.

Tidd, J., Bessant, J. & Pavitt, K. (2006) *Managing Innovation: Integrating Technological, Market and Organizational Change*, 3rd edn. John Wiley & Sons, Hoboken, NJ.

Tjandra I. K. & Tan W. (2002) *Organisational Learning in Construction Firms: The Case Of Construction Firms Operating In Jakarta, Indonesia* – Report on research conducted at National University of Singapore. National University of Singapore, Singapore.

UNCTAD (2000) *Regulation and Liberalization in the Construction Services Sector and its Contribution to the Development of Developing Countries*. Note by the UNCTAD Secretariat. Geneva.

UNCTAD (2009) *World Investment Report 2009*. United Nations, New York.

Vakola, M. & Rezgui, Y. (2000) Organisational learning and innovation in the construction industry. *The Learning Organization*, **7**(4),174–184.

Von Hippel, E. (1988) Economics of product development by users: the impact of 'sticky' local information. *Management Science*, **44**(5), 629–624.

Walker D.H.T. & Johannes D.S. (2003) Preparing for organisational learning by HK infrastructure project joint venture organisation. *The Learning Organisation*, **10**(2),106–117.

Wells, J. (2001) Construction and capital formation in less developed economies: unravelling the informal sector in an African city. *Construction Management and Economics*, **19**, 267–274.

Winch, G.M. (1998) Zephyrs of creative destruction: understanding the management of innovation in construction. *Building Research and Information*, **26**(4), 268–279.

World Bank (2007) *Building Knowledge Economies: Advanced Strategies for Development; World Bank Institute, Knowledge for Development Program*. The World Bank, Washington, DC.

World Commission on Economic Development (WCED) (1987) *Our Common Future*. Oxford University Press, Oxford.

Zawdie, G. & Langford, D. (2002) Influence of construction-based infrastructure on the development process in sub-Saharan Africa. *Building Research and Information*, **30**(3): 160–170.

Innovation and Value Delivery through Supply Chain Management

Derek H.T. Walker

6.1   Introduction

Innovation through Supply Chain Management (SCM) can take many forms and can be viewed from different perspectives. The purpose of this chapter is to provide insights into how the concept of SCM can be applied innovatively, viewed from the perspective of a value-adding process for providing the promised output to the stakeholders and the value contribution to the organisation's sustainability as an enterprise. The question of whether to settle for in-house or external sourcing of the necessary resources is a pivotal procurement decision. If the latter option is chosen, then a host of alternative strategies are open for consideration, and some of these could offer rich opportunities for innovation.

Green (1999a) warns of the dangers associated with the construction industry blindly accepting management innovations such as lean construction, SCM and associated techniques imported from the Japanese car manufacturing industry. From a critical theory perspective (Habermas and Outhwaite, 1996; Saul, 1997), workforce adopting these innovations can become disempowered, paying the price of organisational flexibility through increased employment tenure uncertainty and reduced bargaining power with their employers, as well as being increasingly subjected to an attitude of managerial prerogative at the expense of their autonomy. This chapter presents an innovative way of thinking about a supply chain to reframe concepts of value that can be delivered through projects. The Organisation for Economic Co-operation and Development's five criteria for evaluating aid projects - *efficiency, effectiveness, relevance, sustainability,* and *impact* (OECD, 2007: 37) are adopted in this chapter to conceptualise 'value'. These criteria are relevant to construction projects, although they are rarely known or even understood by clients. The construction industry is familiar with *efficiency* values such as time, cost and fitness for purpose (quality), and these have been used to measure success for a long time (de Wit, 1988). However, more recent project management (PM) literature has focused on the need for *effectiveness* to be considered as prime importance (Cooke-Davies, 2002). *Relevance* has also become prominent as a success criterion with increasing attention now being given to responding to a broader range of stakeholders than a builder's client or even building users/tenants (Dainty *et al.*, 2003; Nogeste, 2004; Bourne and Walker, 2005). *Sustainability* has also become particularly important (Graham and Smithers, 1996; Khalfan *et al.*, 2006), as has *impact* assessment. *Impact* is a concept that is seldom considered because the focus of management has hitherto been largely on delivering a project as a product, and not on delivering service benefits through corporate expertise, competence and knowledge that enhance capabilities for delivering value in the form of constructed facilities.

An organisational sustainability perspective of value will be adopted for this chapter. Sustainability, in this context, refers to the organisation's role to deliver benefits to its core stakeholders – i.e. its clients, its employees and the society that it engages with in its business activities. Thus, a triple bottom line (3BL) approach to value becomes the operating paradigm. The 3BL approach assumes that organisations are obliged to consider not only financial returns on investment but also a social and environmental dividends (Elkington, 1997). A Latin American supra-organisation, established to radically transform SCM, provides the context of the case studies. Recourse is made to literature on the theory of the firm, outsourcing and SCM and the concept of value chains to underpin the discussion in this chapter. Therefore, the aim of this chapter is to challenge existing construction management practices through insights deriving from leading edge PM research. To this end, this chapter provides an overview of procurement trends in outsourcing that have led to concepts of SCM being more readily accepted within the construction and project management related fields. This is followed by a section on the value temporary organisations bring, and how that relates to project procurement, which in turn is followed by sections discussing seriatim outsourcing and SCM, and the emerging SCM view and what innovation it could offer as a procurement strategy. A case study from the logistics and shipping field is then presented as an illustration of SCM as a means of delivering business transformation. It concludes with a discussion of the implications of adopting the emerging SCM view for the construction industry.

6.2   Organisational Value

Construction projects are predominantly delivered by organisations. Organisations assign teams of people such as design consultants, builders, sub-contractors and materials suppliers, drawn from a wide range of companies/firms. These teams focus upon a construction project to become in many ways a single temporary organisation (Lundin and Söderholm, 1995). This organisation's value contribution extends far beyond providing design services or components, or the assembly of components. It is worth reflecting upon the fundamental value that any organisation contributes, because this lies at the heart of this chapter – providing a 'value' perspective on innovation through SCM. Even more central to individual participating firms in a project is the issue of sustainable competitive advantage that would enable them to thrive in business, moving from one project to the next, delivering real value to stakeholders. It is therefore important to consider what advantage firms perceive they would gain from being part of a managed supply chain, as this would provide a useful guide in the determination of appropriate strategies for managing this supply chain.

6.2.1   Sustainable Competitive Advantage

The concept of sustainable competitive advantage provides a suitable starting point for reflection. Porter (1985) is a widely cited authority on the theory of sustainable competitive advantage, the core tenet of which advocates:

> Competitive advantage grows fundamentally out of value a firm is able to create for its buyers that exceeds the firm's cost of creating it. Value is what buyers are willing to pay, and superior value stems from offering lower prices than competitors for equivalent benefits or providing unique benefits that more than offset a higher price… Inter-relationships among business units are the principal means by which a diversified firm creates value… (Porter, 1985: 3).

Porter (1985) also advocates that firms may procure cost advantage through being able to produce at a lower cost than their competitors, or provide a highly customised product/service and therefore become the preferred supplier, or alternatively, they may offer a unique product or service that their competitors cannot deliver. This means that firms need a particular bundle of assets that they can deploy to provide what is valued by their customers. In this respect, Barney (1991) refers to the resource-based view (RBV) of the firm, noting that organisations need to develop their assets and resources in order to be competitive. These assets include tangible assets, such as money (strong balance sheet), productive resources, such as equipment and people to do the work, competence (i.e. the ability to deliver what is required) and knowledge and skill resources (systems put in place to access information and knowledge needed), that people skilled in their use can effectively deploy. The aim of companies from the RBV perspective is to try to build and deploy their resources more *efficiently* than their competitors to gain cost advantage. Alternatively, they could use their bundle of resources more *effectively* to either provide a customised or unique offering, depending upon the client's needs. However, in managing a project supply chain, there would be a cost imperative driven by cost efficiency as well as an effectiveness imperative driven by coordination, timing and relationship considerations. There may also be an unspoken or assumed expectation of relevance to the customer, supply chain partners or the organisation managing the supply chain, that the supply chain is delivering what is needed, and that the long-term impact would be positive and deliver value.

Managing a project with a supply chain may be the consequence of a desire for repeat custom that would warrant sustainable business. However, the long-term view of business would presuppose that the criteria of efficiency, effectiveness, relevance, sustainability and impact relate closely to having a sustainable competitive advantage. Efficiency and effectiveness affect the *financial* bottom line; relevance and impact affect the *social* bottom line; and sustainability should be seen not only in terms of sustainable business practice that would enable organisations to serve their stakeholders continually, but also in terms of the sustainability of the physical environment.

The RBV was later supplemented by a greater emphasis on knowledge management and the role of people bringing social capital assets to the firm through their skills, attributes and behaviours, their network connections and their potential enthusiasm and affective (want-to) commitment (Meyer and Allen, 1991), which was seen to have human resource management implications (Wright *et al.*, 2001). This wider view of a firm's competence as being strongly mediated by its management of human capital and knowledge was a natural progression of the RBV that better explained how this aspect of resources could be deployed (Sanchez, 2004). Another extension of the RBV is recognition of the dynamic nature of competition and of turbulence in the business and economic environment. This recognition led to the perceived gap in the RBV theory of the firm being addressed through the concept of dynamic capabilities (Teece *et al.*, 1997; Teece, 1998, 2001; Eisenhardt and Martin, 2000; Wang and Ahmed, 2007). This recognised that firms are subject to turbulence and shocks, and part of their ability to respond to this uncertainly is to develop the capability to be agile, reflexive and responsive. This called for a shift from being highly prescriptive in developing strategies through the adoption of a command and control strategic approach, to a position where an emergent strategic approach is adopted (Mintzberg *et al.*, 1998). This approach places emphasis upon a broad brush strategic direction being identified and planned, using milestones with shorter-term emergent crafting strategies (Mintzberg, 1987), by navigation towards milestones using a combination of detailed planning (Olsson, 2006; Olsson and Magnussen, 2007), or by creatively muddling through problems and challenges (Lindblom, 1959; Cates, 1979; Lindblom, 1979; Hällgren and Wilson, 2007).

Firms deliver value in a variety of ways. The perception of this value may be mediated by the preferred strategic stance of the client, or of the coordinating firm managing a supply chain. This can be summarised as providing:

1. *Knowledge, competence and dynamic capabilities* to engage with a client and/or coordinating manager of the supply chain to develop an efficient, effective, relevant and sustainable project design that provides positive impact towards delivering the benefits expected by the client/project owner;
2. *Resources* required to deliver the outputs to the expected five criteria – efficiency, effectiveness, relevance, sustainability and impact; and

3. *Required engagement* with, *and commitment* to the client/project owner
    in dynamically responding not only to 'normal' turbulence associated
    with uncertain ventures, such as construction projects, but also to other
    exigencies that may emerge as a result of a potentially changing eco-
    nomic and business environment over the life of the project.

The first potential value relates to knowledge, information and capabilities
at the front end of the project life cycle. Williams and Samset (2010) stress
the importance of project team involvement at the front end of projects, to
ensure that all viable options are available for the team to be able to deliver
a meaningful and realistic (i.e. sustainable and relevant) project concept
(business justification and overall design). This project concept has to be
developed prior to commitment to further project design development.
Williams and Samset (2010) point out that often major cost/time overruns
of projects from the original budget represents unrealistic initial budget
estimates. They also argue that recognising this potential and requesting
several viable alternatives including the zero option (do not do the project)
challenges the creativity of those involved at this stage. This challenge arises
out of scant information being available at the project concept stage. Also,
because traditional procurement paths exclude the contractor and other
supporting members at the project delivery phase, vital knowledge about
practicalities and sustainable design solutions are not usually available
without closer involvement of the eventual supply chain members. This
leads to unwise project designs that may not be relevant or sustainable,
being locked in, so that significant re-work costs may be incurred (Love
et al., 2000, 2009). The contractor, if involved early enough at the front-end
stage of a project, can contribute significant 'buildability' knowledge
(Sidwell and Mehertns, 1996; McGeorge and Palmer, 2002), which can
similarly add value and reduce waste. Coordinating contractors that have
this type of capacity and competence are often part of a learning alliance in
construction and other manufacturing industries involved in projects (Love
and Gunasekaran, 1999) to create learning chains (Maqsood *et al.*, 2007). It
can be seen from the above that learning, knowledge and insights can add
significant value to the front end of projects.

Firms within a supply chain deliver value through their capacity to offer
specialised resources such as materials, plant and equipment or people. These
may be highly specialised, in that each firm would rely, to some extent, on
competitive advantage for the supply of a customised or rare service. Part of
the value they offer is the 'care why' type knowledge (von Krogh, 1998; von
Krogh *et al.*, 2000). This is knowledge about the relevance and significance
associated with a relationship between firms, which also relates to levels of
mutual commitment, so that there is a possibility to know what each partner
needs to know and their being able and willing to share that knowledge.
Merely having the required resources is not enough to deliver optimised
value; there also needs to be commitment and social capital exchanges.
Nahapiet and Ghoshal (1998) identify social capital as having a structural
dimension (sharing of network links), a cognitive dimension (knowledge and
insights, shared codes and language), as well as relational dimensions (trust,

commitment, obligations and identification). Thus, social capital has both explicit and implicit value. The value of commitment and engagement is important because information is often scarce about what is required and what is possible, feasible or relevant at the front end of projects. Therefore, wider engagement with specialisation experts within the supply chain often contribute significant value to a project (Klakegg et al., 2010). This perspective of what value is needed from SCM to effectively deliver projects or construction, clearly sees construction as a lot more complex and complicated than merely mechanistically assembling components. It involves sophisticated integration at the delivery stage where many unforeseen events throw the best-made plans into turmoil. Thus, systems integrators need access to a supply chain that not only provides the goods and services but also delivers assets that are rare and difficult to imitate – knowledge, effective commitment and intangible social capital. Without these assets being willingly deployed, projects can lapse into dysfunctional chaos as compared to the self-correcting responsiveness of well-functioning committed teams who can cope and mutually adjust (Hällgren and Maaninen-Olsson, 2005).

6.2.2   The Cost of Transacting Business

For every economic transaction involving procurement of resources for construction project work, there are usually transparent and hidden costs. These costs are explained in terms of Transaction Cost Economics (TCE). According to Coase (1937) and Williamson (1985; 1991; 1993), TCE relates to the cost of doing business, as every transaction between parties typically involves cost. This can be explicit and easily measured, such as a tendering cost involving advertising, search costs for determining where to advertise the tender, filtering or short listing, evaluation and tender award costs. There can also be a lot of implicit costs expended such as training, learning from mistakes in the award or through governance provision for oversight, assessment progress payment evaluation and a host of other costs. TCE also involves the cost of establishing relationships, trust and the like. It can be appreciated that tendering for the lowest cost can yield the lowest contract award cost, but may entail many pre-tender expenses and heavy post-tender award costs for contract administration, monitoring and evaluation and even conflict resolution related costs. Thus, the lowest cost tender award may easily result in the lowest value result, particularly if the scope of work is difficult to determine and/or the initial value to be obtained relates to knowledge and advice about potential options and solutions. This has profound implications for management of the supply chain, because the total cost, *including transaction costs*, should be considered. As suggested above, the value that organisations contribute includes what they do to lower transaction costs. SCM provides potential opportunities in this respect. Organisations therefore tend to generate value through:

- knowledge, skills, expertise and insights, particularly at the front end of projects;

- social capital, which includes extended networks and knowledge and linkages to provide products or services through *ad hoc* or more formal arrangements;
- products or services;
- commitment, enthusiasm and support; and
- potential savings in transaction costs based on open and mutual negotiations about the ways and means of providing trust and credibility mechanisms that are more effective and efficient than might otherwise be the case under 'normal' trading.

6.3   Value Generation and SCM

At the heart of a construction firm's procurement process lies a fundamental best value decision – a make-or-buy decision that must be taken to sell goods or services that generate more value to the organisation than it expends in terms of a variety of resources. This best value decision is a TCE value inspired decision. Traditionally, most of the literature on procurement (particularly that from the late 1980s to mid-1990s) assumes this decision is mainly a cost/time competitive advantage one (NBCC, 1989; Masterman, 1992; Australian Industry Commission, 1996; Kumaraswamy, 1996; Kelly *et al.*, 2002; Hughes, 2006). Latterly, and leading this debate during the last part of the twentieth century, this focus has shifted towards collaborative and relationship-based approaches through partnering, for example (Larson, 1995; Matthews *et al.*, 1996; Fellows, 1997; Lendrum, 1998; Thompson and Sanders, 1998). In this respect, a more recent focus has emerged based on various stakeholder's conception of what value is exchanged rather than what resources are exchanged (Kelly *et al.*, 2002; Male, 2002; Langford *et al.*, 2003). Overlapping this focus has been an interest in how this happens in construction supply chains (Dainty *et al.*, 2001; Palaneeswaran *et al.*, 2003; Pryke, 2004; Khalfan and McDermot, 2006); how e-portals or enterprise resource planning (ERP) systems (Chan and Swatman, 2000; Chan *et al.*, 2009) achieve integration; and how other means of electronic co-operation both reduce transaction costs and increase efficiency (Croom *et al.*, 2000; Akkermans *et al.*, 2003; Croom, 2005). In addition, there has been an emerging interest in value chains rather than supply chains, particularly from knowledge being a key resource being exchanged and value added (Maqsood *et al.*, 2007).

6.3.1   The Buy or Outsource Decision

Fill and Visser (2000) argue that the buy or outsource decision is triggered by a need to deliver resources from within the organisation (a make decision) or to source this requirement by using the resources from an external organisation (a buy decision). Buy decisions involve sub-contracting a service (e.g. design) or product and service (e.g. the supply and installation of air-conditioning). Apart from outsourcing services such as a consultancy, other options are in use today. This is called in-sourcing, where formally internal expertise is spun off as an external entity, but remains located in-house with project teams working as before, but in a more flexible and cost effective way to the host organisation. The spun off entity is free to take on other work, either as part of its business or on a relationship-contracted basis, usually with minimal transaction costs (McKenna and Walker, 2008). There are clearly numerous options for combining in-house and external combinations for sourcing required resources. Most construction projects require both make and buy decisions. Certainly, much procurement in the construction industry revolves around materials (e.g. ready mix concrete), equipment (e.g. crane hire) and sub-contractors (e.g. mechanical or electrical specialists). Generally these days, contractors tend to be system integrators rather than doing much of the construction work themselves (Winch, 2003). This means that the construction contractor's organisational skill set and competence rests with effectively managing a chain of goods, equipment and services.

6.3.2   Supply Chain Management

London and Kenley (2001) provide an insight into the general industrial organisation's use and development of SCM and how that has influenced the construction industry's adoption and adaptation of SCM. They show influences emerging from the automotive manufacturing industry (Womack et al., 1990), which triggered an interest in lean production (Womack and Jones, 1994, 2000) through to the construction industry where, in particular, Koskela (2000) was active in linking SCM with waste reduction and the effective transformation of construction inputs into completed built facilities. In addition, contributions from Rowlinson and McDermott (1999) and London and Kenley (2001) also trace how strategic construction procurement links to SCM and to the formation of strategic alliance forms.

SCM has been described as a process of coherently integrating all supplier inputs into product/service delivery with a general privilege of gains in competitive advantage for improved cost, time and quality efficiency measures (Sadler, 2007). An efficiency focus is centred upon the transaction; meaning, getting the lowest cost for the project, and a more reliable time delivery of project inputs in the case of construction projects. This is achieved through better capacity of negotiated price based upon bulk purchases, lower transaction costs for reducing the total numbers of suppliers and lowering storage, potential waste and handling costs by Just in Time (JIT) approaches. At the same time, efficiency gains can be achieved through JIT, lean production techniques and many of the approaches reportedly used in manufacturing, automotive and aerospace (Green, 1999b; Michaels, 1999). Part of these gains relate to reducing the number of subcontractors and the use of Keiretsu-like clusters of sub-system assembly integrators, such as those in Japan (Womack et al., 1990); and is particularly well developed in the automotive manufacturing industry (Dyer, 2000). These form an effective group to supply components involving partnered suppliers to co-operate on design and assembly in a way that minimises waste, increases smooth scheduling of

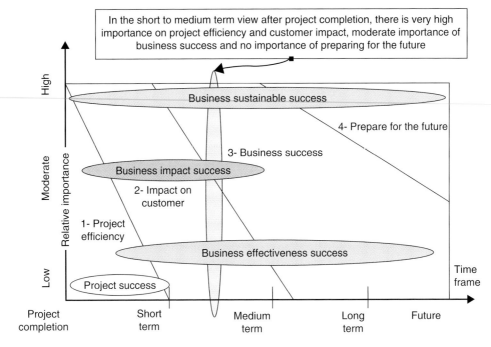

Figure 6.1   Temporal views of success (adapted from Shenhar *et al.*, 2001).

assembly, and provides knowledge exchanges about potential problems or advantages of specific components' properties when combined in assemblies. These types of efficiency gain provide a strong argument for the use of SCM in construction. Effective SCM can therefore be seen to be useful in providing a positive long-term business impact rather than narrow efficiency project objectives (Shenhar *et al.*, 2001). However, this view is challenged by Green (1999a,b), as discussed earlier, and by James (2006) in an article about a range of industries using SCM ranging from supermarket retailers through to the oil and gas sector. These authors caution against unethical use of SCM as a mechanism for coercion by continually reducing costs through profit margin competition.

Figure 6.1 illustrates a series of success factors across various time perspectives. From this, efficiency is associated with project success, whereas gaining positive impact on the customer is associated with business impact success, which is a combination of efficiency and effectiveness. SCM across a business that provides outsourcing TCE benefits can therefore be seen to contribute to long-term business success. SCM can also contribute to business sustainability success by using the knowledge and social capital generated through truly relational SCM that draws firms into business transformations, and prepare them for future success. This is discussed in the next section. However, cautionary suggestions relating to the conduct of SCM (cited above) are salient in context with SCM motivation. Where an outsourcing agent's guiding principle or sole purpose is cost or time reduction, the result in the long term can be exploitation, intimidation and coercion, which could produce a dysfunctional supply chain. Outsourced entities in that chain may then be forced to cut corners or make short-term decisions that could have a detrimental impact, or prevent long-term sustainability due to 'cannibalisation' of weaker links in the chain (James, 2006).

The issue of alliances forming part of an improved and integrated supply chain in a construction context have been extensively explored. For example, research based upon the National Museum of Australia by Walker and Hampson (2003a,b,c) identified how collaboration and co-operation between the client, design team and major system integrators (a significant part of the complete supply chain) provided exceptionally successful project results (Walker and Lloyd-Walker, 2010). Alliances have also been shown to facilitate better relationships between participants (Davis, 2006), and where innovation has been enhanced through alliances (Rowlinson and Walker, 2008). This level of co-operation has been shown to be effective and sustainable in several other industries. Porter (1998), for example, discusses the concept of clusters as loose or more formal industry player groupings, which form a major part, if not all of the entire supply chain. These collaborate and cooperate not only to make business transactions but also to engage in specialised knowledge exchange that allows the chain to be efficient, effective, have a positive impact and create new products on a sustainable basis. Porter (1998) further cites examples from the Californian wine industry, the Italian leather fashion industry and German chemicals industry. The advantages posited through these clusters are similar to those claimed to be achievable through SCM. Other forms of alliance collaboration that contribute to SCM literature in construction have been discussed under the rubric of 'FAs', by which best value is agreed upon from the perspective of the purchaser and provider of goods and services, using a relationship focus based on trust and commitment. Khalfan and McDermot (2006) demonstrate through case studies how this approach can provide innovation and innovative thinking, so that previously unconsidered or poorly considered solutions to problems can be explored, evaluated and adopted. This kind of transformation of the SCM concept takes its form from a managerialist form of efficiency and effectiveness in supply chain integration and optimal management of resources (Sadler, 2007), geared to addressing relevance, sustainability and impact issues.

Fisher (1997: 109) provides some recommendations about how to devise an ideal SCM strategy. The first step is deciding whether the product/service is functional or innovative. Functional products require an efficient process, whereas innovative products require a responsive process. In the construction context, standard mass-like products, such as estate housing, warehousing, small factory workshops – referred to in the USA as 'cooky-cutter' projects – are generally functional, although some customisation is inevitable. Innovative projects – called 'bespoke' in the UK, are highly customised, specialised and require non-routine problem solving throughout the project phases. Fisher (1997: 109) recommends mapping the characteristics of the actual supply chain, so that, if the supply chain is managed as an innovative product, which has its primary focus on responsiveness, then managing it to be primarily efficient will close out all kinds of options, reduce or inhibit knowledge transfer and create a spiral of cost/timing cutting to the detriment of long-term efficiency, long-term business impact or long-term sustainability.

Taking this idea forward, integrated information and data exchange can be viewed as an example of a functional versus responsive strategy. Chan and Swatman (2000) discuss how the use of computer systems has moved on from electronic data interchange (EDI) to e-commerce, particularly with the use of web portals. The level of e-commerce for functional products could be envisaged as being highly focused on reducing waste (particularly paperwork transactions), enhancing financial and scheduling information transfer and other product-specific technical information transfer. A high priority could be to integrate the supply chain through a common or well-linked enterprise resource planning (ERP) system. That could be contrasted with an innovation-oriented product. Here the focus would be a lower priority on mechanistic ERP e-commerce, although that would not be ignored, and a higher priority on access to external information bases, social networking software and other kinds of applications that allow and encourage people to exchange knowledge and perceptions. Jewell and Walker (2005), for example, illustrate how a community of practice social networking software application was successfully used by a UK constructor Carillion for knowledge exchange. A community of practice (CoP) can therefore be seen an informal or formal organisational structure with facilities that enables members with common interest to share knowledge and perceptions (Wenger *et al.*, 2002). Peansupap and Walker (2005) show how CoP can be facilitated by groupware software applications for personalised knowledge exchange. This brief illustration shows how SCM can be effectively sustained through the appropriately matched use of information communication technology.

6.4   Emerging Supply Chain Management Issues

This section discusses emerging conceptual SCM issues pertaining to innovation.

6.4.1   Value Chains

The conceptual shift from SCM to value chain management (VCM) has been advocated as being logical. Figure 6.1 highlighted that efficiency at the project level is a tactical issue, while preparing for or creating a future is a strategic issue. However, successful organisations move beyond merely managing supply chain intelligence, knowledge and information to actively co-create value with their array of suppliers and clients. This is particularly salient in volatile and turbulent conditions, which is highly characteristic of projects at the front end when scoping and strategic design decisions are being made. As Normann and Ramirez (1993: 66) note, 'Increasingly, successful companies do not just add value, they reinvent it. Their focus of strategic analysis is not the company or even the industry but the value-creating system itself ...' In this innovation-oriented view of the need to deliver projects, importance is placed on value co-generation providing a more holistic mental model of what each supply chain member (including the client and end-users) can possibly offer. Normann and Ramirez (1993: 69) also make several salient points about the new logic of value that applies to VCM. First, where value occurs through harnessing a complex constellation rather than a linear supply chain, the business goal is to encourage and mobilise clients and customers to take advantage of creating value for them by liberating them from potential constraints for their ability to freely access value as they see it. They give examples of the multinational furniture company IKEA that mastered the innovation of pre-packaged modular furniture that customers could assemble themselves in the configuration that suits their needs. In the context of construction, this means greater involvement of the client at the briefing and front end of projects. This principle is applied in several ways, for example, by sophisticated clients engaging with visualisation tools to explore various possible options, or construction contractors involving sub-contractors and suppliers in modelling progress through a project to identify possible delivery and/or installation clashes. This approach uses 4D design modelling (Fisher *et al.*, 1997; Fischer and Drogemuller, 2009). Significant recent advances in BIM also allows this interaction with the supply chain (Aranda-Mena *et al.*, 2009), to promote a value chain perception of what would otherwise be referred to as the supply chain.

Following this theme, Normann and Ramirez (1993: 69) argue that as potential value contributions become more complex and involve more actors within a network, the firm's principal role becomes one of systems integrator by reconfiguring the range and types of inputs and transforming relationships between members of a value chain. From a construction context, innovative construction contractors are increasingly leading, or being within a consortium leading large-scale build-own-operate-transfer (BOOT) projects and alliance infrastructure projects. Within these contexts, the construction contractor therefore becomes a catalyst for considering the technical feasibility of options and their relationships with design teams and other consultants to be system integrators. This has evolved from innovative contractors providing 'buildability' advice (Sidwell and Mehertns, 1996; McGeorge and Palmer, 2002) to being at the heart of BOOT schemes (Hall *et al.*, 2000; Walker and Hampson, 2003b).

Finally, if the model of co-production of value chains is accepted, then sustainable competitive advantage must be accepted as being at the core of that value chain. A construction contractor that understands how this chain may deliver sustainable value can then mobilise those value-generating entities in a coherent way to achieve the value objectives. This means that innovative contractors need to be able to map value within value chains. This has been done in other industries; in retailing for example, Kim and Mauborgne (2000: 131) produced a matrix that had on one dimension, six utility levers (including environmental friendliness, convenience characteristics) – i.e. properties of how the product/service could be perceived to be of value; with a second dimension containing six stages of a buyer experience (e.g. purchase, delivery, maintenance, etc). They explain how a product/service can be mapped on a series of cells within these coordinates in the matrix. This approach can be adapted to provide innovative solutions to novel problems. Just being able to map project features against delivery stages could expose and unearth hitherto unconsidered options. This kind of mapping needs input knowledge about the value chain to construct such maps. Walters and Lancaster (2000: 162) show a value chain model for retailing but again this can be adapted to other industries, including construction. Therefore, it is acknowledged that as a firm's core competence includes knowledge of their current and potential place within a value chain, this needs to be managed from a strategic and tactical perspective, as knowledge about competitors, collaborators, clients and environments could also be perceived as an innovative company's prime asset.

6.4.2   Knowledge Chains

Reconceptualising supply chains to value chains leads to a further re-conceptualisation – knowledge chains. Taking this view can be seen as valuable. A supplier or subcontractor, or indeed client, possesses valuable knowledge about a supply chain member's ability to be flexible, adaptable or even able to accommodate JIT delivery, or their history of getting on with others in the value chain. Supply chain members may also hold key knowledge about how their offerings could be better used by others. This knowledge then becomes critical for innovations in processes, products or relationships. For example, in a review of the knowledge management literature, Nonaka *et al.* (2006) re-visit Nonaka and Konno's (1998) concept of *ba*:

> Ba is a shared space for emerging relationships. It can be a physical, virtual or mental space, but all three have knowledge embedded in ba in common, where it is acquired through individual experiences, or reflections on others' experience. For example, members of a product development project share ideas and viewpoints on their product design in a ba that allows a common interpretation of the technical data, evolving rules of thumb, an emerging sense of product quality, effective communication of hunches or concerns, and so on. To participate in ba means to become engaged in knowledge creation, dialogue, adapt to and shape practices, and simultaneously transcend one's own limited perspective or boundaries' (Nonaka et al., 2006: 1185).

The interesting part of the above quote is that supply chains become perceived as knowledge chains in a *ba* sense. Having contractors, consultants, client representatives and those in a supply chain within a *ba* co-located either physically or virtually, can trigger enormously valuable conversations that can lead to value generation. The pre-project design office or the construction office can be seen as a *ba* rather than merely a fixed place; and perceiving this space as a *ba* may facilitate the space to be better poised for improving knowledge exchange, creativity and innovation in both decision making and production. So, what makes this way of looking at SCM provide such an advance? The answer lies in part with the way that knowledge may become sticky and flows slowly (Szulanski, 1996). Szulanski's earlier study (1995) was is based on 271 observations of 122 best-practice transfers in 8 companies, which sought to better understand the different rates of adoption and adaptation of new ideas. This work identified four characteristics of factors that affected knowledge transfer fluidity. The first of these revolved around the nature of knowledge being transferred – some types of knowledge are easier to transfer than others, and some ideas are easier to 'sell' than others. The second revolved around the characteristics of knowledge source – motivation to share knowledge and the perceived reliability and value of that knowledge source affects this. The third revolved around the recipient of knowledge – motivation to accept the knowledge and ability to absorb new knowledge (absorptive capacity). The fourth revolved around the context in which knowledge is transferred, which includes the relationship between source and recipient, and the physical and cultural environmental context.

Therefore, a *ba* that is supportive of containing actors that are committed in a fertile environmental context allows knowledge to be created, transferred and applied more easily than where conditions are less favourable. A knowledge chain view can therefore help prepare the ground for supply chains to increase their absorptive capacity and establish the mindset of actors in that network to be better positioned to exchange valuable knowledge. This can trigger efficiency, effectiveness, impact, relevance and sustainability.

6.4.3   Human Capital Chains

It is important to acknowledge the importance of human capital (HC) and its relevance to VCM. Nahapiet and Ghoshal (1998) see knowledge exchange and knowledge combination as a 'lock and key' concept, needing the opportunity for (and expectation of) parties for an exchange, their motivation to do so, and their compatibility. They present a model of how social capital, through combination and exchange of intellectual capital (knowledge in its many forms), generate new intellectual capital (innovation, knowledge, etc). They include the three dimensions of social capital in their model.

The first dimension of social capital is structural. This relates to network ties, network configuration and organisational structures to facilitate relationships on the network. In the context of construction, the entity that can do this is the supply chain, where participants in this chain can be linked together as a vital part of the vision of the value chain as stimulating intellectual capital through knowledge exchange and enacting innovation to enhance the competitive advantage of the value chain, as well as its constituent members. The second dimension is cognitive, which involves shared codes and languages. This is particularly relevant to project management, as it provides the basis for a clear vision of what the value chain sets outs to achieve in terms that are meaningful for each party (Christenson and Walker, 2004; Steinfort, 2010); and that are efficient, effective, have impact and are sustainable. The third of Nahapiet and Ghoshal's (1998) dimensions of social capital is relational. This comprises trust and commitment in terms of norms, obligations and identification. Trust and commitment are important elements of social capital. Much has been written about trust and Mayer *et al.*, (1995) offers a model of how trust works in relationships. The three principal components are the ability to deliver (as individuals or within organisational constraints), benevolence (positive intentions with no aims to harm in any way) and integrity (being open, honest and consistent in behaviour and practice as espoused and acted upon). Commitment may take several forms. Meyer and Allen (1991) identify three types of commitment: *continuance* is about cost – cost versus the benefit in a highly calculative way; *normative* is about obligation – the duty to do something perhaps as a kind of favour bank, or cultural expectation; and *affective*, which is to do with wanting to do something with aligned objectives, so that motivation is intrinsic. HC can therefore be established through an organisational structure such as a supply chain. However, structural considerations discussed by Nahapiet and Ghoshal (1998) are important here, as is the cognitive dimension; such that there should be recognition of potential contributions of knowledge and experience by members of any value chain. Finally, there is a vital human relationship component of VCM, which is about establishing common (or at least complimentary) elements of culture, norms, behavioural expectations, 'war stories', etc, to bind the chain together in some ways to common goals.

The 'stickiness' of knowledge was focused upon as being a crucial issue in getting the requisite knowledge exchange across a value chain, not only to encourage innovation, but also to make it happen. Absorptive capacity is a fundamental concept in that regard. Zahra and George (2002: 198) provide a conceptualisation of this concept by comparing numerous views of it as published by leading experts. They define absorptive capacity as 'a set of organisational routines and strategic processes by which firms acquire, assimilate, transform and exploit knowledge for purposes of value creation" This is a wider and deeper conceptualisation than previous definitions, including Cohen and Levinthal (1990) who saw absorptive capacity in less dynamic and explicitly systemic terms than Zahra and George (2002). The important additional insights that are relevant to this chapter are that Zahra and George (2002: 192) present a model in which they see experience and knowledge as an input, and competitive advantage as an output. Absorptive capacity therefore has both a potential acquisition and assimilation phase, together with a realisation transformation phase triggered by knowledge exploitation. These elements of the input/process/output model are activation triggers between the input and absorptive capacity process, with social integration mechanisms being critical to allow potential absorptive capacity to be realised by what Zahra and George (2002) call 'regimes of appropriability'. These phases facilitate flexibility of strategy, innovation and superior performance that is characteristic of competitive advantage.

It is now possible to appreciate the critical role that a VCM or SCM approach has on focusing innovation on the delivery of value in construction projects in order to establish and maintain efficiency, effectiveness, positive impact, relevance and sustainability. Readers should consider their experience within any supply chain that may have had aspirations for innovation and sustainable competitive advantage, to question how well the system integrator considered the structural, cognitive and relational dimensions of social capital that Nahapiet and Ghoshal (1998) highlighted; and means for measuring HC, perhaps through a capability maturity model (CMM)? This approach has already been used in the construction context (Manu and Walker, 2006; Maqsood *et al.*, 2007). VCM can also involve measurement, monitoring and improvement of innovation performance using CMM (or similar measurement tools), which if leveraged appropriately, could be a valuable vehicle not just for innovation but also for complete business transformation.

6.5   Case Study of Supply Chain Management Triggering Total Business Transformation

One important purpose of this chapter is to illustrate a practical example of SCM that has had a transformational impact leading to sustainable competitive advantage. A case study example has been purposefully selected that may at first appear incongruous to the construction industry context, but is highly relevant. This case study was initiated in the context of high economic turbulence, the kind of extreme triggers that force innovation and change (Nonaka, 1988; Kotter, 1995). Two main forces triggered an innovative response – the formation of the Great Southern Common Market (Mercosur) and the heightened levels and implications of globalisation for the logistics industry sector within Latin America.

The Mercosur was established during the 1990s, and like the European Union (EU), aimed to better integrate the commercial, political and cultural life of its bloc of members. These include Argentina, Bolivia, Brazil, Chile, Paraguay and Uruguay, with Peru, Venezuela and South Africa as potential future members. An initiative took place in response to the globalisation wave that hit Latin America whilst its political environment was going through drastic changes during the formation of the Mercosur (Arroyo, 2009; Arroyo and Walker, 2004; 2008; 2009). The initiative involved forming a collaborate entity called the Atlantic Corridor, which had as its aim a method of getting together a wide range of logistics supply chain participants to share problems and ideas and to promote efficiency and effectiveness through a community of practice. This turned out to be an effective business transformation vehicle that the construction industry SCM could learn from. In this respect, the research approach involved analysis of extensive documentary evidence drawn from minutes of meetings, corporate intelligence from numerous participants, semi-structured interview data and reflection on insights gained by Arroyo as an embedded actor in the Corridor. The case studies illustrate how efficiency was achieved by bringing about a change in the logistics business to make improvements resulting in cost and time savings. These also brought about effective change by bringing together logistics elements in a way that removed waste and re-work and improved service delivery. They are relevant to this chapter because this approach is similar in many ways to a supply chain delivering construction projects. They had a positive impact because the changes made and explained in these case studies not only enabled companies under threat of being eliminated by their global competitors to transform their business models; the changes made also improved the firm's integration into the global market. Sustainability was achieved by virtue of these firms not only surviving, but also prospering; and that transformation vehicle is still operating today. This case study illustrates how a SCM to VCM mindset change can lead to business transformation, and this could be the greatest innovation participating organisations could imagine.

6.5.1   The Atlantic Corridor Evolution

The globalisation wave that challenged complacency firms that were accustomed to operating in local markets within a protective context, particularly affected the Latin American continent during the late 1980s. South American regional companies had up to that time enjoyed operating within bilateral treaties under a wide array of subsidies and protective regulations that enabled them grow without having to face the competitive pressure of a free market. Foreign companies either found regional entry barriers too high, or were deterred from expanding across a new potential market. However, the creation of the Mercosur offered significant opportunities for business transformation during the early 1990s; and this resulted in Mercosur shipping and logistics firms experiencing a potential threat by global (and more resourceful) world-class shipping companies, port operators and logistic operators. The Mercosur market became strategic to these global firms that were determined to expand into Latin America to maintain their global competitive advantage. Consequently, regional firms suddenly experienced a severe and change-spurring shock to their business logic.

An innovation that emerged from this turmoil was the formation of a form of strategic community of practice (CoP) that became known as the Atlantic Corridor. The genesis of the idea came from several politicians in Brazil, who initially established 'integrating forums' in Brazil and expanded these into a CoP to include members from Argentina and neighbouring countries. The idea of the 'integrating forums' was to initially establish an on-line community at various key points in Brazil and Argentina – linking industrial hubs with consumer centres along the continent. The aim was to achieve, through an improved transportation logistics infrastructure, higher productivity, and thus make participants accomplish greater economies of scale. While this concept appears at first to relate to physical issues such as rationalisation of facilities and identifying expansion and reconfiguration potential, the problems to be solved included political, legislative, managerial and social issues. Members join the Corridor at one of a number of participation levels, ranging from involvement in forum meetings as an observer, through to gaining full access to a database of past and current problems and solutions and relevant posted documents accessible through the Corridor's intranet. This intranet was developed by linking member's own intranets; and it was only recently that this was feasible with the achievement of a critical mass of participants. The Corridor was thus initiated as a regional reaction to the global threat by an increasing number of national and regional companies (both logistic and logistic-related firms) that decided to join forces to face (by then) an uncertain and threatening business environment. These operators came from a wide array of sectors within the shipping and logistic world: ocean and river transportation companies; marine and river port operators; air-station operators; trucking and railway companies; warehousing and distribution operators; freight-forwarding firms; air-cargo freighting operators; consultancy and surveying services; law and insurance services; exporting and importing companies; ship-building and naval repair firms; traders and brokers; etc.

The Corridor provides an example of a business transformation project involving extensive knowledge transfer and trust building as core elements. Arroyo (2009) examined the Corridor's ability to first generate trust, and then to facilitate knowledge transfer among stakeholders. How the Corridor facilitated efficiency for its members, and how it facilitated effectiveness, relevance, positive impact and sustainability, is discussed in the following sub-sections through a summary of five case study results derived from Arroyo's study (2009).

6.5.2   The Atlantic Corridor Facilitating Efficiencies

This subsection provides an account of part of the Corridor's activities and is based on a paper originally presented at a New Zealand conference (Arroyo and Walker, 2004).

Corridor participants gathered fortnightly in their respective video-conference forums (e.g. in the Brazilian remote north-eastern region, or in the remote parts of the Argentine Patagonia region), where they would submit their logistic problems or business limitations for discussion to seek advice on how to quickly resolve these. The problems were wide and diverse, some of which included:

- the lack of a reliable buoying system along the Amazon River;
- the difficulty in interpreting the Buenos Aires customs legislation that affected Brazilian exports;
- the snow intensity along the Andes mountains that impeded the continuous flow of trade between Argentina and Chile;
- the lack of reliable feeder services linking Buenos Aires and Patagonian ports;
- the monopolistic barge services between Manaos and Peruvian river terminal;
- delays from changing rail gauge at the border between Argentina and Brazil;
- draft limitations to reach acceptable economies of scale to reach Bolivia by river; and
- the lack of suitable cranes at this or that port.

Members discussed their problems, proposed solutions, decided upon strategies to solve problems, recommended actions at any level in whichever country the problem lay, and did whatever was necessary to overcome these drawbacks. Member groups were connected on-line and carried out their studies and assessments, which affected their businesses from an operational, financial and commercial perspective. Participants also had a wide exposure of related activities that posed problems to trade and transport. Group membership included:

- ship owners and shipyard operators;
- operators of wide variety of transport medium, including ocean and river terminals, airports, airlines, air stations, barges, truckers, railway operators;
- support logistic groups including insurance underwriters; lawyers, chartering brokers, export and import brokers, custom brokers, bankers, members of provincial and municipal governments, consulting companies, universities; and
- law enforcement agencies, frontier police and coast guard services.

This example represents a complete paradigm shift in thinking about co-operation and collaboration. Many of these participants were loosely involved in supply chains – though early in the formation of the Corridor CoP; they did not tend to see their initiative as a supply chain, or even value chain innovation. Generally, they were reacting to a hostile and threatening environment, and trying through co-operation and collaboration to help solve problems. Sometimes, solutions were shaped by advice that pertained to an individual firm; other times the solution involved collaboration along a supply chain. The main focus of these included efficiency, removing cost and time waste, and to a large extent exposing messy situations to help them resolve systemic issues in order to focus on effectiveness. Some current similarities can be drawn with the concept of 'project alliancing' used in Australia (Walker and Hampson, 2003b; Department of Treasury and Finance Victoria, 2010) and elsewhere (Xu et al., 2005), where projects are being realised in highly uncertain and complex situations.

6.5.3   The Atlantic Corridor Facilitating Business Transformation

As part of the remit of the Corridor, Arroyo was one of several facilitators whose role was to facilitate the formation of collaborative ventures. These took the form of alliances, joint ventures, mergers and formation of stand-alone business entities, with re-configured arrangements that led combining firms to form a vehicle for complete business transformation. Arroyo (2009) details five case studies. The first case study featured an ocean shipping company's joint venture (JV) linking organisations with a different combination of structure, nationality, culture, technology and vision. The JV comprised two regional firms and two extra-regional companies. Each of these firms followed different strategies in the way they got together, how they agreed on the fundamental aim of facilitating outsiders to become regional operators, and how to become more international and global in future. The overarching aim from the outset was to avoid incurring heavy investments. This case study strongly relates to issues of the role of co-operation, leadership, knowledge and corporate strategy in driving the business transformation.

The second case involved a multimodal transportation business transformation, through the way that it connected its operations in ports of the Mediterranean Sea in Europe to end destinations in South America's West Coast. More specifically, this project entailed the set-up of a land bridge between west Mediterranean ports (Spain, France and Italy) and the industrial pole of the Santiago-Valparaiso area through the port of Buenos Aires in Argentina by combining sea, road and rail transportation on a door delivery mode, that is, to the customer's 'door'. This case study demonstrated the important role of successfully combining traditional knowledge, physical assets and management to achieve differentiation and gain competitive advantage. The third case study refers to the way the longest and most strategic railways in Argentina passed on vital knowledge about how foreign companies could reengineer themselves and become a competitive asset, not only at regional level, but also in a global dimension. Without collaboration, co-operation and leadership by partners in this case study, business transformation would have not occurred. This case study illustrates the logistic complementation of knowledge and leadership between the various agents involved that was pivotal in allowing the transformation project to successfully evolve. The fourth case study embodied a Buenos Aires up-river operator that realised the urgent need to react to the globalisation process by becoming regional and gaining world-class competitive advantage to face threats posed by emerging global maritime terminal operators' structure, technology, management and investment. This case study provides a typical reactive model, where two regional players enter into a JV to face global companies that might otherwise easily outperform them individually. The case study further sheds light on aspects of leadership, co-operation and collaboration, strategy and knowledge.

Each of the four case studies refers to short-term tactical transformation project actions. The fifth case study concerned long-term business transformation, and related to the waterways supply market evolution, and the way project stakeholders had to deal with the particular business context faced (either in a purely transactional management mode or transformational style), along with the many constraints that emerged. The response by the players to the business environment, particularly radical changes in focus on SCM towards VCM, seemed to have played a more decisive role than the management styles deployed by the different actors. The specific project focus comprised the construction and operation of fully dedicated barge convoys to carry iron ore supplies from Brazil to Argentina, commencing in 2003 phased over 6 years. The first phase comprised the development and acceptance of an innovative logistics design solution and its realisation through the construction of new barges and leasing of existing barges. The project had in-built replication possibilities to roll out a series of similar projects.

Two driving forces for this joint venture form of alliance become clear from this case study: one is the global dimension of the primary stakeholder (steel manufacturing company), and the other, the complexity of the logistic operation in itself that made for the traditional knowledge of strategic relevance. One might assume that the steel company, holding a business presence in so many countries around the world, and having access to countless databases and logistic experiences to feed from (and benchmark with), would not regard the Corridor as a potentially valuable development tool. However, not only did the primary stakeholders interact during the roundtable forums, exchanging knowledge until eventually giving birth to a knowledge-based logistic design, but also the secondary stakeholders took part in the process and were interested to form part of what was about to take place. However, even though a number of functional aspects, such as financial engineering, insurance, port operations and barging services were often contracted by the steel company on diverse geographical settings, the knowledge specificity and viscosity of the logistic design, along with its unique geographic and operational particulars embodied by the waterways, seem to have found a fertile ground for these various actors to get together and become interested in one another's needs, potential and proposals.

Figure 6.2 illustrates the formation of a JV using a PM perspective of project phases based on the research findings and the role of the Corridor in establishing these JVs.

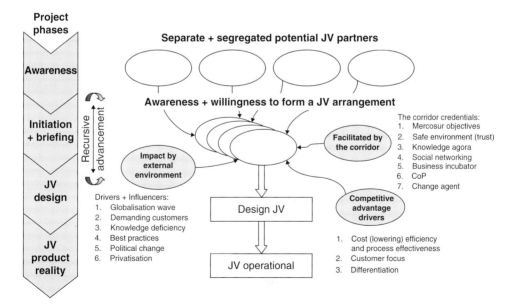

Figure 6.2   The Corridor's role in JV formation (Arroyo, 2009).

Table 6.1   Re-thinking SCM to VCM.

Moving From	To	Notes and Comments
A functional view of SCM	An innovation view of SCM with a focus on value	Other than for 'cookie-cutter' projects, most construction projects are substantially 'bespoke' and so clients welcome innovation to achieve an improved value and benefits outcome.
An efficiency agenda for SCM	Specifically an agenda of being effective, having a positive impact, focussing on value relevance and sustainability.	Clients of construction projects and member in the supply chain generally welcome ideas and expertise for improving value delivered as well as technical expertise
Focus on financial value	Broader conceptualisation of value to include improved knowledge transfer, social capital generation and potential business transformation.	As the Arroyo (2009) work demonstrates, unexpected value can be developed from collaboration in JVs, alliances or other cooperative arrangements as well as the development of new business models.

The Corridor project business transformation process first passed through an awareness gestation phase, and then through to initiation and briefing, where various forms, combinations of roles and ways of collaboration were discussed and decided upon. This was followed by the actual design of the JV. Arroyo (2009) focuses on the end of the awareness phase, as it recursively slipped into the initiation and briefing phase and then into the start of the design of JVs, and how they may be formed. Figure 6.2 also identifies how the external environment impacted upon this process, particularly, how competitive advantage helped shape the drivers for change, and how the Corridor facilitated the shaping of JVs and alliances. At the initial stage, the separate parties joined together under the facilitation and auspices of the Corridor. This entity was sanctioned by advancing the Mercosur ideals, and it was established as a safe (trustworthy) place in which representatives of member firms could meet, exchange knowledge and solve problems. This helped the engaged firms to form a CoP. Later on, more intimate groups coalesced into JV type arrangements, and so the Corridor also had a role as a business incubator and change agent. The competitive advantage achieved was reflected in terms of cost efficiency, customer focus and ability for the JV to be differentiated from potential competitors.

6.5.4   Discussion

The case studies discussed illustrate how a highly innovative concept evolved out of a SCM context in a highly volatile and turbulent economic environment, with extreme challenges of physical distribution within the supply chain. The Corridor concept is important to this chapter because it offers another way of viewing a supply chain or value chain. It reinforces the value of a facilitating vehicle to help trust, commitment and knowledge sharing, which are critical for sparking radical as well as incremental innovation. The examples in Section 6.5.2 present typical short-term efficiency-based innovation that such an entity can readily deliver, whereas Section 6.5.3 presented some longer-term innovations that were initiated through the formation of joint ventures to develop a supply chain end-to-end solution and to facilitate improved efficiency and sustainable value. These ideas relate to re-framing a supply chain regarding its primary purpose, how it can become more effective, efficient, relevant, and have a positive sustainable impact.

The example drawn from the Latin American logistics industry provides an interesting and salient model for reflection, the transition, rationality and concepts of which are summarised in Table 6.1 to illustrate this re-thinking process.

6.6   Conclusion

The chapter focused on five criteria used for evaluating aid projects that have relevance to SCM, specifically *efficiency, effectiveness, relevance, sustainability* and *impact*. The organisational contribution to value was discussed using the RBV of a firm, and what this can bring to the table in any potential SCM innovation. Furthermore, knowledge, skills, experience and social capital are all key assets that could be deployed to enhance value. Whilst this chapter concentrated on value and the value of supply chains, SCM was presented with an explanation of how this might function within a construction industry context. This led to a section on value chains, building upon and extending the seminal work of Porter (1985, 1998), Womack *et al.* (1990) and Womack and Jones (1994), but with a focus on innovation rather than functional products as outputs from such a chain. Case study examples drawn from the empirical work of Arroyo (2009) were discussed. These provided innovative approaches garnered from the Atlantic Corridor, which were considered salient, as they illustrated how an industry sector faced with enormous challenges managed to prevail through a radical transformation in the way it perceived supply chains and value.

The construction industry has been challenged by several government reports to improve its conception of value, and how it can be more innovative and sustainable (Murray and Langford, 2003). What becomes clear from this chapter is that for supply chains in the construction industry to become more efficient, effective, relevant, and have a more positive impact upon sustainability, they will need to re-frame their conception of what SCM is, and draw upon examples from other industries in order to learn from them.

6.7   Acknowledgements

Special acknowledgements are extended to Alejandro Arroyo, Southmark Logistics SA, Buenos Aires, Argentina, for edited excerpts from his Doctoral Thesis (Section 6.5).

References

Akkermans, H.A., Bogerd, P., Yucesan, E. & van Wassenhove, L.N. (2003) The impact of ERP on supply chain management: Exploratory findings from a European Delphi study. *European Journal of Operational Research*, **146**(2), 284–301.

Aranda-Mena, G., Crawford, J. & Chevez, A. (2009) Building information modelling demystified: Does it make business sense to adopt BIM? *International Journal of Managing Projects in Business*, **2**(3), 419–433.

Arroyo, A. & Walker, D.H.T. (2008) *Business transformation through an innovative alliance*. In: ed. Walker D.H.T. & Rowlinson, S., *Procurement Systems – A Cross Industry Project Management Perspective*. Taylor & Francis, Abingdon. UK. 423–444.

Arroyo, A.C. (2009) The role of the Atlantic Corridor Project as a form of strategic community of practice in facilitating business transformations in Latin America. PhD thesis, School of Property, Construction and Project Management. RMIT University, Melbourne.

Arroyo, A.C. & Walker, D.H.T. (2004) *A Latin American Strategic Community of Practice Experience*. 18th ANZAM Conference, Dunedin, New Zealand, 8–11 December, ANZAM: 11 pages on CD-Disk proceedings.

Arroyo, A.C. & Walker, D.H.T. (2009) A Latin American strategic organisational transformation project management experience: the motivation to transform business. In: ed. Bredillet, C. & Middler, C., *European Academy of Management EURAM: Renaissance & Renewal in Management Studies*, 11–14 May, Liverpool. 20.

Australian Industry Commission (1996) *Competitive Tendering and Contracting by Public Sector Agencies*. Report, Australian Government Publishing Service, Melbourne. 48.

Barney, J. (1991) Firm resources and sustained competitive advantage. *Journal of Management*, **17**(1), 99–120.

Bourne, L. & Walker, D.H.T. (2005) Visualising and mapping stakeholder influence. *Management Decision*, **43**(5), 649–660.

Cates, C. (1979) Beyond muddling: creativity. *Public Administration Review*, **39**(6), 527–532.

Chan, C. & Swatman, P.A. (2000) From EDI to internet commerce: The BHP steel experience. Internet research. *Electronic Networking Applications and Policy*, **10**(1), 72–82.

Chan, E.W.L., Walker, D.H.T. & Mills, A. (2009) Using a KM framework to evaluate an ERP system implementation. *The Journal of Knowledge Management*, **13**(2), 93–109.

Christenson, D. & Walker, D.H.T. (2004) Understanding the role of 'vision' in project success. *Project Management Journal*, **35**(3), 39–52.

Coase, R.H. (1937). The nature of the firm. *Economica.*, **4**, 386–405.

Cohen, W.M. & Levinthal, D. (1990) Absorptive capacity: A new perspective on learning and innovation. *Administrative Science Quarterly*, **35**(1), 128–152.

Cooke-Davies, T. (2002) The 'real' success factors on projects. *International Journal of Project Management*, **20**(3), 185–190.

Croom, S., Romano, P. & Giannakis, M. (2000) Supply chain management: an analytical framework for critical literature review. *European Journal of Purchasing & Supply Management*, **6**(1), 67–83.

Croom, S.R. (2005) The impact of e-business on supply chain management. *International Journal of Operations & Production Management*, **25**(1), 55–73.

Dainty, A.R.J., Briscoe, G.H. & Millett, S.J. (2001) Subcontractor perspectives on supply chain alliances. *Construction Management & Economics*, **19**(8), 841–848.

Dainty, A.R.J., Cheng, M.-I. & Moore, D.R. (2003) Redefining performance measures for construction project managers: An empirical evaluation. *Construction Management & Economics*, **21**(2), 209–218.

Davis, P.R. (2006) The application of relationship marketing to construction. PhD thesis, School of Economics, Finance and Marketing. RMIT University, Melbourne.

de Wit, A. (1988) Measurement of project success. *International Journal of Project Management*, **6**(3), 164–170.

Department of Treasury and Finance Victoria (2010) *The Practitioners' Guide to Alliance Contracting*. Department of Treasury and Finance, Melbourne. 161.

Dyer, J.H. (2000). How Chryster created an American Keiretsu. Harvard business review on managing the value chain. *Havard Business Review*, MA. **74**, 61–90.

Eisenhardt, K.M. & Martin, J.A. (2000) Dynamic capabilities: What are they? *Strategic Management Journal.*, **21**(10/11), 1105–1121.

Elkington, J. (1997) *Cannibals with Forks*. Capstone Publishing, London.

Fellows, R.F. (1997) *The Culture of Partnering*. In: ed. Davidson C.H. & Meguid, T.A., *CIB W-92 Procurement Systems Symposium 1997, Procurement – A Key To Innovation*, 18–22 May 1997, The University of Montreal, CIB, *1*, 193–202.

Fill, C. & Visser, E. (2000) The outsourcing dilemma: a composite approach to the make or buy. *Management Decision*, **38**(1), 43–50.

Fischer, M. & Drogemuller, R. (2009) *Virtual design and construction*. In: Newton, P., *Technology, Design and Process Innovation in the Built Environment*. Taylor & Francis, Abingdon, UK. 293–318.

Fisher, M.L. (1997) What Is the right supply chain for your product? *Harvard Business Review*, **75**(2), 105–116.

Fisher, N., Barlow, R., Garnett, N., Finch, E. & Newcombe, R. (1997) *Project Modelling in Construction: Seeing is Believing*. Thomas Telford, London.

Graham, P. & Smithers, G. (1996) Construction Waste minimisation for Australian residential development. *Asia Pacific Journal of Building & Construction Management*, **2**(1), 14–19.

Green, S. (1999a) The dark side of lean construction: exploitation and ideology. In: ed. Tommelein, I.D. & Ballard, G, *Seventh Annual Conference of the International Group for Lean Construction (IGLC-7)*, 26–28 July, Berkeley, CA. 12.

Green, S.D. (1999b) The missing arguments of lean construction. *Construction Management and Economics*, **17**(2), 133–137.

Habermas, J. & Outhwaite, W. (1996) *The Habermas Reader*. Polity Press, Cambridge.

Hall, M., Holt, R. & Graves, A. (2000) Private finance, public roads: configuring the supply chain in PFI highway construction. *European Journal of Purchasing & Supply Management*, **6**(3–4), 227–235.

Hällgren, M. & Maaninen-Olsson, E. (2005) Deviations, ambiguity and uncertainty in a project-intensive organization. *Project Management Journal*, **36**(3), 17–26.

Hällgren, M. & Wilson, T.L. (2007) Mini-muddling: learning from project plan deviations. *Journal of Workplace Learning*, **19**(2), 92–107.

Hughes, W. (2006) *Contract Management*. In: Lowe, D. & and Leiringer, R., *Commercial Management of Projects. Defining the Discipline*. Blackwell Publishing, Abingdon, UK. 344–355.

James, D. (2006) Slave to the supply chain. *Business Review Weekly*, **October**, 52–53.

Jewell, M. & Walker, D.H.T. (2005) *Community of Practice Perspective Software Management Tools: A UK Construction Company Case Study*. In: Kazi, A.S., *Knowledge Management in the Construction Industry: A Socio-Technical Perspective*. Idea Group Publishing, Hershey, PA: 111–127.

Kelly, J., Morledge, R. & Wilkinson, S. (2002) *Best Value in Construction*. Blackwell Publishing, Oxford.

Khalfan, M.M.A. & McDermot, P. (2006) Innovating for supply chain integration within construction. *Construction Innovation: Information, Process, Management*, **6**(3), 143–157.

Khalfan, M.M.A., McDermott, P. & Kyng, E. (2006) Procurement Impacts on construction supply chains: UK experiences. In: McDermott P. and Khalfan, M.M.A., *Symposium on sustainability and value through construction procurement, CIB W092 – Procurement Systems, CIB Revaluing Construction Theme*, 29 November–2 December, The Digital World Centre, Salford, UK. 449–458.

Kim, W.C. & Mauborgne, R. (2000) Knowing a winning business idea when you see one. *Harvard Business Review*, **78**(5), 129–138.

Klakegg, O.J., Williams, T., Walker, D.H.T., Andersen, B. & Magnussen, O.M. (2010) *Early Warning Signs in Complex Projects*. Project Management Institute, Newtown Square, PA.

Koskela, L. (2000) An exploration towards a production theory and its application to construction. PhD thesis, VTT Technical Research Centre of Finland. Helsinki, Finland, Helsinki University of Technology.

Kotter, J.P. (1995) Leading change: Why transformation efforts fail. *Harvard Business Review*, **73**(2), 59–67.

Kumaraswamy, H.M. (1996) *Construction Dispute Minimisation*. In: ed. Langford D.A. & Retik, A., *The Organisation and Management of Construction – Managing the construction project and project risk*, vol. 2. E&FN Spon, London. 447–457.

Langford, D.A., Martinez, V. & Bititci, U. (2003) Best value in construction: Towards an interpretation of value from the client and construction perspectives. *Construction Procurement*, **9**(1), 56–67.

Larson, E. (1995) Project Partnering: Results of Study of 280 Construction Projects. *Journal of Management in Engineering*, **11**(2), 30–35.

Lendrum, T. (1998) *The Strategic Partnering Handbook*, 2nd edn. McGraw-Hill, Sydney.

Lindblom, C.E. (1959) The science of 'muddling through. *Public Administration Review*, **19**(2), 79-88.

Lindblom, C.E. (1979) Still muddling, not yet through. *Public Administration Review*, **39**(6), 517–526.

London, K.A. & Kenley, R. (2001) An industrial organization economic supply chain approach for the construction industry: a review. *Construction Management and Economics*, **19**(8), 777–788.

Love, P.E.D. & Gunasekaran, A. (1999) Learning alliances: a customer-supplier focus for continuous improvement in manufacturing. *Industrial & Commercial Training*, **31**(3), 88–96.

Love, P.E.D., Mandal, P., Smith, J. & Li, H. (2000) Modelling the dynamics of design error induced in rework in construction. *Construction Management and Economics*, **18**(5), 567–574.

Love, P.E.D., Edwards, D.J., Irani, Z. & Walker, D.H.T. (2009) Project Pathogens: The Anatomy of Omission Errors in Construction and Resource Engineering Project. *IEEE Transactions On Engineering Management*, 56(3), 425–435.

Lundin, R.A. & Söderholm, A. (1995) A theory of the temporary organization. *Scandinavian Journal of Management*, 11(4), 437–455.

Male, S. (2002). Building the business value case. In: ed. Wilkinson S., Kelly, J. & Morledge, R., *Best Value In Construction*. Blackwell Science, Oxford. 12–37.

Manu, C. & Walker, D.H.T. (2006) Making sense of knowledge transfer and social capital generation for a Pacific island aid infrastructure project. *The Learning Organization*, 13(5), 475–494.

Maqsood, T., Walker, D.H.T. & Finegan, A.D. (2007) Extending the knowledge advantage: Creating learning chains. *The Learning Organization*, 14(2), 123–141.

Masterman, J.W.E. (1992) *An Introduction to Building Procurement Systems*, 2nd edn. E & FN SPON, London.

Matthews, J., Tyler, A. & Thorpe, A. (1996) Pre-construction project partnering: Developing the process. *Engineering, Construction and Architectural Management*, 3(1–2), 117–131.

Mayer, R.C., Davis, J.H. & Schoorman, F.D. (1995) An integrated Model of Organizational Trust. *Academy of Management Review*, 20(3), 709–735.

McGeorge, W.D. & Palmer, A. (2002) *Construction Management New Directions*, 2nd edn. Blackwell Science, London.

McKenna, D. & Walker, D.H.T. (2008) A study of out-sourcing versus in-sourcing tasks within a project value chain. *International Journal of Managing Projects in Business*, 1(2), 216–232.

Meyer, J.P. & Allen, N.J. (1991) A three-component conceptualization of organizational commitment. *Human Resource Management Review*, 1(1), 61–89.

Michaels, L.M.J. (1999) The making of a lean aerospace supply chain. *Supply Chain Management*, 4(3), 135–145.

Mintzberg, H. (1987). Crafting strategy. *Harvard Business Review*, 65(4), 66–75.

Mintzberg, H., Ahlstrand, B.W. & Lampel, J. (1998) *Strategy Safari: The Complete Guide Through the Wilds of Strategic Management*. Financial Times/Prentice Hall., London.

Murray, M. & Langford, D.A. (2003) *Construction Reports 1944–98*. Blackwell Science Ltd, Oxford.

Nahapiet, J. & Ghoshal, S. (1998) Social capital, intellectual capital, and the organizational advantage. *Academy of Management Review*, 23(2), 242–266.

NBCC (1989) *Strategies for the Reduction of Claims and Disputes in the Construction Industry – No Dispute*. National Building and Construction Council, Canberra.

Nogeste, K. (2004) Increase the likelihood of project success by using a proven method to identify and define intangible project outcomes. *International Journal of Knowledge, Culture and Change Management*, 4, 915–926.

Nonaka, I. (1988) Creating Organizational order out of chaos: Self-renewal in Japanese firms. *California Management Review*, 30(3), 57–73.

Nonaka, I. & Konno, N. (1998) The concept of 'Ba': Building a foundation for knowledge creation. *California Management Review*, 40(3), 40.

Nonaka, I., von Krogh, G. & Voelpel, S. (2006) Organizational knowledge creation theory: evolutionary paths and future advances. *Organization Studies*, 27(8), 1179–1208.

Normann, R. & Ramirez, R. (1993) From value chain to value constellation: Designing interactive strategy. *Harvard Business Review*, 71(4), 65–77.

OECD (2007) *Sourcebook for Evaluating Global and Regional Partnership Programs Indicative Principles and Standards*. Organisation for Economic Co-operation and Development, Development Assistance Committee (DAC): Washington, DC. 148.

Olsson, N.O.E. (2006) Management of flexibility in projects. *International Journal of Project Management*, **24**(1), 66–74.

Olsson, N.O.E. & Magnussen, O.M. (2007) Flexibility at different stages in the life cycle of projects: An empirical illustration of the freedom to maneuver. *Project Management Journal*, **38**(4), 25–32.

Palaneeswaran, E., Kumaraswamy, M. & Ng, T. (2003) Targeting optimum value in public sector projects through 'best value'-focused contractor selection. *Engineering Construction and Architectural Management*, **10**(6), 418–431.

Peansupap, V. & Walker, D.H.T. (2005) Diffusion of Information and communication technology: A community of practice perspective. In: Kazi, A.S., *Knowledge Management in the Construction Industry: A Socio-Technical Perspective*. Idea Group Publishing, Hershey, PA. 89–110.

Porter, M.E. (1985) *Competitive Advantage: Creating and Sustaining Superior Performance*. The Free Press, New York.

Porter, M.E. (1998) Clusters and the new economics of competition. *Harvard Business Review*, **76**(6), 77–90.

Pryke, S. (2004) Twenty-first century procurement strategies: Analysing networks of inter-firm relationships. *RICS Foundation Paper Series*, **4**(27), 1–38.

Rowlinson, S. & McDermott, P. (1999) Selection Criteria. Procurement systems. In: Rowlinson S. & McDermott, P., *A Guide to Best Practice in Construction*. E&FN Spon, London. 276–299.

Rowlinson, S. & Walker, D.H.T. (2008) Case study – Innovation management in alliances. In: Walker D.H.T. & Rowlinson, S., *Procurement Systems. A Cross Industry Project Management Perspective*. Taylor & Francis, Abingdon, UK. 400–422.

Sadler, I. (2007) *Logistics and Supply Chain Integration*. Sage. Thousand Oaks, CA.

Sanchez, R. (2004) Understanding competence-based management: Identifying and managing five modes of competence. *Journal of Business Research*, **57**(5): 518–532.

Saul, J.R. (1997) *The Unconscious Civilization*, 1st edn. Free Press, New York.

Shenhar, A.J., Dvir, D., Levy, O. & Maltz, A.C. (2001) Project success: A multidimensional strategic concept. *Long Range Planning*, **34**(6), 699–725.

Sidwell, A.C. & Mehertns, V.M. (1996) *Case Studies in Constructability Implementation*. Construction Industry Institute Australia, Adelaide.

Steinfort, P. (2010) Understanding the antecedents of project management best practice-lessons to be learned from aid relief projects. PhD thesis, School of Property, Construction and Project Management. RMIT University, Melbourne.

Szulanski, G. (1995) *Appropriating Rents from Existing Knowledge: Intra-Firm Transfer of Best Practice*. Institut Européen d'Administration des Affaires (France), INSEAD – The European Institute of Business Administration.

Szulanski, G. (1996) Exploring internal stickiness: Impediments to the transfer of best practice within the firm. *Strategic Management Journal*, **17**(Winter special Issue), 27–43.

Teece, D., Pisano, G. & Shuen, A. (1997) Dynamic capabilities and strategic management. *Strategic Management Journal*, **18**(7), 509–533.

Teece, D.J. (1998) Capturing value from knowledge assets: The new economy, markets for know-how, and intangible assets. *California Management Review*, **40**(3), 55–79.

Teece, D.J. (2001) Strategies for Managing Knowledge Assets: The Role of Firm Structure and Industrial Context. In: Nonaka I. & Teece, D., *Managing Industrial Knowledge - Creation, Transfer and Utilization*. Sage, London. 125–144.

Thompson, P.J. & Sanders, S.R. (1998) Partnering Continuum. *Journal of Management in Engineering*, **14**(5), 73–78.

von Krogh, G. (1998) Care in knowledge creation. *California Management Review*, **40**(3), 40–54.

von Krogh, G., Ichijo, K. & Takeuchi, H. (2000) *Enabling Knowledge Creation*. Oxford University Press, Oxford.

Walker, D.H.T. & Hampson, K.D. (2003a) Enterprise Networks, Partnering and Alliancing. In: ed. Walker D.H.T. & Hampson, K.D., *Procurement Strategies: A Relationship Based Approach*, Chapter 3. Blackwell Publishing, Oxford. 30–73.

Walker, D.H.T. & Hampson, K.D. (2003b) Procurement Choices. In: ed. Walker D.H.T. & Hampson, K.D., *Procurement Strategies: A Relationship Based Approach*, Chapter 2. Blackwell Publishing, Oxford. 13–29.

Walker, D.H.T. & Hampson, K.D. (2003c) Project Alliance Member Organisation Selection. In: ed. Walker D.H.T. & Hampson, K.D., *Procurement Strategies: A Relationship Based Approach*, Chapter 4. Blackwell Publishing, Oxford. 74–102.

Walker, D.H.T. & Lloyd-Walker, B.M. (2010) *Profiling Professional Excellence in Alliance Management, Draft Report*. Alliance Association of Australasia, Sydney. 48.

Walters, D. & Lancaster, G. (2000. Implementing value strategy through the value chain. *Management Decision*, **38**(3), 160–178.

Wang, C.L. & Ahmed, P.K. (2007) Dynamic capabilities: A review and research agenda. *International Journal of Management Reviews*, **9**(1), 31–51.

Wenger, E.C., McDermott, R. & Snyder, W.M. (2002) *Cultivating Communities of Practice*. Harvard Business School Press, Boston, MA.

Williams, T. & Samset, K. (2010) Issues in front-end decision making on projects. *Project Management Journal*, **41**(2), 38–49.

Williamson, O.E. (1985) *The Economic Institutions of Capitalism: Firms, Markets, Relational Contracting*. The Free Press, New York.

Williamson, O.E. (1991) Strategizing, economizing, and economic organization. *Strategic Management Journal*, **12**(Special Issue), 75–94.

Williamson, O.E. (1993) Calculativeness, Trust, and economic organization. *Journal of Law and Economics*, **36**(April), 453–448.

Winch, G.M. (2003) *Managing Construction Projects*. Blackwell Publishing, Oxford.

Womack, J.P. & Jones, D.T. (1994) From lean production to the lean enterprise. *Harvard Business Review*, **72**(2), 93–103.

Womack, J.P. & Jones, D.T. (2000) *From Lean Production to Lean Enterprise. Harvard Business Review on Managing the Value Chain*. Review H. B. Harvard Business School Press, Boston, MA. 221–250.

Womack, J.P., Jones, D.T. & Roos, D. (1990) *The Machine that Changed the World – The Story of Lean Production*. Harper Collins, New York.

Wright, P.M., Dunford, B.B. & Snell, S.A. (2001) Human resources and the resource based view of the firm. *Journal of Management*, **27**(6). 701–721.

Xu, T., Bower, D.A. & Smith, N.J. (2005) Types of collaboration between foreign contractors and their Chinese partners. *International Journal of Project Management*, **23**(1), 45–53.

Zahra, S.A. & George, G. (2002) Absorptive capacity: A review, reconceptualisation, and extension. *Academy of Management Review*, **27**(2), 185–203.

Part II

Process Drivers

Strategic Management in Construction

Jack S. Goulding

7.1   Introduction

Porter (1979) described how changes in the business environment can often require companies to reconsider their strategic options and positioning in the marketplace regarding 'traditional' economic theories of competition. In this respect, from a strategy development perspective, 'perfect' competition assumes free access to new participants and the availability of 'perfect market information'. However, the extent to which these principles are understood and applied can have a direct impact on business success, especially in periods of rapid change. Organisations are therefore increasingly concerned with providing strategic direction (Morgan, 1990), and these requirements are traditionally achieved through a structured framework often described by the term business strategy (BS). This is often used to define the pattern of decisions needed to determine organisational goals and objectives – the precise definition of which is influenced by perceptions and approaches taken by a range of authors (Ansof, 1968; Andrews, 1987; Davenport and Short, 1990; Porter, 1985). Therefore, as the business environment often involves change, the real challenge for organisations is to organise and align corporate assets to strategic capability in order to achieve corporate goals within the context of their operating market sector. This by default tends to pervade organisational systems, procedures and resources (Andrews, 1987) and consequently, requires strategic decision-makers to fully understand these needs, especially how they can be delivered in order to create strategic advantage, maximise business opportunities and improve process effectiveness (Prahalad and Hamel, 1990; Lee and Dale, 1998; Remenyi and Sherwood-Smith, 1998; Lee and Sai On Ko, 2000; Porter, 2008).

Construction Innovation and Process Improvement, First Edition.
Edited by Akintola Akintoye, Jack S. Goulding and Girma Zawdie.
© 2012 Blackwell Publishing Ltd. Published 2012 by Blackwell Publishing Ltd.

This chapter outlines the core dynamic processes and drivers associated with strategic management in construction business domains. In this respect, it identifies how strategy can be purposefully linked to business imperatives in order to leverage innovation opportunities and gain strategic advantage.

7.2   Construction Sector Dynamism and Drivers

The construction and built environment is widely acknowledged as being subject to high levels of fragmentation (Emmerson, 1962; Banwell 1964; Latham, 1994; Egan, 1998), the corollary of which has continued to adversely affect construction performance and productivity. Contemporary 'change' initiatives have tried to improve performance by focusing on time, quality or cost elements; but Kagioglou *et al.* (2001) noted that most of these problems were often process-related rather than product-related. This resonates with process improvement issues and opportunities cited in extant literature (Davenport, 1993; Hammer and Champy, 1993; Soares and Anderson, 1997; Chan and Land, 1999). Therefore, unprecedented levels of change (Maloney, 1997) has meant that many organisations have had to continually adapt to change by adopting new approaches to business. Typical movements have included 'quality and process management' issues (Hammer, 1990; Davenport, 1993; Al-Mashari and Zairi, 2000), through to the strategic use of ICT to generate strategic advantage (Betts, 1992; Betts and Wood-Harper, 1994; Tan, 1996; Goodman and Chinowsky, 1997; Venegas and Alarcón, 1997; Rezgui and Zarli, 2006).

The changing nature of construction activity and its operating environment is continually evolving, the dynamism and propensity of which is growing at an increasing pace. For example, the sector is influenced by internal and external social, economic, political and technological drivers, which directly affect the way construction organisations view strategy, and the way they conduct business with their stakeholders (Figure 7.1). Other factors include increasing levels of privatisation and globalisation, and a greater diversification of core construction-related business activities (Betts and Ofori, 1992). This diversification required some organisations to reaugment their business units into Architectural, Engineering and Construction (AEC) companies in order to provide a 'one-stop-shop' of bespoke client-orientated services. Other forms of diversification have included joint venturing and vertical/lateral integration, which have helped organisations to develop enhanced synergies of scale and supply to garner extended business benefits (Hillebrandt and Cannon, 1990).

Figure 7.1 identifies the relationship of strategy considerations to the construction supply chain and external governing forces. These forces include environmental, socio-political, economic and technological drivers, the nuances and importance of which are pivotally important to construction organisations, as understanding these relationships (and interplay) ostensibly governs the way strategy is developed and augmented. For example, environmental drivers have a propensity to affect the marketability of products and services, which often influences demand, and socio-political

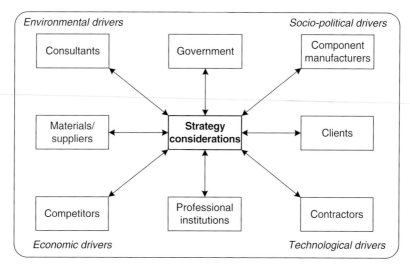

Figure 7.1   Construction supply chain environment.

drivers tend to affect clients' thinking and needs, which often causes organisations to alter their current thinking and/or working practices. However, economic drivers tends to influence the profits and pricing structure of construction organisations, and technological drivers tends to shape the way through which projects are procured and managed, which often mean that new skills are needed to meet these changes. Therefore, when organisations are considering the development of strategy, they will need to openly evaluate these variables in a cogent, robust and defendable way (so they can be defended at Board level). This approach is fundamentally important, as the development and subsequent positioning of strategy will need to not only align to these dynamic market forces and supply chain variables, but also be flexible and agile enough (using modelling options) to be one-step ahead of their competitors. Thus, the impact of stakeholder analysis on strategy should take a balanced view of outcomes and perspectives (Walker, 2000), in order to create a strategy with a clear focus and readily identifiable objectives and deliverables, along with a clear means for measuring outcomes.

7.3   Business Processes Redesign

The process approach of analysing business operations has developmental links with Systems Thinking, Industrial Engineering and the Quality Movement. Historically though, perhaps the origins of process as a discipline should be attributed to Taylor (1911), as this work focused on process improvements using the science of work-study. However, the central argument and tenet of process relates to the evolution of organisational structures and management approaches (Taylor, 1911; Mayo, 1933; Smith, 1759). More recently, the process improvement agenda has been augmented through redesign initiatives (Hammer, 1990; Davenport, 1993) in order to help organisations:

- identify core processes that directly contribute to organisational success;
- assess the current appropriateness and effectiveness of these processes;
- examine the importance and significance of these process areas;
- identify areas for change, using benchmarking and best practice approaches as appropriate;
- assess the effectiveness of process changes made to leverage innovation benefits.

The concept of Business Process Redesign (BPR) is an approach that can help organisations to fundamentally reorganise the way they operate and conduct business in order to facilitate radical improvements (Hammer and Champy, 1993). Variants of these process-based initiatives have also been termed 'Process Innovation' and 'Business Process Reengineering'. BPR therefore aims to challenge organisations to make radical improvements (Petrozzo and Stepper, 1994), along with the required changes necessary to implement change (Davenport, 1993). However, notwithstanding the nuances of nomenclature, the term 'business process' relates to the analysis of an organisation's core business activities, and the term 'redesign' focuses predominantly upon the methods and techniques associated with the business processes involved. Thus, process-related improvement initiatives are intrinsically linked to organisational strategy – the augmentation of which can help organisations optimise their processes to respond to global competition, changes in cyclical economies and changes associated with technological developments (Hammer, 1990). Therefore, organisations are increasingly focusing on processes (not tasks) in order to structure and deliver their work activities. In this context, the term 'process' can be seen as a series of linked activities that take an input activity and transform it into an output deliverable (Johansson et al., 1993). On this theme, the application of a process viewpoint tends to examine process inputs, and asks whether 'value' is being added at each step of transformation to produce these outputs; with typical questions being 'can I add more value for less effort?', 'if so, how and where?' The adoption of this approach can help organisations streamline their processes in order to deliver value-laden activities, over a shorter timescales, with more competitively priced products. However, it is important to note that organisations wishing to follow a BPR approach should attempt to focus on the inter-dependencies within and across the processes within an organisation, and not purely on functional areas alone (Hammer and Champy, 1993).

It should also be acknowledged that whilst most companies will probably have processes embedded into their current working practices, many of these processes may not have been designed or developed using a BPR approach (verbatim); the consequence of this may mean that some or all of these processes may have to be completely redesigned from first principles to fully leverage these benefits. For example, Davenport (1993) identified five key phases for applying BPR, specifically:

1. identifying processes for redesign;
2. identifying the change levers;
3. developing process visions;
4. understanding existing processes; and
5. designing and prototyping new processes.

In this respect, 'identifying processes for redesign' involves determining the aims, objectives and strategic mission of the organisation, and 'identifying the change levers' involves identifying and acknowledging that the organisational infrastructure will most probably change in order to facilitate the new process structure. This is one of the most important areas, as it includes the behavioural characteristics of employees within the organisation, who more often than not will be averse to change (Lewin, 1947). The 'developing process visions' phase includes the articulation of new process structures and metrics for evaluation in order to meet the process vision needed. This also includes the impact this may have on the supply chain, which is of particular importance in the construction industry. The 'understanding existing processes' involves the evaluation of existing processes and practices to appreciate how things work, especially the inter-connections. However, the main issue here is to determine what is wrong, and what is needed to rectify the situation, so that these issues are not repeated. Finally, the 'designing and prototyping new processes' involves detailing the knowledge within these processes, in particular, how this should operate in order to maximise business impact.

7.4   Business Strategy

The term 'business strategy' has been defined by several seminal authors, the terminology and coverage of which embraces such remits as diversification and acquisition policies (Channon, 1978), through to leveraged systems augmented to promote structure and growth (Newcombe, 1990). However, strategy can be seen to be much more than this. For example, it includes ways and methods of doing things to achieve objectives (Andrews, 1987), using controlled and managed approaches (Price and Newson, 2003), to deliver corporate objectives. Therefore, it is important that companies consider strategic development issues in line with the market forces and drivers identified in Figure 7.1, as successful companies more often than not:

1. tend to understand change;
2. often adopt a systems approach to manage and control change;
3. usually pursue competitive advantage as an investment decision;
4. understand risk well; and
5. endeavour to provide innovation to clients as an embedded solution (Perkowski, 1988).

7.4.1   Strategy Development

From an historical perspective, the School of Organisational Theory (ca. 1900) emphasised the importance of labour in business, and the Scientific Management and Classical School (1910–1920) focused on tasks and coordination. However, the concept of corporate strategy and the need to forecast future business direction was not made until the arrival of the Group Dynamics and Bureaucracy Schools of thought (ca. 1940). During this period, Fayol (1949) was one of the first main proponents of this cause, out of which long-term strategy started to develop and evolve. Advancements in strategy development from this period onwards up to the 1960s tended to converge on forecasting trends through leadership and decision making, then in the 1970s, the arrival of Systems Theory replaced the term 'long-term strategy planning' with 'strategic planning', the concept of which focused more closely on the elements of competition and the relationship of this to expanding markets. The arrival of Contingency Theory in the 1980s then led to research in organisational structures and processes, and Porter (1985) was one of the main pioneers of strategic decision making in competitive environments during this period. Therefore, developments have evolved from the strategy formulation side of the process, through to a greater focus on the implementation issues. This allowed the psychological, sociological and political domains to be better integrated into the strategic management arena, and allowed a greater amount of synthesis to occur (Robson, 1997). The current vogue in BS development is now centred on understanding the concepts and principles that govern competition. This involves 'systems thinking', 'chaos theory' and embraces issues relating to the 'learning organisation' (Huber, 1991; Senge, 2006; Stacey, 2010) and Stakeholder Perspective Management (Love and Holt, 2000).

Strategy development can therefore be seen as a core way through which organisational success can be garnered. In this respect, Porter (1979) developed a Five Forces Model to examine interrelated forces, with a particular emphasis on securing competitive advantage. These five forces included:

1. threat of New Entrants;
2. supplier Power;
3. buyer Power;
4. threat of Substitute Products; and
5. jockeying for Position.

This approach can readily be applied to strategic management within the construction domain, as it enables construction organisations to analyse product and location segments of the construction market through these five forces in order to identify areas for business performance improvement (to secure competitive advantage). From this model, the *Threat of New Entrants* force relates to the ease with which new firms can enter and compete in the market. For example, consider the position of a large construction organisation that had heavily invested in a niche area over several years to secure market dominance; a new entrant would therefore find it especially difficult to develop capability and maturity in order to compete on level terms; which, by default, would act as a barrier to entry. This category of force also includes economies of scale, product differentiation, large capital capability, and cost drivers aligned to experience curves. The *Supplier Power* force includes the proximity of supply and its relationship to the organisation, where competitive advantage can be secured from powerful supply relationships. Examples of this include situations where few enterprises dominate supply, where there is no competing product, or where the supplier holds a threat of forward integration over the buyer (vis-à-vis potential acquisition). The *Buyer Power* force includes situations where products are price sensitive but not quality sensitive, and includes positions where a threat of backward integration (securing ownership of the supply chain) over suppliers exists. For example, several large construction organisations are able to exploit their relationships with materials suppliers and specialist subcontractors in order to secure competitive advantage, which has proven to be particularly fortuitous with large construction organisations. The *Threat of Substitute Products* force is where there is potential to use other products or services in lieu of traditional ones. This is predominantly prominent when addressing the price-performance-quality trade-off, especially where new materials or products have proven benefits (cost or otherwise) against conventional solutions. This force therefore enables organisations to micro-analyse these differences with a view to securing competitive advantage. Finally, the *Jockeying for Position* force is used to analyse the marketplace of the competition. In this respect, it is argued that the greatest scope for securing competitive advantage exists where:

1. the competitors are of equal size;
2. there is slow market growth;
3. similar (undifferentiated) products or services exist; and
4. high fixed costs prevail.

This Five Forces Model therefore provides construction organisations with a mechanism for creating several strategic opportunities (Tatum, 1988).

Within a business environment, senior managers are therefore increasingly concerned with investigating the processes involved in creating strategic direction (Morgan, 1997). This direction is often channelled through a structured framework often described by the term 'Business Strategy' or 'Corporate Strategy', where this represents a formal framework that clearly defines the business goals and objectives of the organisation (Ansoff, 1968; Porter, 1985). Research on Business Strategy (BS) formulation tends to embrace a wide range of theories and perceptions, the nuances of which often invite ambiguity and misinterpretation. However, the precise definition of strategy in this remit is more difficult to determine, evidenced by perceptions and methodological approaches taken by a range of authors (Ansoff, 1968; Porter, 1985; Mintzberg and Quinn, 1991; Robson, 1997; Ward and Peppard, 1996). Therefore, from a taxonomy perspective, Business

Strategy within this chapter should be interpreted in the context defined by Porter (1985), as this viewpoint specifically identifies the link between competitive opportunities and high-level processes (Davenport, 1993). Thus, hereafter, the BS can be seen as a pattern of decisions made within a company to determine organisational goals and objectives. Therefore, the core raison d'être of developing the BS should be to establish appropriate policies and plans needed to achieve strategic direction. This includes embracing market forces, internal processes and intellectual capital (employees) and, in most cases, the consideration of shareholders, customers and the wider community. The fundamental details of the BS should consequently focus on how a company competes and positions itself in the marketplace by focusing resources and competence into strategic advantage (to deliver the corporate vision/mission). In this context, it is imperative that key drivers and enablers affecting success are fully understood, especially concerning BS formulation, implementation and subsequent evaluation.

The BS can therefore be seen as a long-term policy that expressly satisfies the needs of the core business imperatives, and not short-term operational issues. The guiding principles and focus should thus concentrate on structuring the pattern of decisions needed to perform and excel within the business environment. This requires considerable business acumen, as these decisions will often shape the way in which the business evolves (and the way it is perceived in the wider community), which can affect the character and image of the company. In essence, strategic decisions help to position the company in the operating environment, allowing it the capacity to mobilise its strengths to fulfil its corporate goals. The central functions within the company must therefore be aligned to these needs, which include systems and procedures, sales and marketing, manufacturing and production, and all organisational resources (human and physical) to support these processes. The BS should therefore be forward-looking, and wherever possible, embrace the following key attributes:

- *Include non-financial aspects* (as focusing purely on financial issues can often prevent the organisation from capturing equally important issues such as quality, corporate perception, culture, employee motivation, etc);
- *Be measurable* (as strategies that cannot be measured or quantified are relatively meaningless – goals must therefore be set, and subsequently measured and evaluated):
- *Be focused and distinctive* (as a broad or indistinct approach can often limit success);
- *Be Innovative and flexible* (as a clear strategy can help direct corporate energies more effectively, whereas flexibility enables organisations to better respond to change).

From a strategy development perspective, the process of creating a BS is normally kept confidential, as it tends to contain commercially sensitive information. Some strategists also argue that release of this information should also be withheld from employees in order to avoid internal resistance or minimise leaks to competitors (Andrews, 1987). Within the context of construction, there are many small and medium-sized enterprises (SMEs), which are currently trying to focus on continuous improvement initiatives in order to not only survive in the marketplace, but also endeavour to secure competitive advantage. Some of these initiatives have focused on the development, integration and automation of ICT, whereas others have focused on developing new products and services following the Five Forces model advocated by Porter (1979). However, business benefits can only normally be secured if the corporate strengths of the organisation are purposefully aligned to market opportunities. Therefore, BS formulation envelops many issues, not least, identifying the company's strengths and weaknesses, and the resources available to service any perceived needs. This process is often referred to as the SWOT Analysis (Strengths, Weakness, Opportunities and Threats). Consequently, strategy development has to balance processes and perceptions with aspirations, ethics, culture and organisational competence. Andrews (1987) classified these issues into four key areas, specifically Market Opportunity, Corporate Competence and Resources, Personal Values and Aspirations, and Acknowledged Obligations to Society. However, strategy formulation should also consider the company's capacity to meet the needs of the market in order to match opportunity with corporate capability (Ward and Peppard, 2002). The formulation of the BS must therefore critically examine all these factors, not least organisational strategic plans and initiatives, business processes, technology issues and corporate resources.

7.5   Business Performance Assessment

The process of assessing business performance includes a number of important areas, from the identification of the organisation's current position and standing in the marketplace, through to the identification of performance gaps with corresponding strategies and measures to address these.

7.5.1   Benchmarking and Gap Analysis

Several definitions of benchmarking are available in the extant literature. However, the fundamental philosophical approach of benchmarking is to analyse existing activities, processes or practices within the company against one or more external organisations in order to measure fundamental differences. Whilst this is typically undertaken with similarly sized companies in the same market sector, there are advantages to be secured from benchmarking against smaller (or larger) organisations in non-cognate areas. The main concept of benchmarking is to determine from the outset the precise areas that are going to be evaluated, along with the measures and metrics used. This is important, as it helps to establish the precise rubrics and purpose of this exercise, so that clarity of purpose can be objectively measured against specific performance criteria. For example, benchmarking can be directed at both micro and macro levels. The micro level may, for example, involve a detailed analysis of a small technical issue, whereas at the macro level this may include issues of a more general nature, including managerial processes, key functional areas, etc. Either way, the core raison d'être of undertaking benchmarking and gap analysis should be to purposefully identify 'best practice' and actions needed to match or exceed this in order to gain competitive advantage.

Performance analysis is a tool that can be used for applying and measuring performance targets, the procedure of which is normally undertaken on mission-critical activities. This technique is also known as Gap Analysis or Variance Analysis. Gap Analysis is widely used in the construction, manufacturing and financial sectors, and can be a valuable tool for controlling resource driven activities over time (to predict trends), the mechanics of which can often be applied to a variety of different process areas. Whilst these areas usually embrace production-related activities, it can also extend to include skills development (Van Daal *et al.*, 1998; Goulding and Alshawi, 2002); intellectual capital (Joia, 2000); gaps in ICT (Ward and Daniel, 2005); gaps in culture (Ward and Peppard, 2002), or external environment issues (Harrison and Pelletier, 2000). The application of this approach is to measure performance achievement against predetermined targets, the purpose of which is to analyse all gaps in performance. The difference between the current level of achievement and the desired target is known as the performance gap (or opportunity gap) – an example of which can be seen in Figure 7.2.

Figure 7.2 identifies three main areas for performance improvement, specifically: Area 'X', Area 'Y' and Area 'Z'. The performance target achievement levels for each of these areas is represented by the dashed line. From this, it can be seen that the greatest area for improvement rests with Area 'Y' (G2), as the opportunity gap identified is greater than G1 or G3. Therefore, if Area 'Y' was classed as a critical BS objective, then resources should be prioritised to this area, in order to address performance deficiencies and maximise performance gains.

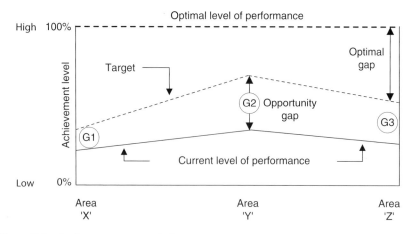

Figure 7.2   Performance gap analysis.

7.5.2   Evaluation

Whilst it is important to understand the concepts and issues of strategy, it is equally important to understand the importance of evaluation, as this is one of the most important aspects of strategy development, and is often misunderstood and/or applied. Evaluation is not just something that needs to be completed retrospectively, it must be planned for well in advance. Things that are needed include issues such as 'What needs to be measured?', 'Why?', 'How?', 'Using what criteria?', etc. Once these decisions have been made, a set of 'measurable' indicators can then be established for subsequent analysis. However, the evaluation process should specifically target areas of particular significance, especially areas that directly contribute to the strategic objectives of the company. It should therefore be linked to the objectives and policies of the company, so that priority areas and desired outcomes can be established and appropriate measures set against predefined criteria for subsequent evaluation. This typically involves the determination of gaps between the desired levels of achievement and the current position of the company in order to appraise the type of activity undertaken, determine the level and suitability of expertise available, assess the appropriateness and effectiveness of core business processes, identify production or financial outcomes, or assess the prevailing level of organisational culture within the company (Rapert et al,, 1996; Warszawski, 1996; Ward and Daniel, 2005).

At this juncture is should be acknowledged that business performance issues can often be adversely influenced by a number of factors. These issues tend to embrace sociological, technological and managerial matters relating to corporate structures, and the interrelationship of people to processes. From a commercial perspective, business performance almost invariably tends to refer to fiscal performance. In this respect, financial performance measures are often used to ensure that resources are deployed appropriately and effectively, with typical measures and metrics being associated with ratio analysis, derived from company balance sheets and profit and loss accounts. However, the use of financial measures alone should be discouraged, as using these in isolation does not always create a balanced view of corporate performance (Andrews, 1987; Mintzberg and Quinn, 1991; Lee and Sai On Ko, 2000). However, when considering the term 'performance', from a definition perspective, this can be interpreted in several different ways. Therefore, in order to provide clarity, the following definitions are presented:

- *Performance Objective*: a critical success factor for achieving the mission/vision;
- *Performance Goal*: a target level or objective;
- *Performance Measure*: a qualitative or quantitative measure of performance;
- *Performance Measurement*: the process of measuring or assessing progress/outcomes;
- *Performance Management*: the use of performance measurement.

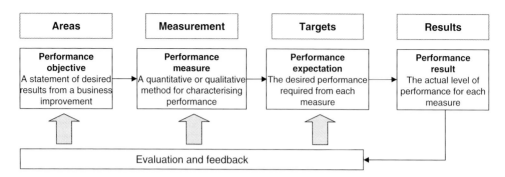

Figure 7.3   Business performance and improvement assessment approaches.

The areas for evaluation should form a feedback loop for reflection, so that the areas (performance objectives), measurement (performance measures), targets (performance expectations) and results (performance results) can be analysed holistically. This relationship can be seen in Figure 7.3.

When it comes to the process of evaluation, it is acknowledged that it is possible to evaluate almost anything and everything within an organisational setting. However, intrinsically, it is more important to understand what areas are needed, why these areas are important, and how much this evaluation process will cost. For example, evaluation areas within a construction organisation may include:

- functional or process areas;
- level of integration/interoperability;
- communication improvements;
- response times or time savings;
- reduced levels of data redundancy/duplication;
- fiscal savings/reductions;
- strategic advantage/market share;
- innovation/new business benefits;
- client satisfaction.

Once the areas for evaluation have been identified, it is important to decide on the criteria that is going to be used for this, for example "What criteria can we evaluate X?', 'Will this be the same for Y?' Typical criteria includes 'success', 'efficiency' and 'effectiveness' (how well, etc). In this respect, evaluation criteria can be either quantitative, qualitative or a combination of both. The main issue to understand here is the overall rationale and thought processes behind the need to undertake this evaluation. For example, 'What do we mean by success?' Success is typically associated with something new and often embraces change or innovation. Success can also be considered subjective, as one person's view of success will often be different from another's. However, success may also be quantified, for example, X% improvement in area Y, or £X profit, or X% increase in turnover, etc. The more complex the criteria, the more involved the analysis will often be, with a corresponding increase in time and cost. Efficiency is another example: 'What do we mean by efficiency?' This tends to be used when something performs optimally, but altruistically, in certain circumstances this may not be entirely effective. Thus, evaluation can be considered as a process of asking and answering three main questions:

1. Will it meet the needs of the organisation?
2. Will it engage proven approaches (methods) to satisfy internal and external audit (if needed)? and
3. Will it provide the organisation with the 'right' results needed (e.g. reports, tables, charts, etc)?

The next section presents a construction case study for discussion. Emphasis is placed on identifying some of these strategic development issues from a practical development and implementation perspective.

7.6   Strategy Development within Construction

Many large construction organisations are often divided (and subdivided) into groups or subsidiaries for logistical or operational reasons, the majority of which tend to operate under the overall control of one parent (or holding) company. In some cases, companies are given complete autonomy from the holding company, the procedure of which enables them to set group-specific business strategies independent of the parent company's BS. This provision can allow a greater degree of flexibility to exploit niche markets, without the necessary burden of being tied into the parent company's BS and corporate infrastructure.

The following section presents a strategy development case study for one large construction organisation based in the UK.

7.6.1   Strategic Challenge

The company had an operating remit covering the AEC market sector, the work of which was predominantly undertaken in the UK (albeit with some overseas activity), with an annual turnover of over £700m, supported by 3,500 technical, managerial and operational staff. The holding company was operationally divided into discreet groups, which, for the purpose of anonymity, are hereafter labelled Group A, Group B and Group C, the representation of which can be seen in Figure 7.4.

From Figure 7.4, three separate groups (Group A, Group B, Group C) are presented, each with their corresponding business strategies. The remit and extent of each group's BS is represented by a solid-line circle, and the parent company's BS is represented by a dashed-line circle. The collective business strategies of each of these three groups can be seen to fulfil the parent company's BS through the overlap in the central core. In this area, the remits of each of the respective group's BS can be considered similar or identical to

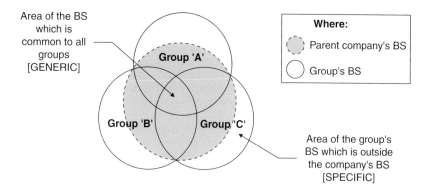

Figure 7.4   Case study business strategy structure.

each other from a needs perspective, and can be thus termed 'generic' (common to all), whereas, areas that fall outside this central core can be said to be specific to that group's needs.

The main purpose of this case study was to highlight the core drivers and barriers affecting the strategic deployment of strategy from a holding company perspective. In particular, it aimed to identify how strategic decisions were holistically managed from inception through to review. This company therefore needed a formal mechanism for assessing the effectiveness of their BS in order to purposefully leverage corporate resources and competence (Andrews, 1987; Mintzberg and Quinn, 1991). In this respect, they needed a new approach to measure their overall performance from a holding company perspective, and not just from a series of combined strategies collated at group level.

7.6.2   Proposed Solution

The first stage of this study was to assess the scale and complexity of the company organisational structure, along with the functional relationship attached to each of the three froups. This was needed to identify the scope and remit of the task, in order to conceptualise the overall scale and complexity of the main process issues involved for developing the proposed solution. Twelve interviews were conducted with all key decision makers within each of the three groups, in order to elicit key information regarding strategy formulation and core critical success factors (CSFs). From this, a case study approach (Yin, 1994) was used to develop a formal framework with actors that 'modelled' what currently happened in practice, against what should happen (best practice). The domain area for this investigation encompassed the strategic use of ICT only, specifically, the core strategic ICT process that supported group and corporate strategy. In this respect, in order to remove organisational bias, all findings were independently tested and validated with a similar-sized company and operating business remit to this company. Only findings from the first phase of this work

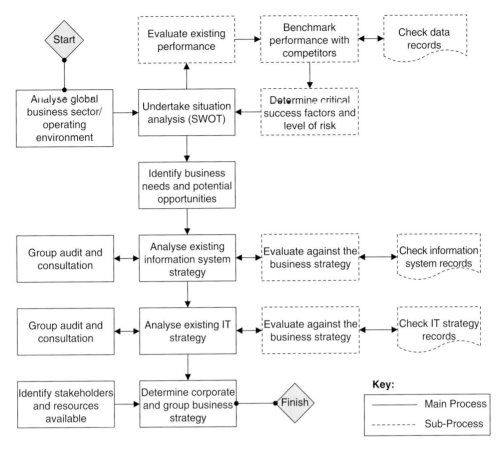

Figure 7.5   Process attributes for phase ZERO.

(Phase ZERO), together with a strategic dependency model, are presented for discussion.

Figure 7.5 represents Phase ZERO of this work, specifically a process representation highlighting the need for strategic ICT in the case study company. The starting point of Phase ZERO is commenced with the process box titled 'Analyse Global Business Sector/Operating Environment'. This allows users to identify the organisation's primary business market, the precursor of which requires a full investigation into the principal factors that can often influence outcomes. Specifically, this process box requires the identification of all current (and future) market issues and forces that could influence outcomes (directly or indirectly). This requires the process box 'Undertake Situation Analysis (SWOT)' to be commenced, as these details can help identify the position and status of the company (from a competition perspective). At this point, a detailed SWOT analysis should be undertaken to identify the key strengths, weaknesses, opportunities and threats facing the company. This is a major activity, and should be separated into readily identifiable and manageable sub-processes – the details of which are identified in 'Evaluate Existing Performance', 'Benchmark Performance with Competitors', 'Check Data Records', and 'Determine Critical Success Factors and Level of Risk' sub-process boxes. The 'Evaluate Existing Performance' sub-process box is used to collate historic key business performance measures achieved (from the previous evaluation period). Typical areas of analysis often include items that can be easily measured and quantified, for example, turnover, direct sales, return on capital employed, earnings per share, tender success percentages, etc. This information is then directed to the 'Benchmark Performance with Competitors' sub-process box, where comparisons are undertaken. At this juncture, it was acknowledged that the process of undertaking comparisons using benchmarking should only be attempted with similar sized and structured construction organisations and, if possible, using companies with a similar level of ICT process maturity. At this stage, the 'Check Data Records' sub-process box is accessed to compare the internal and external data records – the details from which are subsequently sent to the 'Determine Critical Success Factors and Level of Risk' sub-process box. Users are then able to identify all the CSFs needed to move the company forward (in terms of performance and competitiveness). At this stage 'risk' is also evaluated in terms of achieving the identified CSFs, and the corresponding effect this may have on the BS. When all sub-processes activities have been completed, the resultant data is then sent to the 'Identify Business Needs and Potential Opportunities' process box for analysis.

The 'Identify Business Needs and Potential Opportunities' process box is used to outline the business needs and remit of the strategy. Any potential opportunities identified by the benchmarking exercise are discussed at this point. Emphasis is therefore placed on focusing resources and energies effectively, particularly in areas of strength, or perhaps towards new market opportunities. However, these considerations must be tempered with company experience, the amount of risk attached and the type and level of skills available. This information forms the basis of the business need, which (as a prerequisite) requires the exact Information Systems (IS) requirements to be determined, which can be achieved through the 'Analyse Existing Information System Strategy' process box.

The activities undertaken in the 'Analyse Existing Information System Strategy' process box aims to identify the exact needs and expectations required by the BS from an Information Systems Strategy (ISS) perspective. Particular emphasis is therefore placed on assessing whether the existing ISS can deliver the BS. This requires detailed investigation into the existing organisational ISS and support mechanisms, and can be achieved using two sub-process activities, specifically 'Evaluate Against the Business Strategy' and 'Check Information System Records'. The 'Check Information System Records' sub-process box is used to identify all existing processes, procedures and systems currently deployed in the company. This information is then sent to the 'Evaluate Against the Business Strategy' sub-process box for validation and checking purposes. At this point, all IS systems and procedures are formally matched to the requirements of the BS. Group input from the 'Group Audit and Consultation' helps to shape the overall needs of the holding company regarding its overall ISS needs. Any deficiencies or identified needs are subsequently prioritised and sent to the main process box 'Analyse

Existing Information System Strategy' for final analysis. This information is later codified and directed to the 'Analyse Existing IT Strategy' process box for alignment and matching purposes. The 'Analyse Existing IT Strategy' process box is used to critically examine the current systems and procedures in place to deliver the ISS and BS. Emphasis is therefore placed on whether the ITS is able to support the demands and expectations of the ISS, contemplating the existing organisational infrastructure and level of ICT process maturity. This investigation is also supported by two sub-process areas, specifically 'Evaluate Against the Business Strategy' and 'Check IT Strategy Records'. The 'Check IT Strategy Records' sub-process box is used primarily to audit and document the ITS. Emphasis is therefore placed on how well the existing strategy supports the current business operations. However, the focus of attention will need to concentrate on the future demands determined by the 'Analyse Existing IS Strategy' process box. These findings are then sent to the 'Evaluate Against the Business Strategy' sub-process box for evaluation and alignment purposes. In this remit, the ITS is formally matched to the needs and expectations of the BS. Any deficiencies (or areas of need) in the ITS are identified at this stage, the results from which are then directed back to the main process box 'Analyse Existing IT Strategy' for further analysis. When all information has been collated and analysed in the 'Analyse Existing IT Strategy' process box, these findings are then disseminated to all groups (and subsidiaries) through the 'Group Audit and Consultation' process box. This enables consultation with the wider corporate audience, which therefore allows stakeholders to identify their current (and future) position regarding IS and ICT support measures and systems needed. From a corporate perspective, the holding company will need to gain understanding and support from stakeholders, particularly where the strategic alignment of IS and ICT resources are concerned. This undertaking therefore extends into the 'Identify Stakeholders and Resources Available' process box, where all these findings should be analysed. At this point, corporate executives should be in a position to make a formal statement of intent, conscious of the competing forces, drivers and proposed direction of the company – these details therefore naturally embrace corporate resources. The final stage of Phase ZERO is used to determine the corporate BS, which is undertaken in the 'Determine Business Strategy' process activity box. This procedure enables the key BS CSFs to be established and agreed, the findings from which are then disseminated to all groups. All decisions and requirements are formally documented at this point, in the form of a Phase Review Report. The details contained in this report then determine whether there is an intrinsic need for the strategic deployment of ICT within the company. Users are then in a position to exit this process model.

7.6.3   Discussion

Phase ZERO enables the case study company to critically analyse their core ICT business processes in line with market forces, governing conditions and immediate competitors. The next stage of this investigation required the business strategy relationships and dependencies to be determined for

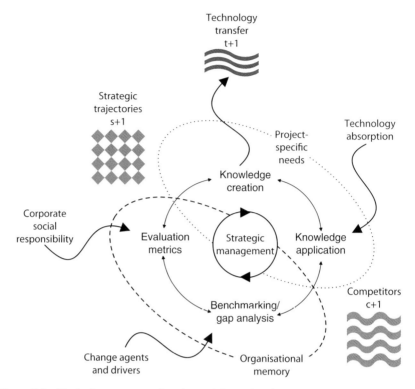

Figure 7.6 Strategic management cycles and dependencies.

the holding company and each of the three groups. This was needed in order to understand, at Board level, how these three strategies interrelated, what was shared and how information flowed between them. It was also important to capture the tangential support structures and mechanisms that made these units function, particularly from a knowledge sharing perspective. This relationship was mapped into a dependency schema, the details of which can be seen in Figure 7.6.

From the outset, it was considered important to understand the strategic behaviour of construction firms (Langford and Male, 2001) and to acknowledge that performance improvement within this sector often extends beyond project specific measures (Smyth, 2010). Furthermore, from a strategy perspective, it was also important to understand competition in relation to business units (Kao *et al.*, 2009) and business/project processes (Gann and Salter, 2000), in order to appreciate the importance of finding the right 'fit' between the construction environment and company specialism (Pries and Janszen, 1995). It was also acknowledged that the implications of learning and transfer should be fully understood (Belling *et al.*, 2004). In this respect, Figure 7.6 identifies the Strategic Management cycle and dependencies for this company. The central core 'Strategic Management' is depicted as a revolving process, through which all strategic decisions are analysed. This is supported by 'knowledge creation', knowledge application', 'benchmarking/gap analysis' and 'evaluation metrics' activities, the hub of which controls how tacit and project specific knowledge is used. This is of particular importance to construction organisations, as it forms the main conduit for organising, embedding and actioning knowledge, especially at project and organisational levels (Senaratne and Sexton, 2011), and how knowledge-pull can deliver innovation through knowledge management (Maqsood et al., 2007). The 'competitors c+1' and 'strategic trajectories s+1' activities can be considered as being symbiotically linked, as the strategic trajectories are directly linked to competitor strategies and pricing structures. In many respects, the trajectories can therefore be seen as strategic objectives, which act as signposting options, which is a useful way of predicting outcomes (Wickham and Wickham, 2007). This can be further enhanced by using probabilistic models, as these can help decision makers predict the likelihood of success of engaging option 1 (s1), option 2 (s2), etc against identified variables, for example, competitors, market niche, project mix, etc. The 'technology absorption' and 'technology transfer t+1' activities are mechanisms through which technology transfer and innovation opportunities are captured, assessed and embedded with the company. In this respect, it was deemed particularly important to understand how innovation could be implemented in construction (Slaughter, 1998; Winch, 2003; Sexton and Barrett, 2004; Ruddock and Ruddock, 2009), as this often engages organisational factors within and across organisational boundaries which, from this company's perspective, involved interaction at both group and corporate levels. Finally, the 'change agents and drivers' and 'Corporate Social Responsibility' (CSR) activities are the formal channels through which change and CSR are managed. For example, CSR embraces the organisation's corporate conscience, and the change agents and drivers include such issues as the organisation's ability to apply transformational change measures (cognisant of external perception). Thus, the extent by which this needs to be proactive or reactive needs to be carefully considered (Nadler et al., 1995), as this can be used as a lever to encourage innovative transformation (Fenton and Pettigrew, 2000).

7.7   Conclusion

This chapter introduced the importance of strategy in the AEC sector in relation to business strategy theories and construction sector drivers. This highlighted the need for organisations to have a strategy with a clear focus and readily identifiable objectives and deliverables. This is especially important where the effects of competition, market positioning and strategy are overtly prominent. In this respect, it was noted that strategy formulation should therefore consider a company's capacity to meet the needs of the market, by matching opportunity with corporate capability. This process should critically examine the issues most likely to affect outcomes – specifically strategic plans and initiatives, business processes, technology, organisational culture and the prevailing level (and type) of corporate resources. On this theme, BPR was introduced as an approach that can often help organisations fundamentally reorganise the way they conduct business operations from a process perspective in order to garner improvements. This was followed by a discussion on performance analysis, especially how this can be used to set and measure performance targets using appropriate measures and metrics. Strategy formulation issues were then presented for discussion through a construction case study exemplar. Particular emphasis was placed on determining this company's capacity to meet the needs of the market, by aligning group corporate capability and resources to strategic plans. This emphasised the strategic challenges facing one company in relation to the strategic deployment of strategy from a holding company perspective, and noted the need to identify functional and organisational areas matched to deliverables with predefined critical success factors. Research findings presented a Phase ZERO process attribute map for the case study company, together with a strategic management dependency schema. The process attribute map identified high-level process activities associated with matching business decisions to corporate strategy, and presented the governing dependencies and interrelationships, and how these can be effectively managed in order to procure defendable business solutions. The strategic management dependency schema highlighted the internal and external dynamics of strategic management from a holistic perspective. In particular, it also identified the importance of embracing project specific needs with organisational needs, especially cognisant of governing forces, competition and strategic trajectory positioning.

In summary, it is particularly important to acknowledge that the construction industry can benefit from understanding how strategy (strategic positioning), skills (intellectual capital), ICT (alignment to strategy) and process (understanding) all interrelate. Understanding these relationships can help companies focus on their particular areas of strength and, if appropriately augmented, could be used to garner strategic advantage.

References

Al-Mashari, M. & Zairi, M. (2000) Revisiting BPR: A holistic review of practice and development. *Business Process Management Journal*, **6**(1), 10–42.

Andrews, K.R. (1987) *The Concept of Corporate Strategy*. Irwin Inc, Illinois.

Ansoff, H.I. (1968) *Corporate Strategy: An Analytic Approach to Business Policy for Growth and Expansion*. Penguin Publishers, Harmondsworth.

Banwell, H (1964) *Report of the Committee on the Placing and Management of Contracts for Building and Civil Engineering Works*. HMSO, London.

Belling, R., James, K. & Ladkin, D. (2004) Back to the workplace: How organisations can improve their support of management learning and development. *Journal of Management development*, **23**(3), 234–255.

Betts, M. (1992) How strategic is our use of information technology in the construction sector. *International Journal of Construction Information Technology*, **1**(1), 79–97.

Betts, M. & Wood-Harper, T (1994) Reengineering construction: A new management research agenda. *Journal of Construction Economics and Management*, **12**, 551–556.

Betts, M. & Ofori, G. (1992) Strategic planning for competitive advantage in construction. *Construction Management and Economics*, **10**(6), 511–532.

Chan, P.S. & Land, C. (1999) Implementing reengineering using information Technology. *Business Process Management Journal*, **5**(4), 311–324.

Channon, D.F. (1978) *The Service Industries: Strategy, Structure & Financial Performance*. MacMillan, London.

Davenport, T.H, (1993) *Process Innovation: Reengineering Work through Information Technology*. Harvard Business School Press, Boston, MA.

Davenport, T.H. & Short, J.E. (1990) The new industrial engineering: Information technology and business process redesign. *Sloan Management Review*, **31**(4), 11–27.

Egan, Sir J. (1998) *Rethinking Construction: The Report of the Construction Task Force*. Blackwell Science Ltd, Oxford.

Elton, M. (1933) *The Human Problems of an Industrial Civilisation*. MacMillan, New York.

Emmerson, H. (1962) *Survey of Problems before the Construction Industries*. HMSO, London.

Fayol, H. (1949) *General and Industrial Management*. Pitman Publishing, London.

Fenton, E.M. & Pettigrew, A.M. (2000) *The Innovating Organization*. Sage Publications Ltd, London.

Gann, D.M. & Salter, A.J. (2000) Innovation in project-based, service enhanced firms: The construction of complex products and systems. *Policy Research*, **29**(7-8), 955–972.

Goodman, R. & Chinowsky, P. (1997) Preparing construction professionals for executive decision making. *Journal of Management in Engineering*, **13**(6), 55–61.

Goulding, J.S. & Alshawi, M. (2002) Generic and specific IT training: A process protocol model for construction. *Journal of Construction Management and Economics*, **20**(6), 493–505.

Hammer, M. (1990) Reengineering work: don't automate, obliterate. *Harvard Business Review*, **July–August**, 104–112.

Hammer, M. & Champy, J. (1993) *Reengineering the Corporation: A Manifesto for Business Revolution*. Harper Business, New York.

Harrison, E.F. & Pelletier, M.A. (2000) Levels of Strategic Decision Success. *Management Decision Journal*, **38**(2), 107–117.

Hillebrandt, P.M. & Cannon, J. (1990) *The Modern Construction Firm*. Macmillan, London.

Huber, G.P. (1991) Organizational learning: The contributing processes and the literatures. *Journal of Organization Science*, **2**(1), 88–115.

Johansson, H.J., McHugh, P., Pendlebury, A.J. & Wheeler, W.A. (1993) *Business Process Reengineering: Breakpoint Strategies for Market Dominance*. John Wiley and Sons, Chichester, UK.

Joia, L.A. (2000) Measuring intangible corporate assets – Linking business strategy with intellectual capital. *Journal of Intellectual Capital*, **1**(1), 68–84.

Kagioglou, M., Aouad, G. & Cooper, R. (2001) Performance management in construction: A conceptual framework. *Journal of Construction Management and Economics*, **19**(1), 85–95.

Kao, C., Green, S.D. & Larsen, G.D. (2009) Emergent discourses of construction competitiveness: localized learning and embeddedness. *Construction Management and Economics*, **27**, 1005–1017.

Latham, M. (1994) *Constructing the Team*. HMSO, London.

Langford, D. & Male, S. (2001) *Strategic Management in Construction*, 2nd edn. Wiley-Blackwell, London.

Lee, R.G. and Dale, B.G. (1998) Business process management: a review and evaluation. *Business Process Management Journal*, **4**(3), 214–225.

Lee, S.F. & Sai On Ko, A. (2000) Building Balanced Scorecard with SWOT Analysis, and implementing 'Sun Tzu's The Art of Business Management Strategies' on QFD methodology. *Managerial Auditing Journal*, **15**(1/2), 68–76.

Lewin, K. (1947) Frontiers of group dynamics. *Journal of Human Relations*, **1**(1), 5-41.

Love, P.E.D. & Holt, G.D. (2000) Construction business performance measurement: The SPM alternative. *Journal of Business Process Management*, **6**)(5), 408–416.

Maloney, W.F. (1997) Strategic planning for human resource management in construction. *Journal of Management in Engineering*, (**13**)3, 49–56.

Mayo, E. (1933) *The Human Problems of an Industrial Civilization*. Macmillan, New York.

Maqsood, T., Walker, D.H.T. & Finegan, A.D. (2007) Facilitating knowledge pull to deliver innovation through knowledge management: A case study. *Journal of Engineering, Construction and Architectural Management*, **14**(1), 94–109.

Mintzberg, H. & Quinn, J.B. (1991) *The Strategy Process: Concepts, Contexts, Cases*. Prentice Hall International (UK) Ltd, London.

Morgan, G. (1997) *Images of Organization*, 2nd edn, Sage Publishers, London.

Morgan, G (1990), Organizations in Society, Macmillan, London, UK

Nadler, D., Shaw, R.B., Walton, A.E., *et al.* (1995) *Discontinuous Change: Leading Organizational Transformation*. Jossey-Bass, San Francisco, CA.

Newcombe, R. (1990) The evolution and structure of the construction firm. *Proceedings, CIB 90, Joint Symposium on Building Economics and Construction Management*, Sydney, 6 March, 358–369.

Perkowski, J.C. (1988) Technical trends in the E&C business: The next 10 years. *Journal of Construction Management and Engineering*, **114**(4), 565–576.

Petrozzo, D.P. & Stepper, J.C. (1994) *Successful Reengineering*. Van Norstrand Reinhold, New York.

Porter, M.E. (1979) How competitive forces shape strategy. *Harvard Business Review*, **March-April**, 137–145.

Porter, M.E. (1985) *Competitive Advantage: Creating and Sustaining Superior Performance*. Free Press, New York.

Porter, M.E. (2008) *On Competition*, updated and expanded edn. Harvard Business School Press, Boston, MA.

Prahalad, C.K. & Hamel, G. (1990) The core competence of the corporation. *Harvard Business Review*, **68**(May–June), 79–91.

Price., A.D.F. & Newson, E. (2003) Strategic management: consideration of paradoxes, processes, and associated concepts as applied to construction. *Journal of Management in Engineering*, **19**(4),183–192.

Pries, F. & Janszen, F. (1995) Innovation in the construction industry: the dominant role of the environment. *Construction Management and Economics*, **13**, 43–51.

Rapert, M., Lynch, D. & Suter, T. (1996) Enhancing functional and organisational performance via strategic consensus and commitment. *Journal of Strategic Marketing*, **4**(4), 193–205.

Remenyi, D. & Sherwood-Smith, M. (1998) Business benefits from information systems through an active benefits realisation programme. *International Journal of Project Management*, **16**(2), 81–98.

Rezgui, Y. & Zarli, A. (2006) Paving the way to the vision of digital construction: A strategic roadmap. *Journal of Construction Engineering and Management*, **132**(7), 767–776.

Robson, W. (1997) *Strategic Management & Information Systems*. Pitman Publishing, London.

Ruddock, L. & Ruddock. S. (2009) Reassessing productivity in the construction sector to reflect hidden innovation and the knowledge economy. *Construction Management and Economics*, **27**, 871–879.

Senaratne, S. & Sexton, M. (2011) *Managing Change in Construction Projects: A Knowledge-Based Approach*. Wiley-Blackwell, London.

Senge, P.M. (2006) *The Fifth Discipline: The Art and Practice of the Learning Organization*, 2nd edn. Random House Business, New York.

Sexton, M. & Barrett, P. (2004) The role of technology transfer in innovation within small construction firms. *Journal of Engineering, Construction and Architectural Management*, **11**(5), 342–348.

Slaughter, E.S. (1998) *Models of construction innovation. ASCE Journal of Construction Engineering and Management*, **124**(3), 226–231.

Smith, A. (1759) *The Theory of Moral Sentiments*. Adam Smith Institute, London.

Smyth, H. (2010) Construction industry performance improvement programmes: the UK case of demonstration projects in the 'Continuous Improvement' programme. *Construction Management and Economics*, **28**, 255–270.

Soares, J. & Anderson, S. (1997) Modelling process management in construction. *Journal of Management in Engineering*, **13**(5), 45–53.

Stacey, R.D. (2010) *Strategic Management and Organisational Dynamics: The Challenge of Complexity (to Ways of Thinking About Organisations)*, 6th edn. Financial Times/ Prentice Hall, London.

Tan, R. (1996) Information technology and perceived competitive advantage: An empirical study of engineering consulting firms in Taiwan. *Construction Management and Economics Journal*, **14**(3), 227–240.

Tatum, C.B. (1988) Technology and competitive advantage in civil engineering. *Journal of Professional Issues in Engineering, American Society of Civil Engineers*, **114**(3), 256–264.

Taylor, F.W. (1911) *The Principles of Scientific Management*. Harper Bros., New York.

Van Daal, B., De Haas, M. & Weggeman, M. (1998) The knowledge matrix: A participatory method for individual knowledge gap determination. *Knowledge and Process Management Journal*, **5**(4), 255–263.

Venegas, P. & Alarcón, L. (1997) Selecting long-term strategies for construction firms. *Journal of Construction Engineering and Management*, **123**(4), 388–398.

Walker, D.H.T. (2000) Client/customer or stakeholder focus? ISO 14000 EMS as a construction industry case study. *The TQM Magazine*, **12**(1), 18–25.

Ward, J. & Daniel, E. (2005) *Benefits Management: Delivering Value from IS & IT Investments*. John Wiley and Sons, Ltd, Chichester, UK.

Ward, J.L. & Peppard, J. (2002) *Strategic Planning for Information Systems*, 3rd edn. John Wiley and Sons, Chichester, UK.

Ward, J and Peppard, J (1996), Reconciling the IT/Business Relationship: A Troubled Marriage in Need of Guidance, Journal of Strategic Information Systems, Vol. 5, No. 1, pp 37-65

Warszawski, A. (1996) Strategic planning in construction companies. *Journal of Construction Engineering and Management*, **122**(2), 133–140.

Wickham., P.A. & Wickham, L. (2007) *Management Consulting: Delivering an Effective Project*, 3rd edn. Financial Times/Prentice Hall, London.

Winch, G. (2003) How innovative is construction? Comparing aggregated data on construction innovation and other sectors – a case of apples and pears. *Journal of Construction Management and Economics*, **21**(6), 651–654.

Yin, R.K. (1994) *Case Study Research: Design and Methods*. Sage Publications, London.

Risk Management in Planning for Process Improvement

Oluwaseyi Awodele, Stephen Ogunlana and Graeme Bowles

The concept of continuous improvement as a means of improving processes, services and products in a competitive marketplace has long been established in business (Blick *et al.*, 2000; Momoh and Ruhe, 2006). To this end, many guidelines and models for process improvement within individual organisations have been produced. This chapter contends that conventional process improvement models are predominantly too rigid for organisations and their projects within the construction industry given its dynamic and volatile nature. A review of literature relating to business process and improvement models is presented and the need to incorporate risk management to make these more meaningful to the needs of construction organisations is argued. For example, organisations can benefit in a number of ways from investing in process improvement, including improvements in product quality, reduction in the time-to-market of the product, improvements to productivity and increased organisational flexibility and stakeholder satisfaction (Florac *et al.*, 1997; Zahran, 1998). Couple this with the reality of ever challenging and competitive markets; organisations must strive to increase productivity, improve relationships with customers and reduce the time to launch new products. In this respect, Kagioglou, *et al.* (1999) acknowledged the need for improvement to the conventional design and construction process was well reported in the literature, and that the construction sector in the UK has, since the 1930s, voiced a desire to change the way it performs its primary activity – the construction of building and civil engineering works. They reported that a succession of government and institutional reports have examined this activity, particularly the practice of construction management, and noted the need for improvement and change.

Construction Innovation and Process Improvement, First Edition.
Edited by Akintola Akintoye, Jack S. Goulding and Girma Zawdie.
© 2012 Blackwell Publishing Ltd. Published 2012 by Blackwell Publishing Ltd.

In a UK Government investigation, Latham (1994) re-affirmed the conclusions of all other previous studies, such as Banwell (1964) and British Property Federation (1983) among others. The report noted the need for effective processes throughout the construction life cycle, starting from the management of the client brief to the selection of the supply chain participants and eventual construction and decommissioning. The main outcome and recommendation of the Latham report was that it called for significant cost savings by the utilisation and formation of effective construction processes, which could lead to increased performance (Kagioglou et al., 1999). It is imperative therefore, for the construction industry to embrace this concept of process improvement, to improve productivity and efficiency of the sector in those areas that have long been the focus of criticism. Furthermore, Egan (1998) recommended an annual reduction of 10% in construction cost and construction time, and a reduction of 20% per annum of defects in projects. This performance improvement requires significant improvements in the way the construction process is enacted, which implies significant reengineering of the construction process and sub-processes. However, ever since these findings, businesses within the construction sector have been applying new methods of working and technologies to improve productivity and attain quality gains (Amaratunga et al., 2002). Furthermore, efficient implementation of different process improvement efforts is problematic, as much of the process research has concentrated on the development of methodologies for understanding process initiatives. Edwards et al. (2000) opined that these methodologies tend to be prescriptive and formulaic, consisting of a series of steps that lead to redesigned processes and hence theoretically significant performance improvements. In their opinion, the variety of organisational elements affected, when combined with the breadth of definition of process orientation, made it difficult to imagine any single process development methodologies being appropriate. The construction industry has therefore few recognised methodologies or frameworks on which to base process improvement initiatives. This lack of clear guidelines has meant that improvements are isolated and benefits are difficult to co-ordinate or repeat. Even when there are methodologies, Pritchard and Armistead (1999) assert that the same methods are unlikely to be equally successful in all cases.

The foregoing implies that the construction industry is limited in its ability to assess construction processes, prioritise process improvements and direct resources appropriately. In an attempt to address this problem, Sarshar et al. (1998) developed a process improvement framework for the construction industry called Structured Process Improvement for Construction Environments (SPICE). SPICE is based on the Capability Maturity Model (CMM), developed by the US Department of Defense, and widely used in the software industry. Evidence from other sectors show that continuous process improvement is based on many small evolutionary steps, rather than revolutionary measures (Paulk et al., 1993; Saiedan and Kuzara, 1995; Sarshar et al., 1998, 1999). Moreover, these steps do not take the possibilities of surprises in the course of improving a given process into cognisance, thus there is lack of integration of risk management in process improvement effort. This lack of integrated implementation approach to exploiting process improvement has been described as one of the important reasons responsible for failure of Process Improvement (Rohleder and Silver, 1997). Hammer and Champy (1993) estimated that 70% of process improvements and reengineering projects are less than successful. It is imperative therefore to raise the success rate of process improvement exercises by integrating risk management throughout the life cycle of a process improvement effort.

An effective process improvement framework begins with the formulation of one or more detailed process improvement plan(s), with the aim of achieving optimum balance of demands with available resources. The planning process in a business context therefore involves:

- identifying the goals or objectives to be achieved;
- formulating strategies to achieve them;
- arranging the means required; and
- implementing, directing and monitoring all steps in their proper sequence.

In planning for process improvement, there are possibilities that certain predictable events might occur, whose exact likelihood and outcomes are uncertain but could potentially affect the objectives in some way, either positively or negatively. The predictable events whose exact likelihood and outcome are uncertain are what manager's term risk (Grimsey and Lewis, 2002). Their capacities for affecting objectives explain the need for adequate risk management exercises in planning for process improvement. In this chapter, it is argued that business process improvement is a complex, risk-prone, collaborative process that requires a set of coordinated risk management processes. Risk management is presented as a proactive rather than a reactive, positive rather than negative effort, which seeks to increase the probability of process improvement. Process improvement is introduced and the integration of risk management at the planning stage for process improvement is also explained.

8.2   Process Improvement

There are many processes common to most, if not all, construction projects: preparation of a brief, establishing project scope, estimating probable cost, procurement of suitable contractors and suppliers, information management, measuring and analysing performance, training employees, closing out the project, etc. Most important business processes span several functional areas. For example, in construction there are many functional areas that tend to operate as separate units, for example, the estimating department is different from the design department. Improvements in this sector therefore are difficult to achieve, due to the cross-functional processes involved. Rohleder and Silver (1997) remarked that, to truly embrace the process improvement philosophy in this type of organisation, the organisation must make firm commitments to both a change in its thinking, and associated modification to the existing, vertical organisational structure, to a horizontal, process-oriented one. They further asserted that management must provide adequate time and resources for the improvement teams and that almost all the people in the organisation, not just a few at the top, should be actively involved in solving problems, reducing costs and eliminating waste.

Since improvement projects tend to ostensibly be managed through regular portfolio management in many organisations, an improvement project must have a sound business case in order to get firm commitment to the process improvement. The business case used in justifying the process improvement project needs to identify measurable impacts on a variety of performance measures for project organisation and the enterprise as a whole. Moreover, there may be measures of value to the process users, often exemplified in adoption or compliance measures that show the new processes are both used and useful. Both impact and adoption measures will be used to provide motivation for change, compare results of alternative approaches, ensure ongoing value of the ongoing investment and meet a variety of other organisation specific needs. Furthermore, people should not perceive identification of problems or the need for improvement as an indication of negative performance, rather it should be seen as the need to overcome complacency, switching to an attitude of preventing instead of reacting to problems. In addition, because process improvements can often naturally lead to higher productivity, fewer workers may then be needed to accomplish the same output. One may tend to feel that process improvements could then lead to layoffs from work. Robinson (1991) asserts that no layoffs should be made because of process improvements, but where possible, reductions should be achieved through attrition, where excess staff could be used to meet increased levels of demand. Care must be taken during the fact gathering exercise for process improvement. This is because fact gathering for process improvement tends to create fears about job changes and job losses. The skills and the integrity of the professional can go a long way towards reducing these anxieties and gaining cooperation. Skills enable the professional to collect critical, relevant data easily and assure that once collected it will not be lost. However, integrity is more important. It includes making sure that the focus of improvement treats people as a resource to be utilised and not an expense to be cut.

8.2.1   Definitions of Business Process

One source of ongoing confusion is the imprecise use of terminology. Process is a word that means different things to different people. Even for practitioners who are extremely familiar with the concepts of process management, there is still confusion across disciplines (Gulledge and Sommer, 2002). For instance, when software engineers document processes, they are often interested in the relationships amongst static activities. When industrial engineers discuss process, they most often focus on the dynamic linking of activities; for example, process flows. For this reason, the following are some definitions of process in general and in construction.

Johansson *et al.* (1993) defined a process as a set of linked activities that takes an input and transforms it to create an output, and it should add value to the input and create an output that is more useful and effective to the recipient. Davenport (1993) describes a process as a specific ordering of work activities across time and place, and which has a beginning and an end, and clearly identified inputs and outputs. Hammer and Champy (1993) remarked that a business process is a set of activities with one or more types of input that creates a valuable output for a customer. A process according to Earl and Khan (1994) is a lateral or horizontal organisational form that encapsulates the inter-dependence of tasks, roles, people, department and functions required to provide a customer with a product or service. In the same vein, Ahmed and Simintras (1996) defined a process as a system, which interlocks cross-functional flows of resources and deals with tasks that were previously considered as isolated in an integrated way. Evans and Lindsay (2008) define a process as how work creates value for customers. Kagioglou, *et al.* (1999) opined that lack of an established definition of a process in the construction industry has led the industry into using procurement systems rather than looking at the overall process as a whole entity. They asserted that a number of lessons could be learned from the manufacturing sector about the implementation and practical use of a 'process view' within the construction industry.

Whilst there are obvious similarities in the definitions, there seems general agreement that process involves a sequence of inter-dependent and linked procedures which, at every stage, utilise one or more resources (time, energy, money) to convert inputs (labour, material, equipments, etc) into outputs. These outputs then serve as inputs for the next stage, until a known goal or *objective* is achieved. In this respect, several authors on process improvement have argued that the change towards a business process approach is directly linked to a company's strategy, and that it would be impossible to create business processes to support a company's objectives if these were not known beforehand. The business objectives need to be linked directly to business strategic planning. Thus, processes are also, directly or indirectly, related to business strategic planning (Sommerville, 2007). Processes without connection to strategic planning have no reason to exist, because they are consumers of resources that add no value. For an effective management of construction project processes, project managers need to understand, control and improve them. In order to understand a process, Kloppenborg (2009) assert that the process boundaries need to be defined, as this will help prevent occurrence of future scope creep.

Another fundamental characteristic of business processes is that the quality of the process can often directly affect the quality of delivered products. In the manufacturing industry, there is a clear link between process and product quality, because the process is relatively easy to standardise and monitor. Once manufacturing systems are calibrated, they can be run repeatedly to output high-quality products. However, construction activities tend to be more bespoke in nature than manufacturing processes, so the influence of individual skills and experience is significant. In addition, construction is prone to many external factors that can influence the achievement of project objectives, namely adherence to the budget and schedule. There is therefore a need for a holistic improvement methodology to provide the means for generalising and describing the knowledge and experience in a structured way, that is, to some extent, transferable to other situations in terms of roles, tasks and the required skills. This can be achieved through a framework that integrates risk management at the planning stage for process improvement.

8.2.2   Classification of Business Processes

It is important to recognise that not every process in an organisation requires the same effort for improvement. For example, Evans and Lindsay (2008) classified business processes under value-creation and support categories. They argue that organisations need to understand their processes well and should differentiate between value-creation and support categories. This will differ across organisations, depending on the nature of the business. Since value-creation processes add, as the term suggests, value to end products and services, they warrant a higher level of attention than the support processes. Value creation processes are often termed core processes (Evans and Lindsay, 2008), since they are fundamental to businesses in maintaining and achieving sustainable competitive advantage. They drive the creation of products and services, and are critical to customer satisfaction. They can be said to have a major impact on the strategic goals of an organisation. Leading companies identify those value-creating processes that affect customer satisfaction throughout the value chain. On the other hand, support processes are those that are most important in supporting an organisation's core processes, employees and daily operations. They provide infrastructure for value-creation processes but, generally, do not add value directly to the product or service. Having now established an understanding of process, it is also important to reflect on process improvement. Process improvement concerns understanding existing processes and making them better in order to increase product quality and/or reduce costs and development time. According to Sommerville (2007), most of the literature on process improvement has focused on perfecting processes to improve product quality and, in particular, to reduce the number of defects in delivered product.

8.2.3   Process Improvement

According to Rohleder and Silver (1997), a survey of the business landscape shows widespread initiatives in process improvements and reengineering. Process improvement helps an organisation meet long-term goals and objectives. One key goal for all organisations is to meet the demands of customers – both internal and external. Customers needs change over time – whether due to economic factors, new product developments, mergers or acquisitions, expansion or contraction. Continuously reviewing processes for potential improvements and efficiencies enables companies to adapt effectively to their clients' changing needs. However, when engaging in true process improvement, organisations should seek to learn what causes things to happen in a process and to use this knowledge to reduce variation, remove activities that contribute no value to the product or service produced, and improve customer satisfaction. A team examines all of the factors affecting the process: the materials used in the process, the methods and machines used to transform the materials into a product or service, and the people who perform the work. Process improvement does not simply mean adopting or 'bolting on' a particular tool or model of a process that has been used elsewhere. Similar organisations that do similar things will have processes in common, however, there will always be some contextual factors, such as procedures and standards that influence the process. It is important that an organisation should always look on process improvement as being specific to itself, or a part of the larger organisation. Furthermore, process improvement initiatives are continuous and as organisations grow, they need to continuously analyse and refine their processes to ensure they are operating as effectively and efficiently as possible to gain competitive advantage. Kagioglou *et al.* (1999) concluded that construction process improvement/reengineering is an area where the construction industry in the UK needs to consider and adopt so that significant improvements in performance could be achieved.

8.2.4 Project and Process Improvement Failures

The aim of process improvement is to meet and often exceed customer expectation in products and services. Organisations through process improvement seek to improve their efficiency, reduce costs or increase revenues. However, with these good intents, too few of these efforts are completed or even partially successful. Many process improvement efforts have ended in failure. This leads one to question: 'Why do companies ever make such efforts in the first place?' A major reason for this is that when process improvements do succeed, they can have significant economic benefit. For example, Gilbert (1993) cited in Rohleder and Silver (1997), gave examples of huge cost savings resulting when the Electric Boat Division of General Dynamics Corporation improved its contracting process, which resulted in a 50% reduction in the number of process steps, eventually saving the company over $3 million annually. However, many factors can prevent the effective implementation of process improvement and hence restrict innovation and continuous improvement. Irani *et al.* (2000) identified some of these issues as:

- loss of nerve, focus and stamina;
- senior management who are comfortable in their ivory towers;
- lack of holistic focus and settling for minor improvement gains;
- human and organisational issues;
- organisation culture, attitude and skills;
- resources restriction; and
- fear of IT.

Bashein and Markus (1994) asserted that senior management commitment and sponsorship, realistic expectation, empowered and collaborative

Figure 8.1   Process improvement cycle.

workers, strategic context, shared vision, sound management practices and sufficient human and financial resources are positive preconditions for efficient business process improvement. According to Paper (1998), successful process improvement hinges on top management support, customer satisfaction, cross-functional teamwork and a systematic means of solving problems. The greatest obstacles faced by process improvement efforts are:

1. lack of sustained management commitment and leadership;
2. unrealistic scope and expectation; and
3. resistance to change.

8.2.5   Steps in a Process Improvement Model

Sommerville (2007) asserts that process improvement is a cyclical activity that involves three principal stages: process measurement, process analysis and process change (Figure 8.1).

Sommerville (2007) described process measurement as a stage where the attributes of the current project or the product is measured with the aim of improving the measures according to the goals of the organisation involved. Process analysis is a stage in process improvement where the current process is assessed to reveal process weaknesses and bottlenecks. Process models that describe the process, are usually developed during this stage, while process change is the stage where changes to the process that have been identified during analysis are introduced. In this context, processes can be improved in either a continuous or a breakthrough fashion. It is important for all project core team members and experts in a particular process to be thinking of ways they can improve at any time. Kloppenborg (2009) asserts that slow and steady improvement is a good foundation, but that sometimes substantial improvement may be what is needed.

Whichever one, or regardless of the size of improvement desired, many models exist to guide the effort, such as Six Sigma, that uses the define, measure, analyse, improve and control (DMAIC) methodology. In Kloppenborg (2009), the DMAIC methodology is seen as a 15-step process

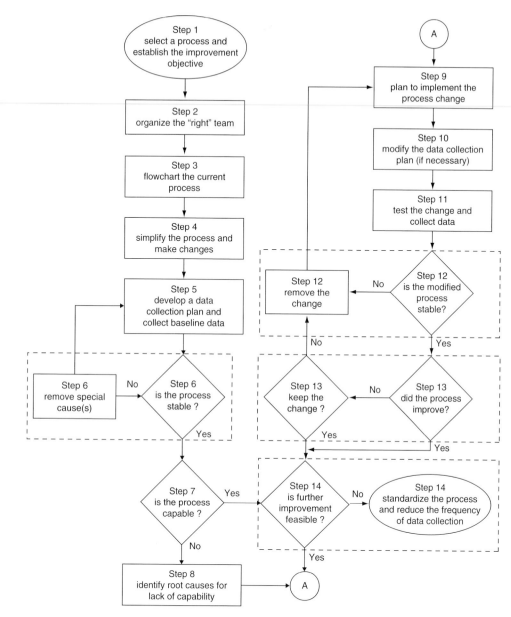

Figure 8.2   Basic process improvement model (*Handbook for Basic Process Improvement*, 1996). Reproduced by permission of the Balance Scorecard Institute.

comprising 5 project phases: define, measure, analyse, improve and control. However, most improvement models are based on the plan-do-check-act (PDCA) cycle, as contained in *Basic Process Improvement Handbook* (1996). This handbook presented a process improvement model, which has two parts:

1. a process simplification segment; and
2. a Plan-Do-Check-Act (PDCA) cycle (Figures 8.2 and 8.3).

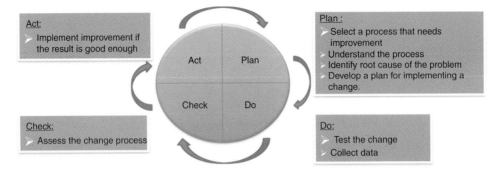

Act:
➤ Implement improvement if the result is good enough

Plan :
➤ Select a process that needs improvement
➤ Understand the process
➤ Identify root cause of the problem
➤ Develop a plan for implementing a change.

Check:
➤ Assess the change process

Do:
➤ Test the change
➤ Collect data

Figure 8.3 Plan-do-check-act cycle for process improvement. From Kloppenborg. *Aise-Contemp Proj Mgmt*, 1E. © 2009 South-Western, a part of Cengage Learning, Inc. Reproduced by permission. (www.cengage.com/permissions).

Table 8.1 Steps and actions in the process simplification segment (handbook for basic process improvement, 1996). Reproduced by permission of the Balance Scorecard Institute.

Steps	Description of actions
1	Select the process to be improved and establish a well-defined process improvement objective. The objective may be established by the team or come from outside.
2	Organize a team to improve the process. This involves selecting the "right" people to serve on the team; identifying the resources available for the improvement effort, such as people, time, money, and materials; setting reporting requirements; and determining the team's level of authority. These elements may be formalized in a written charter.
3	Define the current process using a flowchart. This tool is used to generate a step-by-step map of the activities, actions, and decisions which occur between the starting and stopping points of the process.
4	Simplify the process by removing redundant or unnecessary activities. People may have seen the process on paper in its entirety for the first time in Step 3. This can be a real eye-opener which prepares them to take these first steps in improving the process.
5	Develop a plan for collecting data and collect baseline data. The data will be used as the yardstick for comparison later in the model. This begins the evaluation of the process against the process improvement objective established in Step 1. The flowchart in Step 3 helps the team determine who should collect data and where in the process data should be collected.
6	Assess whether the process is stable. The team creates a control chart or run chart out of the data collected in Step 5 to gain better understanding of what is happening in the process. The follow-on actions of the team are dictated by whether special cause of variation is found in the process.
7	Assess whether the process is capable. The team plots a histogram to compare the data collected in Step 5 against the process improvement objective established in Step 1. Usually the process simplification actions in Step 4 are not enough to make the process capable of meeting the objective and the team will have to continue on to Step 8 in search of root causes.

The process simplification cycle consists of seven steps, and the process improvement is expected to start process improvement with these steps. Depending on the stability and capability of the process, the team may continue on to Step 8 or go directly to Step 14. The PDCA cycle, which is the second part of the improvement model, consists of seven steps (Steps 8 through 14) flowing from the process simplification segment. In the handbook, it is stated that using all 14 steps of the model should increase the team's process knowledge, broaden decision-making options and enhance the likelihood of satisfactory long-term results.

The PDCA cycle is similar to what was presented as plan, do, study and act in Deming's cycle (PDSA) – a tool that was developed and promoted by Edward Deming who worked with Japanese industry after World War II to improve quality. The cycle was named the 'Deming Cycle' by the Japanese sponsors of the work in 1950 (Evans and Lindsay, 2008).

From Table 8.1 it is evident that Step 7 marks the ends of the process simplification cycle. It should be emphasised that even if the data indicates that the process is meeting the objective, the team should consider whether it is feasible to improve the process further before going on to Step 14. In Step 8, which is the beginning of PDCA Cycle, the team should start by identifying what causes the product or service to be unsatisfactory. As the team plans a change, they normally conduct a test and collect data, evaluate test results to find out whether the process has improved, and decide whether to standardise or continue to improve the process.

8.3   Planning for Process Improvement

Companies often redesign business processes to achieve improvements in their performance, such as better service and quality (Lee and Ahn, 2008). Process improvement is a long-term approach to improving organisational performance, with substantially less risks of destroying value when compared to short-term approaches. Planning, according to the *Cambridge Advance Learner Dictionary Online* is the act of deciding how to do something. Kerzner (2009) defines planning as '… the function of selecting the enterprise objectives and establishing the policies, procedures and programs necessary for achieving them… by establishing a predetermined course of action within a forecasted environment.' This assertion advocates that without proper planning, programmes and projects can start-off 'behind the eight ball' and identifies consequences of poor planning to include:

1. project initiation without defined requirements;
2. wild enthusiasm;
3. disillusionment;
4. chaos;
5. search for the guilty;
6. punishment of innocent; and
7. promotion of the non-participants.

Table 8.2   Steps and actions in the plan-do-check-act cycle (handbook for basic process improvement, 1996). Reproduced by permission of the Balance Scorecard Institute.

Steps	Description of actions
8	Identify the root causes which prevent the process from meeting the objective. The team begins the Plan-Do-Check-Act Cycle here, using the cause-and-effect diagram or brainstorming tools to generate possible reasons why the process fails to meet the desired objective.
9	Develop a plan for implementing a change based on the possible reasons for the process's inability to meet the objective set for it. These root causes were identified in Step 8. The planned improvement involves revising the steps in the simplified flowchart created after changes were made in Step 4.
10	Modify the data collection plan developed in Step 5, if necessary
11	Test the changed process and collect data.
12	Assess whether the changed process is stable. As in Step 6, the team uses a control chart or run chart to determine process stability. If the process is stable, the team can move on to Step 13; if not, the team must return the process to its former state and plan another change.
13	Assess whether the change improved the process. Using the data collected in Step 11 and a histogram, the team determines whether the process is closer to meeting the process improvement objective established in Step 1. If the objective is met, the team can progress to Step 14; if not, the team must decide whether to keep or discard the change.
14	Determine whether additional process improvements are feasible. The team is faced with this decision following process simplification in Step 7 and again after initiating an improvement in Steps 8 through 13. In Step 14, the team has the choice of embarking on continuous process improvement by re-entering the model at Step 9, or simply monitoring the performance of the process until further improvement is feasible.

With these possible consequences of poor planning, it is evident that in every endeavour, it is important to prepare a road map as to what needs to be done, who is responsible for it and when it should be delivered. Thus, planning for project improvement can be described as all effort geared towards determining what needs to be done, by whom and by when, that is, in order to achieve desired process improvement, it is also necessary to ascertain which process needs improvement, how these are going to be improved, who will carry out the improvement and when this is expected. Similar to general project planning, where a project management plan is developed, the outcome of process improvement planning is the production of process improvement plan, which forms part of the overall process management plan.

The planning for process improvement stage involves prioritising the key areas for improvement, and identifying the process whose improvement in the project will make the greatest difference. The improvement plan should contain a small number of 'breakthrough' process (i.e. those value-creation process within the organisation) rather than many minor process or support processes relating to parts of the organisation. This justifies the need for adequate knowledge of the processes in the organisation and the need to be able to classify them as either value-creation (core processes) or support processes. Kloppenborg (2009) opines that to manage project processes effectively, project managers need to understand, control and improve them. Similar to the project manager's need to produce a comprehensive project management plan (PMP) for effective management of a project, managing project processes requires detailed planning to establish how the improvement process will progress.

A cursory review of the actions involved in the planning stage for process improvement in the model in Figure 8.3 and supported by Tables 8.1 and 8.2, shows that actions in Step 1 through to Step 10 can be regarded as planning for process improvement stage. Where possible, improvement projects should be allocated to improvement teams who are given complete responsibility to plan and implement improvement actions. Ideally, these teams should comprise staff involved in the particular process at all levels. The improvement plan is reported on a periodical basis to senior management. Thus, for improvements to be worthwhile, they should aspire to achieve sustainable change. This may be through focusing the exercise on identifying and tackling the root causes of the problem, changing the organisation's management system in line with the proposed improvement, embedding changes into process documentation, and incorporating training and development plans where necessary.

8.4   Risk and its Management

The main objective of Project Management is to maintain a healthy balance between the three conventional objectives of any construction project (i.e. cost, time and quality), which is often referred to as the 'iron triangle'. Anything that can threaten the achievement of these objectives is considered as a risk to the project, which can prevent the project manager from meeting cost, time, quality and any other objectives predefined for the project. Risk can affect productivity, performance, quality and budget of a construction project (Kangari, 1995). Risk is therefore a permanent element in every decision-making process, including design and planning decisions and is inherent in every human endeavour, which can prove difficult to deal with. As such, it is important to develop a proper risk management framework, both in theoretical and in practical terms (Wang *et al.*, 2004). Many researchers have defined risk based on their perceptions and the needs or outcomes of their studies. For example, Wideman (1986) and Akintoye and Macleod (1997) defined risk as the likelihood of unforeseen factors occurring, which would adversely affect the successful completion of the project in terms of cost, time and quality. Cooper and Chapman (1987) defined risk as 'the exposure to the possibility of economic or financial loss or gain, physical damage or injury, or delay as a consequence of uncertainty associated with pursuing a particular course of action.' In this respect, Skorupka (2008) noted that the term 'risk' was derived from an Italian verb 'riscare', which means 'to have the cheek to do something'. Risk to the economist focuses on the financial aspects, engineers relate risk to process disruption and cost, the military consider the risk of completing a task, police officers treat risk as threat to the citizens, and employees may see risk as being dismissed from work. It is therefore an imperative to clearly specify the meaning of risk in this respect. The definition given in the Project Management Body of Knowledge Guide (PMBOK, 2004) is therefore adopted in this chapter.

PMBOK Guide (2004) defines project risk as an uncertain event or condition that, if it occurs, has a positive or a negative effect on at least one project objective, such as time cost, scope or quality. Two things are important in this definition. The first is the possibility of loss or gain in any risk situation. Risk is about deviation from a desired target, and loss and gain are possibilities at all times. It is possible to have cost over-run or cost under-run, and it is also possible to have time over-run or under-run. However, when there is positive gain, most people are not concerned, but when adverse effects are experienced (i.e. when the project manager fails to meet the set objectives), people get angry or are unhappy with the outcome. It is now commonly realised that risk and opportunity should therefore go together.

Kloppenborg (2009) suggested two tactics that project managers and teams can adopt in addressing risks. First, any risk that may inhibit success-ful project completion (to the satisfaction of stakeholders, on time, and on budget) needs to be identified, and a plan must be developed to overcome it. Second, a risk that can create a positive effect on a project can be considered as an opportunity to complete the project better, faster and or at lower cost, and a plan should be developed to capitalise upon it. The implication is that risk has the potential for causing loss, often called the downside risk. According to Smith *et al.* (2002), the loss can be financial, time, corporate image, poor quality, etc. There is also the possibility of the event leading to favourable outcomes, whereby things turn out better than what was planned; this is referred to as the upside of risk.

The focus in this chapter is on the downside of risk, that is, the unfavour-able impacts those events can cause on the project objectives when they occur, since the overriding intention of most risk management actions is to minimise potential losses. Given this, risk emanates from the uncertainty associated with pursuing a particular cause of action. Uncertainty derives from the absence of sufficient information about the action and/or the possible out-comes. From an information perspective, it is possible to identify two levels of uncertainty. The first level of uncertainty is experienced when decision makers have sufficient information to make probability judgement about a particular event, even though information may be incomplete (i.e. it is pos-sible to say this event is more probable than not). This is the region of risk management. Risk then is concerned with being able to make probability judgement. A second level of uncertainty is experienced when the informa-tion available to decision makers is so bad or insufficient that it is not possi-ble to make probability judgement. Such a situation is called pure uncertainty. There are procedures for dealing with pure uncertainty in economic analysis, but managing pure uncertainty is outside the scope of this chapter. In the context of process improvement, risks are those factors that can jeopardise the successful implementation of process improvement effort or the realisation of the expected benefits (i.e. efficiency, reduction in cost or increase in revenues). This therefore underscores the need for their management.

8.4.1   The Risk Management Process

Kezsbom and Edward (2001) stated that risk management is an important and integral element of project management. All managers manage risk either consciously or unconsciously, but rarely systematically. Managing risk is a forward thinking act, where individuals or groups take a look at the future of a particular process or endeavour to identify, in a responsible manner, the downsides or upsides of actions or inactions. With this, balanced thinking is achieved, which provides a framework to facilitate more effective decision making. Risk Management is therefore all about maximising opportunity, that is, increasing the probability and impact of positive events, and decreasing the probability and impact of events adverse to the project. While corporate governance can be said to be the 'glue' that holds the organisation together in pursuit of its objectives, risk management provides the resilience. Risk management can therefore be said to be a formal and orderly process of systematically identifying, analysing and responding to risks throughout the life cycle of a project, to obtain the optimum degree of risk elimination, mitigation and/or control. Smith and Merritt (2002) posit that project risk management has become a popular management topic because many organisations now recognise the high cost of dealing with project problems that could have been anticipated. Risk management continues to be a major feature of project management and is becoming increasingly important globally. The popularity of risk management in the management arena can be seen as a product of the following factors:

- global economic problem/budgetary constraints;
- recognition of true cost of underperformance;
- legislations;
- competitiveness of global markets; and
- true payback of risk management.

The PMBOK Guide (2004) defines risk management as the processes concerned with identifying, analysing and responding to uncertainty throughout the project life cycle. It includes maximising the result of positive events and minimising the consequences of adverse events. The body of knowledge asserts 'to be successful, the organisation should be committed to addressing the management of risk proactively and consistently throughout the project.' Dey (1999) suggested that project risk management processes are threefold:

1. identifying risk factors;
2. analysing their effect; and
3. responding to risk.

Winegard and Warhoe (2003) prescribed that the risk management process should follow five consecutive steps:

1. identification of risk;
2. risk assessment and analysis;
3. risk mitigation, i.e. development of risk reduction and reaction to threats;
4. implementation of risk management plan; and
5. review and correction of risk assessment.

PMBOK (2004) provides the project risk management processes through six stages as:

1. *Risk Management Planning*: deciding how to approach, plan and execute the risk management activities for a project;
2. *Risk Identification*: determining which risks might affect the project and documenting their characteristics;
3. *Qualitative Risk Analysis*: prioritising risks for subsequent further analysis or action by assessing and combining their probability of occurrence and impact;
4. *Quantitative Risk Analysis*: numerically analysing the effect on overall project objectives of identified risks;
5. *Risk Response Planning*: developing options and actions to enhance opportunities, and to reduce threats to project objectives, and
6. *Risk Monitoring and Control*: tracking identified risks, monitoring residual risks, identifying new risks, executing risk response plans, and evaluating their effectiveness throughout the project life cycle.

It should be noted that risk management is not just a one-way process but rather an iterative process, as these processes often interact with each other, and also with processes in other knowledge areas of project management. Each process can involve effort from one or more persons (or groups) based on the needs of the project. Each process occurs at least once in every project and occurs in one or more project phases, if the project is divided into phases. It is evident then that risk management needs to be integrated at the planning phase of process improvement, which is the stage at which process improvement areas are identified in order to understand process issues, issues to be dealt with and plans for improvement. Before moving on to how risk management can be integrated into planning for process improvement, it is important to understand the different stages involved in the risk management planning process.

8.4.2   Risk Management Planning

Winning is the science of being totally prepared; and to fail to plan, is planning to fail. These two popular sayings highlight the importance of planning to most enterprises and most often such planning is taken for granted. Kerzner (2009) asserts that the primary benefit of not planning is that failure will then come as a complete surprise rather than being preceded by periods of worry and depression. Risk Management Planning is therefore a crucial part of the risk management process, and in fact determines the success of all the other five processes. To be totally prepared means to be ready to deal with any situation. It therefore can be described as all efforts geared towards deciding how best to approach and conduct risk management activities for a particular project or process. Just as in general project planning where a project management plan is developed, the outcome of Risk Management Planning is the production of a risk management plan (RMP) that forms part of the overall project management plan (PMP). Depending on the size and the complexity of the project, for example, in small projects, risk management may be informal, whereas for large, complex projects, management needs to develop and prepare a written RMP.

When preparing the RMP, managers rely greatly on a good historic database that details and records the attitudes and tolerance of their organisations and the people they have worked with towards risk. Moreover, it is possible that their organisations have pre-defined approaches to risk management. For instance, the organisation may have developed standard templates for risk planning. Then it will be easy for the manager to make use of the template, but with some level of care. This is so because, though it is possible to generalise or adopt a given risk template for all projects in an organisation, some projects may have specific or unique risk elements inherent in them. Therefore, having considered the attitude towards risk as well as the risk tolerance of the organisation, managers need to evaluate project risks through the statement of work (SOW) and examine the work breakdown structure (WBS) for the project to be able to understand the sources of risks.

The roles and responsibilities of the people that will be involved in risk planning could also be assessed using the SOW and WBS. This allows thorough evaluation of the risks associated with sub-tasks separately (Dey et al., 1994). In this respect, a good RMP gives a brief summary of the approach, tools and data sources that are to be used to perform risk management on the project. It should also define the roles of the risk management team members, for every activity in the risk management process, along with their responsibilities. The plan also assigns resources, and contains estimates of costs needed for risk management, and also defines when and how often the risk management process will be performed throughout the project life cycle. Finally, the RMP provides a structure that ensures a comprehensive process of systematic identification of risk to a consistent level of detail and contributes to the effectiveness and quality of risk identification, which is the next process in risk management. To be able to develop the RMP, there is a need for project team members to hold a planning meeting. This meeting is usually attended by the project manager and some selected project team members, depending on the complexity of the project, and by other stakeholders, that is, anyone in the organisation with responsibility to manage the risk planning and execution. In the context of process improvement, this meeting can also involve process team leaders within an organisation.

In summary, if the organisation has pre-defined approaches to risk management, for instance, standard templates for risk categories and definitions of terms, such as levels of risk, probability by type of risk, impact by type of objectives and the probability and impact matrix can then be tailored to a specific project. In the absence of such predefined templates for planning, the project manger will need to create new templates, and discuss these with others who are working or have worked on similar projects. This helps to establish the standard for use on subsequent projects. Risk planning therefore helps managers to know the feasibility of risk management activities in order to decide whether the effort is worthwhile when compared to the resources/time expended against the risk exposure.

8.4.3   Risk Identification

Risk identification is essentially a process for uncovering any risks that could potentially afflict a process. This step is of considerable importance because other processes such as risk analysis and response can only be performed on the identified potential risks. In this context, the process must involve an investigation into all the potential sources of process risks and their consequences. Risk identification must be done iteratively as new risks may be identified as the process of improvement progresses throughout its life cycle. The risk identification process is beneficial to process improvement as it focuses the attention of process managers on strategies for the control and detection of potential risks. This highlights those areas where further efforts are needed. Risk identification is a simple but difficult task, because there are no absolute procedures that may be used to identify risks in a process. Often, managers rely heavily on their experience and on the insight of other key personnel involved in the process. Depending on the process documentation available and the nature of the process, a variety of thought starters can prompt risk discovery. Regarding risk, Smith and Merritt (2002) noted that managers need to focus on the interface between the consultant and the client, between departments of the client organisation, between phases or tasks of a client process, or between geographic areas. They further suggest that the project schedule should clearly show dependencies between tasks, to help pinpoint risk-prone areas. Alternatively, managers may use process maps that show interfaces between processes or tasks. Flanagan and Norman (1993) compared attempts of risk identification in projects with multiple layers of planning, complex vertical and horizontal interactions.

Organisations that keep good records of their past projects or project managers that conduct reviews of their projects at closure, can use this experience of tacit knowledge and lessons learned to garner insights on potential risks. Furthermore, a good project scope statement will detail all assumptions in the project; it is then easier to evaluate uncertainty and determine project assumptions. The outcome of the risk identification exercise is a document called the 'risk register'. This document includes *inter alia*, a list of identified risks, including their root causes and uncertain project assumptions. Often at the risk identification stage, the potential responses to a risk may be identified. These potential responses will also be recorded in the risk register, which becomes a useful input to the risk response planning process.

Thus, in order to choose the right technique for identifying risk, practitioners should consider several factors, such as the organisation's objectives, the nature of the project in terms of sizes, duration and the company's strategies for risk management.

The following are some of the tools and techniques suggested by the PMBOK guide (2004):

- documentation review;
- information gathering techniques;
- checklist analysis;
- assumptions analysis; and
- diagramming techniques.

These can be supported by information gathering techniques for risk identification, which include:

- brainstorming;
- Delphi technique;
- interviewing;
- root cause identification, and
- strengths, weakness, opportunities and threats (SWOT) analysis.

8.4.4   Qualitative Risk Analysis

In the risk analysis stage, managers try to estimate the overall magnitude of the risk and the expected losses. Typically, the risk event drivers and their impact drivers are determined at this stage. Any risk event that cannot be justified through probability of occurrence and impact are automatically dropped from the risk register. Broadly, risks can be analysed either qualitatively or quantitatively, depending on the purpose, required degrees of detail, and data and resources available for analysis. In qualitative analysis, risks are subjectively estimated and ranked in a descriptive manner. Jiang *et al.* (2002) described qualitative risk analysis as the process of prioritising risks for subsequent further analysis or action by assessing and combining their probability and impact. Qualitative estimation can be used for the following purposes:

- as an initial screening activity to identify risks that require more detailed estimation, when it provides sufficient information for decision making; or
- where available data or resources are insufficient for a quantitative estimation.

Because of such analysis, risks can be rated, for example, as high, moderate or low. Usually qualitative risk analysis is a rapid and cost-effective means of establishing priorities for risk response planning, and lays the foundation for quantitative risk analysis. As in risk identification, review of process documentation of past experiences or lessons learned from previous projects can be good sources of information to prioritise the identified risks. With this, an update is made to the risk register. Risks are listed in terms of their probability and impact in matrix form, which the project manager can use to focus attention on those risks with high significance to the project, where responses can lead to better project outcomes. In the risk register, risks may also be grouped by categories showing the root causes of risks or areas of the project that require particular attention or urgent attention and those that can be handled later. From this, some risk items will go straight to the response stage, while some may require further analysis, for example, quantitative risk analysis.

8.4.5   Quantitative Risk Analysis

Quantitative risk analyses are performed on risks that have been prioritised through the qualitative risk analysis process as potentially and substantially influencing the project's competing demands (PMBOK, 2004). Quantitative risk analysis is the process of numerically analysing the effect on overall project objectives (Kloppenborg and Deborah, 2004). The evaluation of both consequences and probability are based on data from a variety of sources, for example, past project records, collected field data, experimental data (including prototype testing), etc. Quantitative risk analysis is often used when the need to predict with confidence the probability of completing a project on time, on budget, at the agreed-upon scope, and/or the agree-upon quality is critical (Kloppenborg, 2009). Dey and Ogunlana (2004) identified some of the quantitative risk analysis tools and techniques currently in use as:

- Statistical Probability Distribution;
- Simple Multi-Attribute Rating Technique;
- Expected Value Technique;
- Sensitivity Analysis;
- Decision Trees;
- Bayes' Theorem;
- Simulation;
- Utility Theory;
- Analytic Hierarchy Process;
- Fuzzy-set Theory;
- Neuro-Fuzzy Networks, and
- Financial Methods.

After the identified risk have been analysed quantitatively, the risk register initiated in the risk identification stage is then updated accordingly, where risks are prioritised according to the level or threat posed, or opportunity offered.

8.4.6   Risk Response Planning

After the risk is estimated, it should be determined whether the risk level is acceptable or not by comparing it with the acceptance criteria determined at the risk management planning stage. Risk response planning therefore involves determining in advance how to respond to each major risk. Minor risks are handled by simply being aware of their potential and dealing with them if and when they occur. Kloppenborg (2009) identify six types of risk response strategies that can be applied to major risks as:

1. avoid
2. transfer
3. mitigate
4. accept
5. research, and
6. exploit.

PMBOK (2004) categorised risk response under four main headings:

- strategies for threats (i.e. those risks that have negative impacts) are avoid, transfer and mitigate;
- strategies for positive risks (i.e. opportunities) are exploit, share and enhance;
- strategies for both threats and opportunities is acceptance; and
- contingent response strategies, which are strategies, put in place for use only if certain events occur.

It must be stated that the approaches are not mutually exclusive, and in most cases, their combination will provide the most efficient solution. It is also important to note that risk response measures should be addressed as part of the initial risk assessment during the planning stage, as many risk response measures may be impossible or costly to implement once the structure has been commissioned. After proper consideration of alternatives for risk response, the most appropriate ones should be selected and implemented. Since new risks could be introduced by the risk treatment, they should also be identified, assessed, treated and monitored. For action plans to work, they must be taken seriously. This means they become another task in the project and require budget, schedule and labour resources, which is no different from other project tasks (Smith and Merrit, 2002). Communication and consultation are important considerations at each step of the risk management process. Effective communication can help ensure that those responsible for implementing risk management on the process under improvement and those with stakes, understand the basis on which decisions are made and why particular actions are required. Since stakeholders have significant impacts on the decisions made, it is important that their perceptions of risk, as well as their perceptions of benefits, are identified and documented, and that the underlying reasons for them are understood and addressed appropriately.

8.4.7   Risk Monitoring and Control

Risk monitoring and control is the process of identifying, analysing and planning for newly arising risks, keeping track of the identified risks and those on the watch list, reanalysing existing risks, monitoring trigger conditions for contingency plans, monitoring residual risks, and reviewing the execution of risk responses, while evaluating their effectiveness (PMBOK, 2004). It is therefore evident that under risk monitoring and control, managers try to focus on the proposed processes and responses proffered for the identified risks. The outputs of the other five stages must be kept under review as things evolve. Changes in the environment, or simply the discovery of better information, may render the original assessment out of date, thereby triggering the need for reassessment. There should be periodic reassessment of risks (i.e. re-measurement) and risk audits. Risk audits examine and document the effectiveness of risk responses in dealing with identified risks and their root causes, as well as the effectiveness of the risk management process. Thus, Kerzner (2009) commented that risk monitoring and control is not a problem-solving technique but, rather, a proactive technique to obtain objective information on the progress to date of reducing risks to acceptable levels. Earned Value Analysis or other methods of project variance and trend analysis (e.g. programme metrics, schedule performance monitoring and technical performance measurement) could be used to monitor overall performance; and deviation from the baseline plan may indicate the potential impact of new threats. It is not generally necessary to begin the whole process over again when this happens, unless the change or deviation is particularly profound, but those parts that are directly affected by changing circumstances must be brought up to date.

8.5   Integrating Risk Management into Planning for Process Improvement

Having understood what the actions are, in terms of risk management and planning for process improvement, and where planning for improvement should be situated, it is important to highlight the need for integration of risk management into planning for process improvement. When management sets a goal, these may sometimes be missed because of foreseeable and unforeseeable circumstances. Risk is predominantly inherent in most human endeavours, including planning for process improvement. It is also important to acknowledge that improving one process may have unintended knock-on effects on other processes. For example, if a property development company makes an improvement in its sales order processing, once that process is improved, the improvement will affect order fulfilment process – the delivery of the product to customers. The improvement in sales order processing may apparently create a backlog in order fulfilment in the production department. Implementing a proper project management approach – effective project risk management – would have addressed such issues as part of the risk planning, and the order fulfilment processes should have been reviewed as an extension of the sales order process. In other words, the initial process (i.e. sales order processing) would have been assessed to determine if making changes to it would be beneficial to the company as a whole, given investments needed in other parts of the company.

Chrissis *et al.* (2003) stated that there is a need for risk management integration into day-to-day decision making, and that by employing a collective set of activities for identifying, assessing, handling, monitoring and communicating risk information, this could uncover potential issues before they can disrupt normal operations. They explain that two areas were targeted for proactive risk management:

1. change as a major source for introducing new risks into a system or process – risk assessment will be performed whenever change to the baseline is proposed; and
2. communication and monthly reviews of significant risks as proactive measures – the presentation of risk information from one programme element may help identify collateral risks in other areas.

Therefore, since planning for process improvement can be conceptualised as a stage, managers seek to propose a plan for change to a system or process, by understanding that the process may need improvement, cognisant of threats and opportunities along the paths to achieving this change. This is also true when the change is finally introduced. It is important, therefore, that risk management is integrated into planning for process improvement. In addition, it is known that risk can affect productivity, performance, quality and budget of a construction project (Kangari, 1995). It implies therefore that at the planning for process improvement stage, where decisions as to 'whether to improve', 'what processes need to be improved', 'when to improve them', 'how they should be improved' and 'alternative plans on the steps to follow in their improvement' are taken, require adequate risk management. This is because organisations need to be able to make informed decisions to help achieve goals or objectives, improve efficiency and procure competitive advantage.

Considering the activities or the types of questions planning seeks to answer, it is evident that after actions have been completed in a process simplification segment (Table 8.1, Steps 1–7), it is at that point (i.e. starting from Step 8) that risk management needs to be incorporated (Table 8.2, Steps 9 and 10). It is the point where questions such as 'Are there any risks associated with the proposed change?' 'What will the change cost?' and 'What are the costs aside from money?' It includes time, number of people, materials used and other factors. 'Which employees will be affected by the change?' 'How will they be affected (positively or adversely)?' 'Who should be responsible for implementing the change?' 'What has to be done to implement the change?' 'Where will the change be implemented?' 'How will the implementation be controlled?' 'At what steps in the process will measurements be taken?' 'How will data be collected?' 'Is a pilot (small-scale) test necessary prior to full implementation of the change?' 'How long will the test last?' 'What is the probability of success?' 'Is there a downside to the proposed change?' Risk management offers answers to all of these and many other questions. Moreover, as the team picks one of the processes to work on, after considering the possible root causes identified in Step 8, they can then develop a plan to implement a change in the process to reduce or eliminate the root cause. The change action developed for implementation can be a source of another potential risk to the organisation in achieving desired objectives (Pipattanapiwong *et al.*, 2003). It is only through deploying project management techniques such as risk management that this potential risk can be exposed and adequately mitigated against '*ab initio*'. Although risk management process can be introduced at any time into process improvement, at a particular point – planning for process improvement – it has the potential to yield the most significant organisational, operational or external benefits to the organisation. It is proposed therefore that risk management processes should be installed at the planning stage for process improvement (Figure 8.4).

8.6   Conclusion

This chapter described process improvement initiatives, along with the need to have clearly articulated mechanisms in place to help organisational performance. In this respect, business process was seen as a series of steps, both at the process simplification stage and at the plan-do-study-act stage. It was argued that risk predominantly emanates from uncertainties associated with pursuing certain causes of action. Therefore, it is important when planning an improvement exercise, to acknowledge that unforeseen events may (and often do) occur, which can adversely affect outcomes in terms of cost, time and quality. The need for risk management integration when planning for process improvement was therefore stressed in this chapter. In addition, organisations need to understand (and should be able to differentiate) their core/value-creation business processes and supportive processes; and greater effort should be given to improving value-creation processes, rather than wasting resources on improving supportive processes that add little or no value to the entire system. There is also a need for effective communication and adequate consultation at every stage in risk management process. This is urgent, as improvement teams, and those affected by the improvement, should understand what decisions are made, and why improvement interventions are required, as this can help create acceptance (i.e. facilitate buy-in) for changes.

Project risk management was also presented, and risk management integration into the planning stage of process improvement was advocated, particularly as an iterative exercise, rather than a single point intervention. It is through this holistic approach that practitioners can critically appraise how work is being performed, to help understand unforeseen events that may impinge on process improvement outcomes. By way of conclusion therefore, practitioners in the construction industry need to be aware that when developing and implementing a plan for process improvement, it is essential to integrate risk treatment strategies in order to be effective. Achievement of goals and objectives of any proposed plan depends on proper integration of

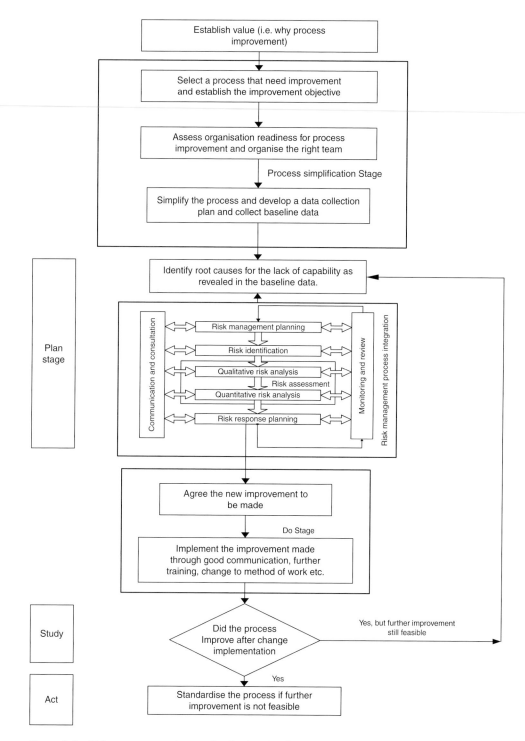

Figure 8.4   Risk management integration in planning for process improvement.

risk management at the planning stage. Only then, can the capital invested by the organisation in process improvement produce the desired results, for example, increased market share, improved customer satisfaction levels, enhanced profitability through cost, time and quality performance, etc.

References

Ahmed, P.K. and Simintras, A.C. (1996) Conceptualizing business process reengineering. *Business Process Management Journal*, **2**(2), 73–92.

Akintoye, A.S. & MacLeod, M.J. (1997) Risk analysis and management in construction. *International Journal of Project Management*, **15**(1), 31–38.

Amaratunga, D., Sarshar, M. & Baldry, D. (2002) Process improvement in facilities management: the SPICE approach. *Business Process Management Journal*, **8**(4), 318–337.

Banwell, H. (1964) *Report of the Committee on the Placing and Management of Contracts for Building and Civil Engineering Works.* HMSO, London.

Bashein, M.L & Marcus, M.L. (1994). Preconditions for BPR success. *Information Systems Management*, **11**(2), 7–13.

Blick, G., Gulledge, T. & Sommer, R. (2000) Defining business process requirements for large-scale public sector ERP implementations: a case study. *Proceedings of the European Conference on Information Systems*, Wirtschafts Universitat, Wien.

British Property Federation (BPF) (1983) Manual of the BPF system for building design and construction. In: ed. Kagioglou, M, Cooper, R. & Aouad (1999), *Reengineering the UK Construction Industry: The Process Protocol.* British Property Federation, London. Available from http://www.processprotocol.com/pdf/cpr99.pdf (accessed 7 March 2011).

Cambridge Advance Learner Dictionary. Available from http://dictionary.cambridge.org/ (accessed 7 March 2011).

Chrissis, M.B., Konrad, M. & Shrum, S. (2003) *CMMI Guidelines for Process Integration and Product Improvement*, Software Engineering Institute (SEI) Series. Addison Wesley Publishing Co, London.

Cooper, D.F. & Chapman, C.B. (1987). *Risk Analysis for Large Projects – Models, Methods and Cases.* John Wiley & Sons, Inc, New York.

Davenport, T.H. (1993), *Process Innovation – Reengineering Work through Information Technology,* Harvard Business School Press, Boston, MA.

Dey, P. (1999), Process reengineering for effective implementation of projects. *International Journal of Project Management*, **17**(3), 147–159.

Dey, P., Tabucanon, M.T. & Ogunlana, S.O. (1994) Planning for project control through risk analysis: The case of a pipeline project in India. *International Journal of Project Management*, **12**(1), February, 23–33.

Dey, P.K. & Ogunlana, S.O. (2004) Selection and application of risk management tools and techniques for build-operate-transfer projects. *Industrial Management & Data Systems*, **104**(4), 334–346.

Earl, M. & Khan, B. (1994) How new is business process redesign? *European Management* Journal, **12**(1), 20–30.

Edwards, C., Braganza, A. & Lambert, R. (2000) Understanding and managing process initiatives: A framework for developing consensus. *Knowledge and Process Management*, **7**(1), 29–36.

Egan, J. (1998) *Rethinking Construction.* Department of the Environment, Transport and the regions, HMSO, London.

Evans, J.R. & Lindsay, W.M. (2008) *The Management and Control of Quality*, 7th edn. South-Western Learning, Canada.

Flanagan, R. & Norman, G. (1993) *Risk Management and Construction*. Blackwell Scientific Publications, London.

Florac, W.A., Park, R.E. & Carleton, A.D. (1997) *Practical Software Measurement: Measuring for Process Management and Improvement*. CMU/SEI-97-HB-003. The Software Engineering Institution, Pittsburgh. PA.

Gilbert, P.A. (1993) The road to business process improvement – can we get there from here? *Production and Inventory Management Journal*. **34**, 80–86.

Grimsey, D. & Lewis, M.K. (2002) Evaluating the risks of public private partnerships for infrastructure projects. *International Journal of Project Management*, **20**, 107–118.

Gulledge, T.R. & Sommer, R.A. (2002) Business process management: public sector implications. *Business Process Management Journal*, **8**(4), 364–376.

Handbook for Basic Process Improvement (1996) Available from http://www.balancedscorecard.org/portals/0/pdf/bpihndbk.pdf (accessed 10 September 2010).

Hammer, M. & Champy, J. (1993) *Reengineering the Corporation: A Manifesto for Business Revolution*. Harper Business, New York.

Irani, Z., Hlupic, V., Baldwin, L.P. & Love, P.E.D. (2000) Reengineering manufacturing processes through simulation modelling. *Journal of Logistics and Information Management*, **13**(1), 7–13.

Jiang, J.J., Chen, E. & Klein, G. (2002) The importance of building a foundation for user involvement in information system projects. *Project Management Journal*, **22**(1), 20–26.

Johansson, H.J., McHugh, B., Pendlebury, A.J. & Wheeler, W.A. (1993) *Business Process Reengineering, Breakpoint Strategies for Market Dominance*. John Wiley & Sons, Ltd, Chichester, UK.

Kagioglou, M., Cooper, R. & Aouad, G. (1999) *Reengineering the UK Construction Industry: The Process Protocol*. Available from at http://www.processprotocol.com/pdf/cpr99.pdf (accessed 7 March 2011).

Kangari, R. (1995) Risk management perceptions and trends of US: Construction. *ASCE Journal of Construction Engineering and Management*, **121**(4), 422–429.

Kerzner, H. (2009) *Project Management A Systems Approach to Planning, scheduling, and Controlling*, 10th edn. John Wiley & Sons, Inc, Hoboken, NJ.

Kezsbom, D.S. & Edward, K.A. (2001) *The New Dynamic Project Management*. John Wiley & Sons, Inc, New York.

Kloppenborg, T.J. (2009) *Project Management A Contemporary Approach*, International student edn. South-Western Cengage Learning. Canada.

Kloppenborg, T.J. & Deborah, T. (2004) Using a project leadership framework to avoid and mitigate information technology project risks. In: ed. Slevin, D.P., Cleland, D.I. & Pinto, J.K., *Innovations: Project Management Research*. Project Management Institute, Newton Square, PA.

Latham, M. (1994) *Constructing the Team*. HMSO, London.

Lee, S. & Ahn, H. (2008) Assessment of process improvement from organisational change. *Information and Management*, **45**, 270–280.

Momoh, J. & Ruhe, G. (2006) Release planning process improvement – An industrial case study. *Software Process Improvement and Practice*, **11**, 295–307.

Paper, D. (1998) Business process reengineering: Creating the conditions for success. *Long Range Planning*, **31**(3), 426–435.

Paulk, M.C., Curtis, B., Chrissis, B. & Weber, C.V. (1993) *Capability Maturity Model for Software*, Version 1.1 Tech Report, CMU/SEI-93-TR-24. Carnegie Mellon University, Pittsburgh, PA.

Pipattanapiwong, J., Ogunlana, S.O. & Watanabe, T. (2003) Multi-party risk management process for a construction project. In: ed. Akintoye, A., Beck, M. & Hardcastle, C., *Public Private Partnership: Managing Risks and Opportunities*, Chapter 16. Blackwell Science, Oxford. 351–367.

PMBOK Guide (2004) *A Guide to the Project Management Body of Knowledge*, 3rd edn. Project Management Institute, Newtown Square, PA.

Pritchard, J. & Armistead, C. (1999) Business process management – Lessons from European business. *Business Process Management Journal*, 5(1), 10–35.

Robinson, G.D. (1991) *Continuous Process Improvement*. Free Press, New York.

Rohleder, T.R. & Silver, E.A. (1997) A tutorial on business process improvement. *Journal of Operation Management*, 15, 139–154.

Saiedan, H. & Kuzara, N. (1995) SEI capability maturity model's impact on contractors. *IEEE Computer*, 28, January, 16–26.

Sarshar, M., Hutchinson, A., Aouad, G., Barrett, P., Minnikin, J. & Shelley, C. (1998) *Standardised Process Improvement for Construction Enterprises (SPICE)*. Proceedings of 2nd European Conference on Product and Process Modelling, Watford.

Sarshar, M., Finnemore, M. & Haigh, R. (1999) SPICE: Is the capability maturity model applicable in the construction industry. *Proceedings of 8th International Conference on durability of Building Materials and Components (CIB W78)*, 30 May–3 June, Vancouver, BC.

Skorupka, D. (2008) Identification and initial risk assessment of construction projects in Poland. *Journal of Management in Engineering*, 24(3), 120–127.

Smith, C.M.C. & Merritt, G.M. (2002) Managing consulting project risk. *Consulting to Management*, 13(3), 7–13.

Sommerville, I. (2007) *Software Engineering*, 8th edn. Addison-Wesley Publishing Co, London.

Wang, S.Q., Dulaimi, M.F. & Aguria, M.Y (2004) Risk management framework for construction projects in developing countries. *Construction Management and Economics*, 22, 237–252.

Wideman, R.M. (1986) Risk Management. *Project Management Journal*, 17(4), 20–62.

Winegard, A. & Warhoe, S.P. (2003) Understanding risk to mitigate changes and avoid disputes. AACE *International Transactions*. The Association for the Advancement of Cost Engineering, Orlando, FL.

Zahran, S. (1998) *Software Process Improvement: Practical Guidelines for Business Success*. Addison Wesley Publishing Co., Reading, MA.

Modern Methods of Construction

Wafaa Nadim

The construction industry is widely acknowledged for its vital economic role and contribution to countries' GDP. In addition to this economic importance, the construction industry also accounts for a high political, environmental and social profile, which is attributed to its key role in providing housing, its impact on the environment, as well as being a major employer (EC, 2006). Over time, construction practices, methods and materials have continuously evolved to fulfil nations' needs (Ngowi *et al.*, 2005). The amount of raw material and natural resources used in construction for creating and operating the built environment is estimated to exceed those consumed in any other sector (EC, 2006).

From a scope perspective, the construction industry is subjected to two definitions, namely a 'narrow' definition and a 'broad' definition. The 'narrow' definition mainly concerns 'on-site' construction activities, whereas the 'broad' definition encapsulates the whole supply chain from quarrying raw material to demolition of the product and the associated professional services (Pearce, 2003). The construction output by value is mainly concentrated in the developed countries (ILO, 2006), with Europe accounting for 30% of global output, followed by the USA (21%) and Japan (20%). However, construction employment is almost the exact reverse of the distribution of output. This is explained by the differences in technology use that substitutes expensive labour in the developed countries (ILO, 2006). Given the scale and impact of construction, from an application perspective, the construction industry is still fraught with problems and challenges, which includes, among others, failure to meet market demands, skills shortages and poor quality of products (Morton, 2002; CIOB, 2006, 2008).

Construction Innovation and Process Improvement, First Edition.
Edited by Akintola Akintoye, Jack S. Goulding and Girma Zawdie.
© 2012 Blackwell Publishing Ltd. Published 2012 by Blackwell Publishing Ltd.

Thus, many initiatives have been launched worldwide to improve the construction industry practices, one of which is through Modern Methods of Construction (MMC).

This chapter explores the need for the construction industry to change, introduces MMC, and exemplifies the different systems therein, with particular emphasis on Offsite Production (OSP). Hence, OSP is explored from a strategic as well as from an implementation perspective, in order to identify the requirements needed to help overcome the industry's inherent problems. The chapter further reports on a study carried out in the UK (2008) to identify and quantify the main OSP knowledge gaps amongst construction professionals. Finally, the chapter concludes by introducing a flexible training and education model based on the quality function deployment (QFD) concept, the remit of which can help quantify and prioritise the construction industry's OSP training, education and business needs.

9.2   The Need for Change

The criticism of the construction industry has been mainly attributed to its underperformance, resulting in poor quality of products, and poor health and safety record in comparison to other industries (Emmerson, 1962; Banwell, 1964; Latham, 1994; Egan, 1998; Koskela, 2000; Morton, 2002; Lambert, 2003; Woudhuysen and Abley, 2004; EMCC, 2005). This is largely due to the peculiar needs of the end product (predominantly a building of some kind), which often necessitate production to take place on site, and therefore tends to be a one-of-a-kind product. Vrijhoef and Koskela (2005) identified and disaggregated the peculiarities of the construction industry at three different levels, namely the product, project/production and industry levels. This peculiarity is further perplexed by the fragmented nature of the industry, which has been identified as a major contributor to the complex nature of construction industry practices, often leading to the so-called 'over the brick-wall/silo approach' (Palotz, 2006). Fragmentation often refers to the increase in the number of firms accompanied with the decrease in their average size. This fragmentation in structure consequently results in low investment in research and development (R&D) and a fragmented delivery process (González, 1999). The ramifications of this are manifested in the disparate and uncoordinated procedures that create an environment for 'risk dumping' and hence, lead to a 'fully-fledged' blame culture (Saffin, 2007), which is a fundamental barrier to improvement (NAO, 2001). This is also evidenced by the inadequate capture of client requirements, lack of integration of information, lack of life-cycle analysis, and poor communication flow amongst project stakeholders (Evbuomwan and Anumba, 1998; Anumba *et al.*, 2008).

Given the challenging nature of the construction industry, many initiatives have been instigated worldwide to effect change, in order for the construction industry to overcome and mitigate the inherent problems, be able to compete globally, and hence maintain its strategic importance (Egan, 1998; NAO, 2001). A summary of these initiatives can be seen in Table 9.1.

Table 9.1 International initiatives to promote change in the construction industry (Nadim, 2009).

Report/Initiative	Year	Context
Australia	1999	Building for growth
Finland	2002	Reengineering the construction process using Information Technology
Ireland	1997	Building our future together
Japan	1998	Future Directions of the construction industry, coping with structural changes of the market
Singapore	1999	Construction 21, Reinventing Construction
South Africa	1997	Creating an enabling environment for reconstruction growth and development in the construction industry
USA	1994	National Construction Goals
UK	1962	Greater integration of the design and construction process
UK	1994	Constructing the Team (Latham report), to review contractual and procurement arrangements
UK	1998	Construction Task Force (Rethinking Construction), Committed leadership, focus on customer, integrated process and teams, quality driven agenda, commitment to people
UK	2004	Construction Client Group CCG, To support private and public clients

The common trend noted amongst the international initiatives to promote the construction industry was the emphasis on meeting client needs, integrating design and construction, and benchmarking performance. These achievements were mainly sought through process improvement (NAO, 2001). However, the future driving forces for change were anticipated to extend beyond merely 'process improvement', to further embrace 'sustainable communities' concepts and emphasising the 'social needs' (RIBA, 2005). In this context, the future driving forces for change are envisioned to encompass:

- climate change and the depletion of natural resources;
- 'building information object modelling and larger industrialisation and standardisation of buildings;
- increasingly demanding and powerful consumers;
- growing global competition to deliver high-quality buildings more effectively and
- demand for intelligent buildings to minimise the increasing energy costs.

However, change can only take place and be successful when governed by a set of ingredients such as change in culture, the introduction of enabling technologies, promotion of new contractual arrangements, and risk management (Adamson and Pollington, 2006; Saffin, 2007).

9.3   Modern Methods of Construction

The idea of improving performance in the construction industry, by learning from other industries such as 'manufacturing', is not a new phenomenon (Womack *et al.*, 1990; Gann, 1996; Egan, 1998; Höök and Stehn, 2008). This was particularly evidenced in the 1960s, when the 'future' dream for improving the construction industry was:

> …a group of white coated, well paid workers, slotting and clipping standard components into place in rhythmic sequence on an orderly, networked and mechanised site to a faultless programme without mud, mess, sweat, or swearing…' (Carter, 1967).

This 'dream' has apparently driven the call to improve the construction industry through the take up of the production concept, which is governing the manufacturing industry (Leabue and Viñals, 2003). This call for change is multifaceted, incorporating: the people involved, the technology used and the processes employed. In other words, it was a call for 'innovation', the mode of which largely depended on the 'degree' of change required from current practices (Van de Ven, 1986; Laborde and Sanvido, 1994; Seaden *et al.*, 2003). Slaughter (1998) distinguished five modes for innovation, which encompassed:

1. *Incremental innovation*: requiring small change to occur;
2. *Modular innovation*: requiring change in concept;
3. *Architectural innovation*: i.e. change in concept which requires change in the link of units;
4. *System innovation*: requiring multiple innovations to occur that are linked together and require major change in components; and finally
5. *Radical innovation*: that requires an entirely new approach, resulting in major change.

From an innovation point of view, MMC is one of the means sought to improve and change the construction industry practices through emphasising the manufacturing concept and the introduction of different levels of mechanisations, hence catering for the different modes of innovation (Slaughter, 1998). In this context, MMC is closely intertwined with the concept of industrialising the construction industry (Tatum *et al.*, 1986; Warszawski, 1999; O'Brien *et al.*, 2000; Gibb, 2001; Morton, 2002). Intrinsically, industrialisation can be seen as a business strategy that transforms the traditional construction process into a manufacturing and assembly process by engaging people, embracing (new) technologies and translating clients' needs into building requirements (Nadim, 2009). Hence, industrialisation aims to improve business efficiency, quality of product, customer satisfaction, environmental performance, sustainability, and predictability of timescales. This makes industrialisation broad-based, rather than merely confined to a particular product and therefore requires engaging people and processes in order to improve the delivery and performance of construction (CABE, 2004; MBS, 2004; Taylor *et al.*, 2004; NAO, 2005; Plus Group and Riverside, 2005; Barker 33 Cross-Industry Group, 2006; Gibb and Pendlebury, 2006).

9.3.1   MMC Classification

MMC is subject to numerous classifications, which range from those defining the type of product produced to those specifying the location of production. The type of products include panelised, volumetric units, hybrid (which combines panelised and volumetric approaches) subassemblies, and components such as floor or roof cassettes, pre-cast concrete foundation assemblies, pre-formed wiring looms and engineering composites (NAO, 2005). Other classifications distinguished location of production such as site-based systems, for example, tunnel from, slip form, jump form, etc (Homein, 2006; NHBC, 2006) (Figure 9.1). Another classification suggested by Gibb and Pendlebury (2006) represents four distinctive levels, which may overlap, namely components sub-assembly, non-volumetric sub-assembly, volumetric preassembly and complete buildings.

MMC may also be regarded as an overarching umbrella, which embraces a variety of building approaches such as Offsite Manufacturing (OSM) and onsite work. Hence, MMC and OSM should not be considered as synonyms (Goodier and Gibb, 2005; Homein, 2006). OSM is a process that incorporates prefabrication and pre-assembly, and involves the design and manufacture of units often remote from the work site and their installation. These units are then transported and assembled at the work site, thus referred to as 'offsite fabrication/production'. From a terminology perspective, Offsite Manufacturing, Offsite Construction (OSC), Offsite Production (OSP), Offsite Fabrication (OSF), factory built assembly, industrialised construction, prefabrication, modular construction, and system building have often been used interchangeably (Gibb and Pendlebury, 2006; NHBC, 2006; BRE, 2007).

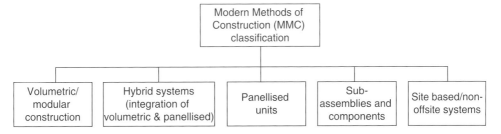

Figure 9.1   MMC classifications.

9.3.2   Offsite Production Evolution and Application

Gibb (1999) and Leabue and Viñals (2003) attributed the development of OSP as a response to sporadic demands for buildings and facilities. These encompassed colonisation in the nineteenth century, commercial developments in the late 1980s, natural disasters, industrial revolution and advances in technology. Further drivers, which called for the wider use of OSP, included the decrease of available skilled labour and the consequent increase in labour costs, change in client demands and expectations, development of digitally controlled manufacturing, and the increased concern about the health and safety of workers. In addition, clients have become more pragmatic with respect to their expectations, requiring better quality products that are delivered faster at a reasonable cost (Gibb, 1999).

OSP has been employed in several construction projects worldwide, predominately in the commercial sector, where the high cost of projects often justifies the associated initial capital investments (Leabue and Viñals, 2003). However, residential projects to a large extent lagged behind with respect to OSP implementation. The most iconic residential buildings featuring OSP are Kisho Kurokawa's 1972 Capsule Tower in Japan and Moshe Safdie's Habitat 67 in Montreal Canada (Sebestyen and Pollington, 2003). The mode of production and delivery of OSP buildings, as well as the mode of intervention and support of the State/Government, is largely governed by historical, geographical and social elements (O'Brien *et al.*, 2000; Leabue and Viñals, 2003, ManuBuild, 2009). While a number of failures were recorded internationally, successful OSP examples were often underpinned by R&D and through collaboration with the industry, as was the case in Sweden and Japan (O'Brien *et al.*, 2000). Therefore, OSP arguably provides a safe, efficient and productive work environment when compared to traditional construction practices (Figure 9.2), which are fraught with problems and challenges (Gibb, 1999). From a sustainability point of view, OSP allow reduced levels of defects, thus resulting in a better quality, less waste, improved health and safety, improved environmental performance of the final product, social benefits (from improved working conditions), less deliveries to site, in addition to greater efficiency in the use of resources – both materials and labour (Gorgolewski, 2003). While

Figure 9.2   Hard and labour intensive work environment (Nadim, 2001).

it is argued that OSP may help reduce the amount of labour needed on the job; Gibb (1999) negated this notion, suggesting that OSP should be seen as part of an effective use of labour rather than a reduction in labour.

Initiatives and calls have been launched worldwide to promote the wider use of OSP in order to reap the benefits (Egan, 1998; Barker, 2003; Leabue and Viñals, 2003; CIDB, 2009). These benefits include meeting market demands, overcoming skill shortages and improving the quality of final product. In response to these initiatives, the UK and Australia, for example, have incorporated OSP in their construction industry vision for 2020 (Hampson and Brandon, 2004; ConstructionSkills, 2008). However, when employing OSP working practices, there is a need to distinguish between an 'offsite produced product' and 'processes of OSP'. A 'house' may be entirely prefabricated in a factory using the traditional procedures and then trans-ported to the site. In this case, the factory would only allow a better work environment by simply moving work from a site to a factory. This does not acknowledge the benefits derived from the production technology, such as using machines, robots and new methods for management (Leabue and Viñals, 2003). Notwithstanding this, the degree of industrialisation used in OSP can differ depending on the enterprise, the region, and type of com-ponent/product produced, where Leabue and Viñals (2003) identified five levels of industrialisation, these being:

1. *Use of simple machinery*: components of the building is produced in a factory and assembled using traditional methods;
2. *Mechanisation*: using specialised tools in order to reduce the depend-ence on labour;
3. *Automation*: replacing labours with machines;
4. *Robotisation*: using sophisticated programmable machines that accom-plish complex and diversified work to allow replacing more labour;
5. *Reproduction*: replacing the rest of the traditional work with new procedures, in order to reduce the number of operations required for producing complex objects so as to bring gains that far exceed mecha-nisation and robotisation.

In order to capitalise on the benefit from employing OSP systems, a 'project-wide' strategy should be developed at an early stage in the project to allow measuring of the effects. This strategy requires that key decisions are made during the concept design and detail design phase prior to the start of the production phase. The earlier the OSP decision is made, the more likely the benefits are to be maximised. The key decisions required encom-pass agreeing on the strategy for employing OSP, OSP applications, logistics for unit installation, and OSP details (Gibb, 1999). However, to allow these decisions to take place, special procurement methods are suggested such as 'strategic partnering', 'two-stage tendering', 'nominated suppliers', 'manage-ment forms of contract', 'design and build', and 'design and manage'. In addition, a number of issues need to be agreed upon and determined to help realise and maximise the benefits of OSP application (Homein, 2006; build-offsite and BAA, 2008), which include:

- a date for design freeze/phased freeze;
- a timetable for delivery of the units;
- notice required by the manufacturer to check that tolerances on site are within the agreed limits;
- penalties for late delivery;
- penalties for delays if the site is behind schedule and unable to accept delivery at the agreed time;
- tolerances and standards that the units will be manufactured to;
- formal procedures for checking the units prior to accepting handover;
- latent defects liability period;
- manufacturers' responsibilities on site during the erection/installation of their units, and their requirements/conditions during that period.

In contrast to the traditional construction process, the amount of work required prior to going to site in OSP projects often increases, as the super-structure and fitting-out can take place in the factory, thereby decreasing the amount of work carried out on site (NHBC, 2006). The amount of work carried out in the factory depends largely on the type of the system employed, which in the case of volumetric systems is larger than the work required for panellised systems for example. Consequently, the amount of work required on site is reduced compared to all other systems. At a strategic level, OSP as a technology differs from other technologies, as it is not realistically feasible to be solely acquired by one single organisation or across a single supply chain. Therefore, in order for OSP to justify the capital investment and hence, be cost effective, it needs to be taken up widely by the construction industry to provide market continuity and stability. This requires the industry to move beyond the notion of a 'learning organisation' (Senge, 1992; Andrew and Ciborra, 1996) to the notion of a 'learning industry' (Eurich, 1990; Rezgui et al., 2004).

9.3.3   Offsite Production Strategies

Industrialisation is a notion that transcends the use of offsite-produced elements and the assembly of buildings to a rather 'visionary' system. This, in addition to the technical aspects, tends to include economic, management and market exploitation aspects. In this respect, the industrialisation process is envisioned to start with market analysis, management and design, followed by marketing of the finished product, which is then distributed, installed and finally maintained (Leabue and Viñals, 2003). This often requires exten-sive communication channels throughout the different stages of the process and so necessitates that different stakeholders are 'on the same side of the fence' and share a common 'language' to accommodate the new work conditions (Carter, 1967).

There are five main distinguished manufacturing strategy models that may be applied in the industrialised construction industry, namely buy-to-order, make-to-order, assemble-to-order, make-to-stock and ship-to-stock. These strategies are governed by a 'decoupling' point that defines the extent to which the process allow capturing customer's requirements, and whether these are real or speculative requirements (Barlow *et al.*, 2003); thus, are dependent on the extent of standardisation/customisation allowed. In the context of standardisation, this typically involves the 'extensive' use of components, methods or processes, in which there is regularity and repetition (Gibb, 1999). This is predominately governed through the design on a modular basis to coordinate the size of factory-made components with the design of buildings, that is, through 'dimensional coordination' (Gann, 1996) and jointing conventions (Carter 1967; Jones, 1967; Tindale, 1967). In order to avoid 'identical standardisation' of products, which does not consider customer preferences (Gibb, 1999), 'high-powered' computer-aided design and digitally controlled manufacturing machinery is often sought to facilitate the provision of mass customisation. This requires flexible production lines to allow produce a wide range of alternative assemblies and final products. Thus, the call is for more emphasis to be placed on the standardisation of interfaces between components – for example, smart components (ManuBuild, 2009) rather than through the standardisation of products. These interfaces are not necessarily physical, but may to be extended to encapsulate managerial/contractual and/or organisational interfaces (Gibb, 1999). However, the level of standardisation/customisation is often dependent on the number of stages that allow customisation along the different stages of the supply chain, such as design, fabrication, assembly and distribution (Barlow *et al.*, 2003), which can affect time and cost of delivery. In this context, pure standardisation does not conventionally allow customisation at any stage of the supply chain. In contrast, pure customisation does allow configuration to meet customer requirements across all stages of the supply chain, but consequently has a relatively high impact on cost and lead-times. Nevertheless, the Japanese construction industry managed to balance production cost reduction and mass customisation through the rationalisation of the manufacturing processes (Toffler, 1970; Noguchi, 2005).

9.4   Open Building Manufacturing – ManuBuild Project

ManuBuild was a European Framework Six Integrated Research Project (2005–2009), partly funded by the European Commission (EC). In an attempt to measure the perception of the construction industry stakeholders in Europe, semi-structured interviews were conducted to capture views with regard to the shift towards OSP and the introduction of 'openness' to the system. The interviews addressed the business models, the information platforms, the level of automation and production systems, the involvement of end users in the construction process, maintenance and refurbishment, the market and the risk involved (Hervàs and Ruiz, 2007). Transcripts from 54 open-ended interviews (carried out in 4 European countries) were analysed to help inform and effect change, and thus help improve the construction industry. The analysis suggested the need to balance the people involved, the technology used and the process underpinning the business. Whilst there is

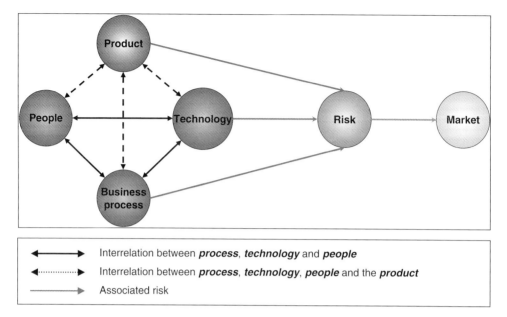

Figure 9.3 boxes:

Figure 9.3   OSP Conceptual Model (adapted from Nadim and Goulding, 2011).

an already acknowledged model to balance people, process and technology (Wysocki, 2004), two extra dimensions were suggested to this relationship, namely the 'product' dimension and the 'market' dimension (Nadim and Goulding, 2011). This relationship was consolidated in a conceptual model to help identify and create a focal point concerning the OSP knowledge needed and the subsequent implementation thereof (Figure 9.3).

Figure 9.3 presents a conceptual model based on four core dimensions (people, product, technology and process) and relationship with risk and the market; and vice versa, i.e. how process, technology and people involved affect and govern the selection of the OSP product in question (Nadim and Goulding, 2011):

9.4.1   The 'Process' Dimension

The process dimension represents the backbone of the OSP business. It therefore requires an appropriate infrastructure and means to allow the OSP business to grow and prosper. This mandate requires processes to be less complicated, with the removal of non value-adding activities – and should clearly demonstrate value for money. In addition, the roles of the different stakeholders need to be well defined and integrated throughout the different phases, from conception to maintenance and demolition. This necessitates improving the organisational structure and awareness of the awarding authorities regarding the new processes and contractual models.

9.4.2   The 'People' Dimension

OSP, like any (new) technology, often involves change, which can attract resistance associated with protectionism and conservatism inherent within the construction industry (Gibb, 1999; Mtech Group, 2008). This makes the people dimension one of the most important drivers, and at the same time, an inhibitor to the wider uptake of OSP. This may be related to a number of factors, not least culture, change required and skill shortages. Hence, multi-skilling could be seen as a possible solution (Cather *et al.*, 2001).

9.4.3   The 'Technology' Dimension

The technology dimension entails the product, the process (Gibb, 1999; Goodier and Gibb, 2005; NAO, 2005), as well as ICT needed to underpin the business vis-à-vis integration of information. This requires information and communication processes be simplified, particularly the process to move from 'made-for-stock' to cater for 'made-to-order'; and the product technology automation to extend beyond the successful implementation evidenced in kitchen and bathroom pods (Pan *et al.*, 2005) to include building material.

9.4.4   The 'Product' Dimension

Whilst a wide range of products/systems are available (Gibb, 1999; Gibb and Pendlebury, 2006; Homein, 2006), the selection tends to depend on several parameters such as cost, quality, design, sustainability and flexibility (NAO, 2005). However, the cost parameter seems to represent the determinant factor for accepting and employing OSP (Gibb, 1999; Warszawski, 1999; Barlow *et al.*, 2003; Leabue and Viñals, 2003; Goodier and Gibb, 2005). However, for successful OSP implementation, the cost element needs to be counterbalanced by 'adaptability', 'customisability', 'flexibility', quality of interfaces, and most importantly how these coalesce with needs, cognisant of the 'multi-generation house' concept (ASID, 2001).

9.4.5   The 'Market' and 'Risk' Dimension

There is an element of risk associated with the process, technology, people and product relationship, the level of which largely determines the market characteristics and market availability for the OSP system in question. The wider uptake of OSP is therefore dependent on the market, and the amount of investment behind it (Gorgolewski, 2003; Venables *et al.*, 2004; Goodier and Gibb, 2005; NAO, 2005; Bill, 2008). In contrast to the traditional construction practice, where is predominantly low capital intensive and tends to use clients' money, the OSP approach often requires intensive capital investment in order to set up and help provide continuity. This capital investment can only be made if there is a consistent and predictable market driven by demand (Barlow *et al.*, 2003; Bill, 2008). Furthermore, this should also take into consideration the trend to build in an existing building environment (i.e. reconstruction, refurbishment, extension), the trend of individualisation, and also the decreasing population evident in most EU countries. This confines OSP risk, as unstable/unpredictable market demand, along with the perceived uniformity of this approach, which is perceived to restrict architectural creativity, especially where alternative (traditional) products of similar qualities are available.

9.5   Offsite Production in the UK Construction Industry

Meeting market demands is one of the major drivers, which instigated the call for the wider use of MMC in the UK. The amount of dwellings needed over a period of 16–25 years starting in 2002/03 was estimated to range from 160,000–225,000 units/year (Barlow *et al.*, 2002; Gorgolewski, 2003). This led to the Office of Deputy Prime Minister setting targets of 25% of all new social housing schemes in the Registered Social Landlords sector to use OSP/MMC in anticipation of reducing the shortfall in housing production by increasing the speed at which industry could build to meet this demand. However, whilst MMC in general and OSP in particular has been the focus of much attention in the UK (Barker, 2003; Venables and Courtney, 2004; NAO, 2005; BRE, 2007), the actual value of the MMC market in the UK in 2004 was £2.2 billion, representing only 2.1% of the total value of the UK construction sector, 3.6% of new build (Goodier and Gibb, 2005). In addition, the UK OSP market accommodates a wide range of OSP products, including foundation, superstructure, envelope, mechanical and electrical services, etc (Ogden, 2007; Buildoffsite, 2008); where the greatest area for potential OSP exploitation is estimated to include kitchen and bathroom pods, external walls, timber frame, and roofs (Pan *et al.*, 2005). Other drivers cited for using OSP in the industry were overcoming skills shortages, ensuring time and cost certainty, achieving high-quality products and minimising on-site activities (Goodier and Gibb, 2005). Whilst the OSP concept is not new (Gibb, 1999), the advancement in technologies together with the increasing complexity of current construction projects introduces a challenging proposition to the wider uptake of OSP.

9.5.1   The UK Construction Industry Skill Problem

Approximately 70–80% of the UK construction workforce have been estimated to have no formal qualifications, with 35% being classified as labourers; compared to 5% in Denmark, 7% in the Netherlands and 17% in Germany (Venables *et al.*, 2004; Clarke and Winch, 2006). This phenomenon suggests that the UK construction industry is largely dependent on the 'lower level skills'. This is further compounded by the claim that 80% of firms within the built environment experience skills problems with their existing workforce (CIC, 2004), along with a failure to attract younger generations to the construction industry (McNair and Flynn, 2006; Stephen and Flynn, 2006). Therefore, without appropriate skills, there is a real risk of decline in competitiveness and reduction in economic growth (Leitch, 2005, 2006).

The introduction of new technologies often requires new ways of working and thinking along with new skills and/or the enhancement of existing skills (Eurich, 1990; Construction Skills, 2008). The term 'skill' was initially used to refer to the capabilities of an individual to undertake a particular task (Taylor, 2005). This interpretation has evolved over time to be used interchangeably with the term 'competence', thus implying attributes that encompass adaptability and flexibility to change within the workplace, in addition to the possession of 'transferable behavioural characteristics (Prahalad and Hamel, 1990). This is particularly important, as the development of competence within organisations often provides access to a variety of markets, and can significantly enhance the customer experience. Therefore, from a skills perspective, new skills are important, especially where new material, equipment, systems, processes and practices are constantly maturing and evolving (Eurich, 1990). However, whilst acknowledging that improving the construction industry cannot solely be dependent upon the provision of new skills *per se*, improvement is unlikely to occur or be effective without the appropriation of skills matching. In this respect, future technological developments in the construction industry suggest that skills are needed to move the workforce away from blue-collar labourers towards white-collar employees (Hauck and Rockwell, 1997; Leitch, 2005; Gurjao, 2006). In this context, future skill requirements are anticipated to be largely 'generic', encompassing soft skills (e.g. problem solving, communication, and the ability to work in teams) as opposed to job specific skills (Nateriello, 1989; Hauck and Rockwell, 1997; Hager *et al.*, 2001; Stasz, 2001; Leitch, 2006).

Skill shortages often occur in businesses operating with less skilled workers, leading to increase in operating costs, and consequently losing business to competitors (DFEE, 2000; Construction Skills, 2006). While serious shortfalls in manual skills are evidenced in the UK construction industry, professionals and managerial skill shortages are argued to be just as acute (Dainty *et al.*, 2005). These shortages are linked to a number of issues, not least failure to modernise recruitment and training, the changing nature of construction markets, the introduction of new technologies, the growth of self-employment, the fragmentation of the industry and associated decline in construction train-ing (Gann and Senker, 1998; Mackenzie *et al.*, 2000). It is further anticipated that an increase in construction demand would be affected by skill gaps in middle and senior management (CIOB, 2008; Bill, 2008). In this context, skills gaps are perceived as being 'holes' in the knowledge and competence of existing staff; leading to lower profitability and productivity, reduced quality output and an 'under par' health and safety record (Construction Skills, 2006), where skilled craft workers are likely to lack technical and practical skills, while managers, administrators and professionals are most likely to lack management and team working skills (DFEE, 2000). Nevertheless, skill gaps seem to be more evident within the 'lower level' occupations rather than within management or professional positions (LSC, 2005).

9.5.2   Multidisciplinary Training and Education

Technological advancements and increased competition often require transforming the skill profile of construction personnel from highly trained individual 'specialists' to broadly trained 'generalists', who are able to work in groups across multi-disciplinary projects making 'cradle-to-grave' decisions, including finance, design, construction, operation and maintenance (Eraut, 1994; Farr and Sullivan, 1996). In addition to the soft skills required, future skills will also need to be aligned with the increased use of ICT in construction processes, along with new products, new materials, etc (Gann and Senker, 1998). In light of the type of skills required, managers and professionals are expected to account for an increased share of employment, whereas skilled crafts are expected to form a declining share (DFEE, 2000). Furthermore, there is likely to be a growing need for customer-focused staff and possible increase in 'less-skilled manual workers when prefabrication techniques become more widespread' (Clarke and Wall, 1998). This notion was also emphasised by Williamson and Bilbo (1999), with respect to administrating construction technologies and the design/construct/manage interface, rather than performing them. Accordingly, change in the construction industry is suggested to take place on two levels, namely on the process and the technology level where the bulk of work is transferred offsite, hence initiating the need for 'offsite' as well as 'onsite' skills (Gurja, 2006). Therefore, due to the continuously changing environment, and the increased amount and complexity of information exchanged amongst construction stakeholders (Jaraiedi and Ritz, 1994), training and education should be seen as a continuous rather than a 'one-off' process, in line with the pace of change (Eurich, 1990).

From a training perspective, inter-disciplinary/multi-disciplinarily training and education allows greater collaboration amongst the different professions, thereby enabling shared understanding, minimising the chance of friction amongst professions, and facilitating a seamless workflow (Gann and Salter, 1999; Wood, 1999; Campbell, 2001; Chan et al., 2002; Slotte and Tynjälä, 2003; Anumba et al., 2008). However, this requires the industry to establish the capabilities needed by (potential) employees (Tener, 1996) and requires commitment through appropriate channels supported by two-way communication between industry and academia (Trauth et al., 1993; Slotte and Tynjälä, 2003).

Numerous benefits can be established through industry–academia collaboration (Lamancusa et al., 1995; Tener, 1996). Slotte and Tynjälä (2003) identified benefits on three different levels, namely at industry, academia and student levels. Industrial benefits can be manifested in the access granted to the broad theoretical knowledge, inquiry and reflection, as well as in the skilled and knowledgeable workforce attained. Academia, on the other hand, would benefit from coming in contact with work practices, thereby improving business awareness, and access to work-based case studies. Finally, students would gain access to lifelong learning opportunities, which would help with their career development through the exposure to real-world factors and interaction with industrial personnel (Lambert, 2003). In other words, industry–academia collaboration could help encapsulate theory, practice and problem solving, which is often referred to as 'theorising practice' and 'particularising theory' (Eraut, 1994; Slotte and Tynjälä, 2003).

9.5.3   An OSP-QFD Collaborative Training and Education Model

The UK construction industry and academia collaboration has long been sought to improve the industry practices through skill development (Tener, 1996; NWDA, 2004; ACBEE, 2007). However, in order to measure the extent to which the UK's industry-academia collaboration considered OSP professional skill requirements, a study was carried out in the UK in 2008 to explore the construction industry and academia OSP perception and future requirements (Nadim and Goulding, 2009). This study suggested that academia is often unaware of the specific and disparate training needs of the industry in general (Alter and Koontz, 1996; Wang, 2003; Kunstler, 2005), and recognised that there was a mismatch between industry and academia with respect to industry skill requirements (and the delivery thereof). This study further noted that uncertainty prevailed within both the construction industry and academia with respect to OSP requirements, and the future of OSP in the UK.

Drawing on the argument that professional and managerial skill shortages represent a major challenge for the wider uptake of OSP, Nadim and Goulding (2010) investigated the UK construction industry professionals' perception with regard to OSP implementation. Whilst there is general agreement that the UK construction industry is ready to embrace OSP practices, findings suggested a number of problems, concerns and uncertainties. These were not merely discipline-specific, rather multidisciplinary, which need to be addressed to fully appreciate and understand OSP practices. The study concluded that major factors contributing to OSP perceived problems in the UK were mainly associated with culture and resistance to change, as well as inadequate current construction processes. The initial increased cost, manifested by the capital investment and alteration of cash flow, and the perceived design inflexibility further compounded the problem. These assertions associated with OSP implementation, necessitates the provision of adequate OSP training and education programmes to enlighten industry stakeholders of the specifics of OSP practices.

In the context of training and education, Sahney et al. (2003) argued that 'education' was a service industry and would thus need to learn from other industries with respect to measuring the quality of its services to the satisfaction of their customers. In this context, 'education' would need to be broken down into 'system components', reflecting the 'manufacturing process' (Jaraiedi and Ritz, 1994; Hwarng and Teo, 2001; Ahmed, 2006; Gonzalez et al., 2008). Therefore, current educational programmes need to be built upon a product platform concept to satisfy customer needs and allow the desired outcomes to be measured. It was suggested that the supply chain management concept be applied to the relationships between industry, university and associations to rapidly provide and continuously update knowledge and competence needed to run businesses in a timely and cost-efficient manner (Shunk, 2002). This requires knowledge to be seen as a transferable commodity, changing the knowledge supply process from 'push' to 'pull', and including academia as an integral partner. In this context, industry needs to define their requirements to help set the exit capabilities for training and education, in order to enable academia to satisfy those needs (Shunk, 2002).

Quality Function Deployment (QFD) is one of the various tools of 'Total Quality Management' (TQM) that supports product/system design through considering and incorporating customer requirements. It was first developed in Japan during the 1960s by Yoji Akao in the ship-building industry vis-à-vis incorporating customer demands/needs into the product development process (Akao, 1990; Cohen, 1995; Brackin and Rogers, 1999; Akao and Mazur, 2003). This is achieved by capturing the voice of customer (VOC) and working 'backwards' towards defining the design specifications (Abdul-Rahman et al., 1999). The main goal of QFD is to translate subjective quality criteria into objective ones that are quantifiable and measurable, which are then used to design and manufacture a product that satisfies the customer (Hauser and Clausing, 1988; CIRI, 2007). QFD therefore supports interdisciplinary work for effective communication and coordination beyond the borders of organisational units and disciplines, and hence integrate the views of the disparate stakeholders (Cohen, 1995; Chan and Wu, 2002; Pietsch, 2002). QFD also allows prioritisation and benchmarking competitiveness to take place. Drawing on Jaraiedi and Ritz (1994), and in the context of OSP training and education, the QFD tool was seen as an approach to help answer the question: 'How to deliver quality training and education programmes and services based on the needs of customers/ industry?' (Figure 9.4).

Figure 9.4 represents an OSP-QFD development concept model for scrutinising and synchronising industry and training/education institutions' polarised views and expectations. It was designed specifically to provoke OSP multidisciplinary dialogue, where Nadim (2009) identified, quantified and prioritised the major OSP knowledge gaps within the UK construction industry. The VOC (the 'What'], was extrapolated from the knowledge gaps captured from the construction industry stakeholders perceptions and sorted according to their importance from a training and education perspective. The areas suggesting 'larger' knowledge gaps (with higher means) would consequently take priority from a training and education perspective. The service elements (SE) (the 'How'], were then identified as the measurable means by which the VOC could be addressed from a training and education perception. In this context, the 'How' represents the voice of the training/ education organisations (Han et al., 2001), as the specific courses that

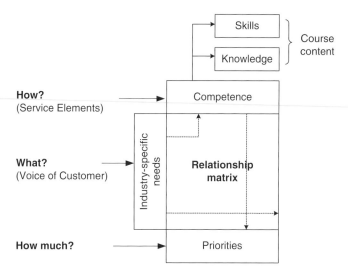

Figure 9.4   OSP-QFD development concept (adapted from Nadim, 2009).

contain topics/knowledge areas that would satisfy customers' expectations/
requirements (Gonzalez *et al.*, 2008). The SEs were derived from qualitative
data analysis and literature, and were categorised under the four main OSP
dimensions of: process, product, people and technology.

From an OSP-QFD implementation perspective, the VOC is located in the
first column on the left-hand side of the matrix, followed by the importance
of each VOC in the adjacent column to the right (Figure 9.5). The SE (the
competence needed) is placed at the top of the matrix. The strength of
relationship between each VOC and SE is determined in the relationship
matrix using three modes of strengths, strong relationship (r = 9), medium
relationship (r = 3) and weak relationship (r = 1), whereas an empty 'blank'
cell suggests no relationship exists between VOC and SE. The SE priorities
('how much'), identify the SEs that fulfil most of the VOC, and should
therefore be considered for further deployment and development. These
priorities are determined based on the weights for each SE using the following formula: *weight* $(w_j = \Sigma_{i=1} d(i) * (ij)$, where d_i is the degree of importance
of the VOC and r_{ij} represents the strength of relationship between VOC *(i)*
and SE *(j)*. The number of SE priorities considered for further deployment,
ultimately depend on the training/education institutions capacity and availability of resources to develop and deliver those training/education priority
programmes.

The first three resultant priorities of the OSP-QFD training and education
model are customer/end user oriented, mainly concerned with the final
product with respect to affordability, life-cycle costing (LLC) and customer
satisfaction. These are then followed by the management side of OSP practices with regard to the supply chain involved and design to manufacture
concept, taking into consideration logistics and assembly, building technology

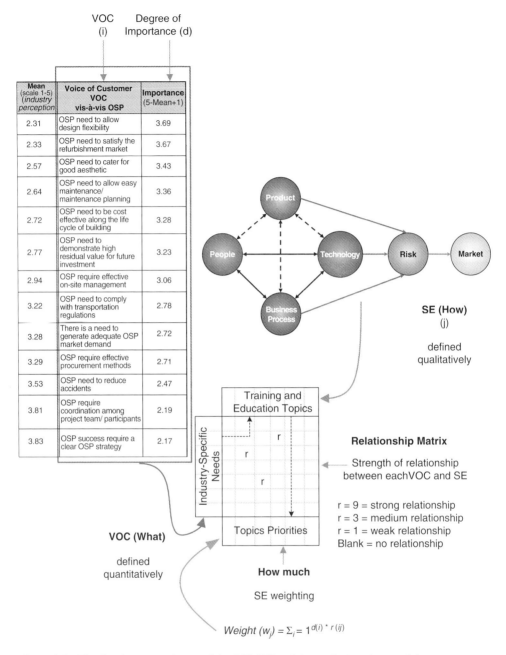

Mean (scale 1-5) (industry perception)	Voice of Customer VOC vis-à-vis OSP	Importance (5-Mean+1)
2.31	OSP need to allow design flexibility	3.69
2.33	OSP need to satisfy the refurbishment market	3.67
2.57	OSP need to cater for good aesthetic	3.43
2.64	OSP need to allow easy maintenance/ maintenance planning	3.36
2.72	OSP need to be cost effective along the life cycle of building	3.28
2.77	OSP need to demonstrate high residual value for future investment	3.23
2.94	OSP require effective on-site management	3.06
3.22	OSP need to comply with transportation regulations	2.78
3.28	There is a need to generate adequate OSP market demand	2.72
3.29	OSP require effective procurement methods	2.71
3.53	OSP need to reduce accidents	2.47
3.81	OSP require coordination among project team/ participants	2.19
3.83	OSP success require a clear OSP strategy	2.17

VOC (i)   Degree of Importance (d)

Product

People   Technology   Risk   Market

Business Process

SE (How) (j)

defined qualitatively

Training and Education Topics

Industry-Specific Needs

r

r

r

Relationship Matrix

Strength of relationship between eachVOC and SE

r = 9 = strong relationship
r = 3 = medium relationship
r = 1 = weak relationship
Blank = no relationship

Topics Priorities

VOC (What)

defined quantitatively

How much

SE weighting

Weight $(w_j) = \Sigma_i = 1^{d(i) * r(ij)}$

Figure 9.5 The development scheme of the OSP-QFD training and education model.

used, environmental performance and energy efficiency. Other priorities included issues such as regulation, customisation, standardisation and risk management.

The OSP-QFD model is particularly flexible, as it allows the development of OSP training and education programmes based on industry needs and priorities in response to the changing environment. This can be used to define and develop OSP training and education programmes in accordance with organisational expertise, capacity and resources. This can also be used to help shape programmes with training/education providers, in order to enhance overall competitiveness through bespoke alignment with industry drivers.

9.6   Conclusion

The major reasons for construction industry problems and challenges relate predominantly to the nature of the one-of-kind product, with unique project characteristics and fragmented industry. These challenges have prompted the need for change, which was initially driven by process improvement. However, future drivers for change will mostly be driven by political agendas, such as climate change and the depletion of natural resources. Despite global technological advances, the construction industry still lags behind regarding the uptake of these advances to improve industry practice. This has largely contributed to the industry's failure to attract young professionals, resulting in skills shortages – a barrier for meeting future demands.

Offsite Production, under the overarching umbrella of Modern Methods of Construction, has been sought as one means of promoting the construction industry to overcome skill shortages and meet market demands. However, the invocation of this will require extensive communication channels across the supply chain, along with a common 'language' and a shared understanding of OSP practices amongst the different stakeholders. In order to achieve such multidisciplinary understanding, an OSP consolidated conceptual model was proposed as a viable solution to help inform, scrutinise and synchronise industry and training/education institutions' polarised views and expectations. It is envisaged that this could help create and secure a shared and collective congruence of understanding amongst construction industry stakeholders with regard to OSP training and education priorities.

References

Abdul-Rahman, K.C.L. & Woods, P.C. (1999) Quality function deployment in construction design: Application in low-cost housing design. *International Journal of Quality & Reliability Management*, 16(6), 591–605.

ACBEE (2007) *Accelerating Change in the Built Environment Education*. Available from http://www.cebe.ltsn.ac.uk/ (accessed 7 April 2011).

Adamson, D.M. & Pollington, T. (2006) *Change in the Construction Industry*. Routledge. Oxford.

Ahmed, S. (2006) QFD application to improve management education at KIMEP. *Issues in Information Systems*, 7(1), 193–198.

Akao, Y. (1990) *Quality Function Deployment: Integrating Customer Requirements Into Product Design*. Productivity Press, Cambridge.

Akao, Y. & Mazur, G.H. (2003) The leading edge in QFD: past, present and future. *International Journal of Quality & Reliability Management*, 20(1), 20–35.

Alter, K. & R. Koontz, J. (1996) Curriculum development and continuing education in project management for the speciality subcontracting industry. *Journal of Construction Education*, 1(2), 66–78.

Andrew, R. & Ciborra, C. (1996) Organisational learning and core capabilities development: the role of IT. *Journal of Strategic Information Systems*, 5, 111–127.

Anumba, C.J., Issa, R.R.A., Pan, J. & Mutis, I. (2008) Ontology-based information and knowledge management in construction. *Construction Innovation*, 8(3), 218–239.

ASID (American Society of Interior Designers) (2001) *Aging in Place: Aging and the Impact of Interior Design*. Available from http://www.asid.org/NR/rdonlyres/837A82C3-48C1-4C92-BFC9-2E37E9C8E1C7/0/aging_in_place1.pdf

Banwell, H. (1964) *Report of the Committee on the Placing and Management of Contracts for Building and Civil Engineering Works*. HMSO, London.

Barker 33 Cross-Industry Group (2006) *Modern Methods of Construction – Executive Summary of Final Report*. Available from http://www.hbf.co.uk/fileadmin/documents/barker/MMC_Final_Draft.pdf

Barker, K. (2003) *Barker Review of Housing Supply – Final Report – Recommendations*. HM Treasury. Available from http://www.hm-treasury.gov.uk/consultations_and_legislation/barker/consult_barker_index.cfm

Barlow, J., Childerhouse, P., Gann, D., Hong-Minh, S., Naim, M. & Ozaki, R. (2003) Choice and delivery in housebuilding: Lessons from Japan for UK housebuilders. *Building Research & Information*, 31(2), 134–145.

Bill, T. (2008) *Caledonian says Crunch will Boost OffSite Sector Building*. Available from http://www.building.co.uk/story.asp?sectioncode=30&storycode=312390 1&c=2 (accessed 22 October 2008).

Brackin, P. & Rogers, G.M. (1999) *Assessment and Quality Improvement Process in Engineering and Engineering Education*. 29th ASEE/IEEE Frontiers in Education Conference, 10–13 November. San Juan, Puerto Rico.

BRE. (2007) *Modern Methods of Construction*. Available from http://www.bre.co.uk/housing/section.jsp?sid=377 (accessed April 2007).

Buildoffsite (2008) *Year Book*.

CABE (2004) *Design And Modern Methods of Construction Review*.

Campbell, M. (2001) *Skills in England 2001: The Key Messages*. Policy Research Institute, Leeds Metropolitan University, Leeds.

Carter, J. (1967) Components and architect. *RIBA Journal*, **November**, 476–477.

Cather, H., Morris, R. & Wilkinson, J. (2001) *Business Skills for Engineers and Technologists*. Butterworth-Heinemann. Oxford.

Chan, E.H.W., Chan, M.W., Scott, D. & Chan, A.T.S. (2002). Educating the 21st century construction professionals. *Journal of Professional issues in Engineering Education and Practice*, **128**(1), 44–51.

Chan, L.-K. & Wu, M.-L. (2002) Quality function deployment: a literature review. *European Journal of Operational Research* **143**(3), 463–497.

CIC (Construction Industry Council) (2004) *Built Environment Professional Services Skills Survey 2003/2004*. London.

CIDB (2009) *Malaysia International IBS Centre*. Available from www.cidb.gov.my

CIOB (2006) *Skills Shortages in the UK Construction Industry: Survey 2006*. The Chartered Institute of Building, Berkshire.

CIOB. (2008) *Skills Shortage in the UK Construction Industry Survey*.

CIRI (2007) Quality Function Deployment. *Product Brief Development Tools*, AUT University. Available from http://www.ciri.org.nz/resources.html. (accessed 2 August 2008).

Clarke, L. & Wall, C. (1998) UK construction skills in the context of European developments. *Construction Management and Economic*, **16**(5), 553–567.

Clarke, L. & Winch, C. (2006) European skills framework? – but what are skills? Anglo-Saxon versus German concepts. *Journal of Education and Work*, **19**(3), 255–269.

Cohen, L. (1995) *Quality Function Deployment: How to Make QFD Work for You.* Addison Wesley Publishing, Ltd, Reading, MA.

Construction Skills. (2006) Sector Skills Agreement. Available from http://www. constructionskills.net

ConstructionSkills (2008) *2020 Vision – The Future of* UK. A scenario based report.

Dainty, A.R.J., Ison, S.G. & Briscoe, G.H. (2005) The construction labour market skills crisis: the perspective of small-medium-sized firms. *Construction Management and Economic,* **23**(4), 387–398.

DFEE (2000) *An Assessment of Skills Needs in Construction and related Industries. Skills Dialogue: A comprehensive summary from employers of skills required in Construction and related Industries.* The Institute for Employment Studies.

Egan, J. (1998) *Rethinking Construction: The Report of the Construction Task Force to the Deputy Prime Minister.* DTI. London.

Eichert, J. & Kazi, A.S. (2007) Vision and Strategy of ManuBuild – Open Building Manufacturing. In: ed. Kazi, A.S., Hannus, M., Boudjabeur, S. & Malone, A. *Open Building Manufacturing: Core Concepts and Industrial Requirements.* Espoo, Finland, Manubuild in collaboration with VTT – Technical Research Centre of Finland.

EMCC (2005) *EMCC dossier on the European Construction Sector.* Available from http://www.emcc.eurofound.eu.int/content/source/eu05017a.html (accessed 6 November 2005).

Emmerson, H. (1962) *Survey of Problems before the Construction Industries.* HSMO, London.

Eraut, M. (1994) *Developing Professional Knowledge and Competence.* Routledge Falmer. London.

Eurich, N.P. (1990) *The Learning Industry. Education for Adult Workers.* ED330812. Carnegie Foundation for the Advancement of Teaching. Princeton, NJ.

European Commission (2006) *Construction Overview. Enterprise and Industry.* Available from http://ec.europa.eu/enterprise/construction/index_en.htm (accessed 29 December 2006).

Evbuomwan, N.F.O. & Anumba, C.J. (1998) An integrated framework for concurrent lifecycle design and construction. *Advances in Engineering Software,* **29**(7–9), 587–579.

Farr, J.V. & Sullivan, J.F. (1996) Rethinking training in the 1990s. *Journal of Management in Engineering,* **12**(3), 29–33.

Gann, D. & Salter, A. (1999) *Research Interdisciplinary Skills for the Built Environment Professionals: A Scoping Study.* Sussex University Science and technology Policy Research Unit (The Ove Arup Foundation).

Gann, D. & Senker, P. (1998) Construction skills training for the next millennium. *Construction Management and Economics,* **16**(5), 569.

Gann, D.M. (1996) Construction as a manufacturing process? Similarities and differences between industrialized housing and car production in Japan. *Construction Management and Economics,* **14**, 437–450.

Gibb, A. & Pendlebury, M. (2006) *Glossary of Terms.* Buildoffsite: Promoting Construction Offsite.

Gibb, A.G.F. (1999) *Offsite Fabrication.* Whittles Publishing Services, Scotland.

Gibb, A.G.F. (2001) Standardisation and pre-assembly – distinguishing myth from reality using case study research. *Construction Management and Economic,* **19**(3), 307–315.

Gonzalez, M.E. (1999) Regulation as a cause of firm fragmentation: The case of the Spanish construction industry. *International Review of Law and Economics,* **18**, 433–450.

Gonzalez, M.E., Quesada, G., Gourdin, H. & Hartley, M. (2008) Designing a supply chain management academic curriculum using QFD and benchmarking. *Quality Assurance in Education*, **16**(1), 36–60.

Goodier, C. & Gibb, A. (2005) *buildoffsite: The value of the UK market for offsite*. Loughborough University, Loughborough. Available from http://www.buildoffsite.org

Gorgolewski, M. (2003) Offsite Construction, SIPs. Building for a future, the independent journal for 'green building' professionals and enthusiasts. *Offsite Fabrication: from Building for a Future Spring*. Available from www.buildingforafuture.co.uk/spring03/offsite_fabrication_SIPs.php

Gurjao, S. (2006) *Inclusivity: The Changing Role of Women in the Construction Workforce*. CIOB, University of Reading, Reading.

Hager, P., Crowley, S. & Melville, B. (2001) *Changing Conceptions of Training Ffr Evolving Workplaces: The Case of the Australian Building and Construction Industry. RCVET working paper 01–03*. Research Centre for Vocational Education and Training. TD/TNC 68.177. University of Technology, Sydney. Research Centre for Vocational Education and Training (RCVET), Sydney.

Hampson, K. & Brandon, P. (2004) *Construction 2020: A Vision for Australia's Property and Construction Industry*. Cooperative Research Centre for Construction Innovation, Brisbane.

Han, S.B., Chen, S.K., Ebrahimpour, M. & Sodhi, M.S. (2001) A conceptual QFD planning model. *International Journal of Quality & Reliability Management*, **18**(8), 796–812.

Hauck, A.J. & Rockwell, Q.T. (1997) Desirable characteristics of the professional constructor: The results of the Constructor Certification Skills and Knowledge survey. *Journal of Construction Education*, **2**(1), 24–36.

Hauser, J.R. & Clausing, D. (1988) The House of Quality. *Harvard Business Review*, **66**(3), 63–73.

Hervàs, F.C. & Ruiz, M.I.V. (2007) Stakeholder requirements for open building manufacturing. In: ed. Kazi, A.S., Hannus, M., Boudjabeur, S. and Malon, A., *Open Building Manufacturing: Core Concepts and Industrial Requirements*. Espoo, ManuBuild in collaboration with VTT.

Homein (2006) *Modern Methods of Construction*. Available from http://www.homein.org/searchresults.jsp?category=7&q=mmc (accessed April 2007).

Höök, M. & Stehn, L. (2008) Applicability of lean principles and practices in industrialised housing production. *Construction Management and Economics*, **26**, October, 1091–1100.

Hwarng, H.B. & Teo, C. (2001) Translating customers' voices into operations requirements: a QFD application in higher education. *International Journal of Quality & Reliability Management*, **18**(2), 195–225.

ILO International Labour organisation. (2006) *Global Distribution of Construction Output and Employment. Construction*. Available from http://www.ilo.org/public/english/dialogue/sector/sectors/constr/global.htm (accessed December 2006).

Jaraiedi, M. & Ritz, D. (1994) Total quality management applied to engineering education. *Quality Assurance in Education*, **2**(1), 32–40.

Jones, A. (1967) Components and the Builder. *RIBA Journal, November*, 485–488.

Kazi, A.S., Hannus, M., Boudjabeur, S. & Malon, A. (2007) *Open Building Manufacturing: Core Concepts and Industrial Requirements*. ManuBuild in collaboration with VTT – Technical Research Centre of Finland. Finland.

Koskela, L. (2000) An exploration Towards a production theory and its application to construction. PhD thesis 296. Espoo. Technical Research Centre of Finland, VTT Publications, Finland.

Kunstler, B. (2005) The hothouse effect: a model for change in higher education. *On the Horizon*, **13**(3), 173–181.

Laborde, M. & Sanvido, V. (1994) Introducing new process technologies into construction companies. *Journal of Construction Engineering and Management*, **120**(3), 488–508.

Lamancusa, J.S., Jorensen, J.E., Zayas-Castro, J.L. & Ratner, J. (1995) *The Learning Factory – A New Approach to Integrating Design and Manufacturing into Engineering Curricula*. ASEE, 25–28 June, Anaheim, CA.

Lambert, R. (2003) *Lambert Review of Business-University Collaboration*. Final report. HMSO, London.

Latham, M. (1994) *Constructing the Team*. HSMO, London

Leabue, D. & Viñals, J. (2003) *Avis Bâtir et innover: Tendances et défis dans le secteur du bâtiment*. Conseil de la science et de la technologie. Gouvernement du Québec.

Leitch, S. (2005) *Skills in the UK: A Long-term Challenge*. Interim report. HSMO, London.

Leitch, S. (2006) Leitch Review of Skills: Prosperity for All in the Global Economy – World Class Skills. Final report. HMSO. TSO, London.

LSC (Learning and Skills Council) (2005) *National Employers Skills Survey 2004: Key Findings*. LSC, UK.

Mackenzie, S., Kilpatricik, A.R. & Akintoye, A. (2000) UK construction skills shortage response strategies and an analysis of industry perceptions. *Construction Management and Economics*, **18**(7), 853–862.

Manubuild. (2009) *Open Building Manufacturing*. Available from http://www.manubuild.net

MBS (2004) *Modern Building Services: Enjoying the Benefits of Prefabrication*. Available from http://www.modbs.co.uk/news/printpage.php/aid/285/Enjoying_the_benefits_of_prefabrication

McNair, S. & Flynn, M. (2006) *Managing an Ageing Workforce in Construction: A Report for Employers*. Department for Work and Pension (DWP).

Mtech Group (2008) What disigners think. In: *Offsite Director 08*. Building Magazine.

Morton, R. (2002) *Construction UK: Introduction to the Industry*. Blackwell Science Ltd, Oxford.

Nadim, W. (2001) *The Work Environment in the Construction Industry: Hard and Labour Intensive*, Mubarak TV Studio Complex Project Image. Cairo, Egypt.

Nadim, W. (2009) Industrialising the construction industry: A collaborative training and education model. PhD thesis. School of the Built and Human Environment. University of Salford, Salford.

Nadim, W. & Goulding, J.S. (2009) Offsite production in the UK: The construction Industry and academia. *Architectural Engineering and Design Management (AEDM)*, **5**(3), 136–152.

Nadim, W. & Goulding, J.S. (2010) Offsite production in the UK: The way forward? A UK construction industry perspective. *Construction Innovation*, **10**(2), 181–202.

Nadim, W. & Goulding, J.S. (2011) Offsite Production: a model for building down barriers - A European construction industry perspective. *Engineering, Construction and Architectural Management*, **18**(1), 82–101.

NAO (National Audit Office) (2001) *Modernising Construction*. Comptroller and Auditor General.

NAO (National Audit Office) (2005) *Using Modern Methods of Construction to Build Home More Quickly and Efficiently*. National Audit Office.

Nateriello, G. (1989) *What do Employers Want in Entry-Level Workers? An Assessment of Evidence*. Trends and Issues No. 12.OERI-RI88062013. Columbia University, New York.

Ngowi, A.B., Pienaar, E., Talukhaba, A. & Mbachu, J. (2005) The globalisation of the construction industry – a review. *Building and Environment*, **40**, 135–141.

NHBC (2006) *A Guide to Modern Methods of Construction*, NF1. NHBC Foundation. Housing Research and Development in partnership with BRE Trust.

Noguchi, M. (2005) *Japanese Manufacturers' 'Cost-Performance' Marketing Strategy for the Delivery of Solar Photovoltaic Homes*. Solar World Congress, Orlando, FL.

NWDA (2004) *North West Development Agency – Construction Sector: Mapping, Strategy and Action Plan*. NWDA.

O'Brien, M., Wakefield, R. & Beliveau, Y. (2000) *Industrializing the Residential Construction Site. Center for Housing Research*. Virginia Polytechnic Institute and State University. Virginia.

Ogden, R. (2007) *Discovering Offsite. Buildoffsite Promoting Construction Offsite PowerPoint Presentation*. Available from buildoffsite. www.buildoffsite.com

Palotz, T.R. (2006) *Eine neue Ökonomie für den Wohnungsbau: Konzepte des Kosten- und Flächensparenden Bauens und Ansätze der Übertragbarkeit*. Books on Demand, Norderstedt, Germany.

Pan, W., Gibb, A. & Dainty, A. (2005) *Offsite Modern Methods of Construction in Housebuilding: Perspectives and Practices of Leading UK Housebuilders*. Loughborough University, Loughborough.

Pearce, D. (2003) *The Social and Economic Value of Construction*. Davis Langdon Consultancy.nCRISP New Construction Research and Innovation Strategy Pancl, London.

Pietsch, W. (2002) QFD dissemination – principles and practice. *1st international Symposium on Quality Function Deployment (KFG)*, Dokyuz Eylül Üniversitesi, Faculty of Business, Izmir.

Plus Group and Riverside (2005) *Modern Methods of Construction in the North West: Breaking the Barriers*. MMCNW Steering Group.

Prahalad, C K. & Hamel, G. (1990) The core competence of the corporation. *Harvard Business Review*, **68**(May–June), 79–91.

Rezgui, Y., Wilson, I.E., Damodaran, L., Olphert, W. & Shelbourn, M. (2004) *ICT Adoption in the Construction Sector: Education and Training Issues*. ICCCBE-X: the Xth International Conference on Computing in Civil and Building Engineering, Bauhaus-Universität, Weimer, Germany, 2–4 June 2004. Professur Informatik in Bauwesen.http://e-pub.uni-weimar.de/frontdoor.php?source_opus=207&la=de

RIBA (2005) *RIBA Constructive Change: A Strategic Industry Study into the Future of the Architects' Profession*. RIBA. Available from http://www.architecture.com/go/Architecture/Debate/Change_2096.html

Saffin, D. (2007) Breaking down barriers to collaboration. *Insite Magazine*. The IT Construction Forum.

Sahney, S., Banwet, D.K. & Karunes, S. (2003) Enhancing quality in education: application of quality function deployment – an industry perspective. *Work Study*, **52**(6), 297–309.

Seaden, G., Guolla, M., Doutriaux, J. & Nash, J. (2003) Strategic decisions and innovation in construction firms. *Construction Management and Economics*, **21**(6), 603–612.

Sebestyen, G. & Pollington, C. (2003) *New Architecture and Technology*. Architectural Press, Oxford.

Senge, P.M. (1992) *The Fifth Discipline: The Art and Practice of the Learning Organisation*. Century Business, London.

Shunk, D. (2002) *Competency-based, Anytime-Any place System*. e-Business and e-Work Conference, 16–18 October. Corinthia Towers Hotel. Prague, Czech Republic.

Slaughter, S.E. (1998) Models of construction innovation. *Journal of Construction Engineering and Management*, **124**(3), 226–231.

Slotte, V. & Tynjälä, P. (2003) Industry-University Collaboration for Continuing Professional Development. *Journal of Education and Work*, **16**(4), 445–464.

Stasz, C. (2001) Assessing Skills for work: two perspectives. *Oxford Economic Papers*, **53**(3), 385–405.

Stephen, M. & Flynn, M. (2006) *Managing and Ageing Workforce in Construction: A Report for Employers*. Centre for Research into the older Workforce. Department for Work and Pensions.

Tatum, C.B., Vanegas, J.A. & Williams, J.M. (1986) *Constructability Improvement Using Prefabrication, Preassembly, and Modularisation*. Stanford University. Department of Civil Engineering. Technical Report No. 297. Construction Industry Institute. Constructability Task Force.

Taylor, A. (2005) What employers look for: the skill debate and the fit with youth perceptions. *Journal of Education and Work*, **18**(2), 201–218.

Taylor, P.L., Morrison, S., Ainger, C. & Ogden, R. (2004) *Design and Modern Methods of Construction Review*. The Commission for Architecture and the Built Environment (CABE).

Tener, R.K. (1996) Industry-University Partnership for Construction Engineering Education. *Journal of Professional issues in Engineering Education and Practice*, **122**(4), 156–162.

Tindale, P. (1967) Components and the nation. *RIBA Journal*, **November**, 478–481.

Toffler, A. (1970) *Future Shock*. Bantam Books. New York.

Trauth, E.M., Farwell, D.W. & Lee, D. (1993) The IS Expectation Gap: Industry versus Academic Preparations. *MIS Quarterly*, **17**(3), 293–307.

Van de Ven, A.H. (1986) Central problems in the management of innovation. *Management Science*, **32**(5), 590–607.

Venables, T., Barlow, H. & Gann, D. (2004) *The Housing Forum: Manufacturing Excellence*. UK capacity in offsite manufacturing. Innovation Studies Centre, Tanaka Business School, Imperial College.

Venables, T. & Courtney, R. (2004) *Modern Methods of Construction in Germany: Playing the Offsite Rule*. Global Watch Mission Report, DTI.

Vrijhoef, R. & Koskela, L. (2005) *Revisiting the Three Peculiarities of Production in Construction*. IGLC-13 Lean Construction Theory, Sydney.

Wang, J.T.J. (2003) Curricular planning of upgrading the practical and professional competence of students in Technological Colleges. *International Conference on Workforce Education and Development*, 17–19 November. Taipei, Taiwan.

Warszawski, A. (1999) *Industrialized and Automated Building Systems*. 2nd edn. E&FN Spon, London.

Williamson, K.C. & Bilbo, D. (1999) A Road Map to an effective Graduate Construction Program. *Journal of Construction Education*, **4**(3), 260–277.

Womack, J.P., Jones, D.T. & Ross, D. (1990) *The Machine that Changes the World*. Rawson Associates, Maxwell Macmillan, New York.

Wood, G. (1999) Interdisciplinary working in the built environment education. *Education + Training*, **41**(8), 373–380.

Woudhuysen, J. & Abley, I. (2004) *Why is Construction so Backward?* Wiley-Academy, Chichester, UK.

Wysocki, R.K. (2004) *Project Management Process Improvement*. Artech House Inc, Boston, MA.

10

Construction Innovation through Knowledge Management

Charles Egbu

10.1   Introduction

The business environment of today is uncertain, unpredictable and competitive. Organisations conduct businesses in this second decade of the twenty-first century facing fiercer economic and market conditions than ever before. To survive, remain in business and profit-maximise, organisations need to be competitive and draw on organisational assets that distinguish them from others. More than ever before, there is a growing recognition and acceptance, in competitive business environments and project-based industries, that knowledge is a vital organisational and project resource that gives market leverage and contributes to organisational innovations and project success. Barney (1991) noted that sustained competitive advantage is obtained through capabilities and resources that are valuable, rare, non-imitable and non-substitutable. Such capabilities can be seen in the knowledge and experiences of individuals who are the workforce of organisations. Two decades later in 2011, after Barney (1991), the role of knowledge (individual and organisational knowledge) in contributing to organisational competitiveness and innovation is even greater. However, today, many organisations are at a crossroads. They are having to shed their workforce to remain in business. By so doing, they run the risk of depleting their valuable knowledge base, which is so important for organisational innovations and competitiveness. It is therefore important that organisations are able to manage their knowledge assets in an environment of uncertainty and economic turbulence. If they do not do so effectively, they may end up having organisational memory loss and serious depletion of knowledge assets, which would be detrimental to competitiveness and organisational innovations and renewal.

Construction Innovation and Process Improvement, First Edition.
Edited by Akintola Akintoye, Jack S. Goulding and Girma Zawdie.
© 2012 Blackwell Publishing Ltd. Published 2012 by Blackwell Publishing Ltd.

Knowledge management (KM) has truly emerged as a vital activity for organisations to preserve valuable knowledge and exploit the creativity of individuals that generates innovation. Egbu (2005) argued (in Anumba *et al.* (2005) that knowledge management is important for a number of reasons. First, it is important because the rise of time-based competition as a marketing weapon requires organisations to learn quickly. Second, it is important because of the globalisation of operations and because of the growth in number of mergers and takeovers where multiple organisations share knowledge in a collaborative forum. In project-based industries, such as the construction industry, the situation is even more complex. The activities of construction organisations are often characterised by short-term working contracts and diverse working patterns. Knowledge management is important in this context, because it brings together diverse knowledge sources from different sections of the demand and supply chains achieving cross-functional integration. Given this, an understanding of how organisations manage knowledge assets for improved innovation is important.

This chapter presents the role of KM in the acquisition, storage and use of tacit and explicit knowledge in an organisation. It also discusses how knowledge management impacts upon innovations in project based environments, along with the challenges facing project based organisations in managing knowledge through effective knowledge management practices, particularly in turbulent economic and market conditions.

10.2   Knowledge and Knowledge Management – Context and Definition

Few would contest the fact that people conceptualise knowledge according to their subjective interpretations. The debate about the meaning of knowledge is a pastiche of abstract ideas, which is too substantial to approach in this chapter. However, it is useful to explore the epistemological ideas of some authors briefly to establish a better conceptual understanding of knowledge in different contexts. Egbu and Botterill (2001) postulate that there is a complex dialectic between those who define knowledge as a scientific truth that exists independently of human action, and those who argue that knowledge is socially constructed. Knowledge has been traditionally thought of as a static concept, something that exists independently of human beings. Western philosophical preoccupations with the idea of knowledge as truth failed to account for essential dimensions in epistemology. Positivism took this view further by asserting that scientific truths exist objectively in the world, and that knowledge is a random combination of these truths. Epistemologists of this perspective have repeatedly neglected the human element of knowledge. In contrast, there are those who assert that knowledge is socially constructed and is completely determined by social structures. In this respect, knowledge can be seen as a process that is context-specific. This social constructionist perspective has Marxist undertones, and has been further revised by a view that human action determines knowledge (Habermas, 1984). Thus, social interaction is the principal motor of knowledge and human beings are responsible for conditioning their own environment (Nonaka and Takeuchi, 1995). Davenport and Prusak (1998) define knowledge as consisting of truths, beliefs, perspectives, concepts, judgements, expectations, methodologies and know-how, and originates in the minds of 'knower's'.

Defining knowledge management precisely can therefore be somewhat problematic. Nowhere in the literature on knowledge management is there a single unified meaning of the concept. Alvesson (1993) (cited in Despres and Chauvel, 1999: 110) argue that knowledge management is clearly on the slippery slope of being intuitively important but intellectually elusive. Many authors have attempted to explain certain elements of KM, specific to their own academic domains. Much of the ambiguity associated with KM is rooted in these authors' epistemological beliefs. Knowledge management could arguably be viewed as any process or practice of creating, acquiring, capturing, sharing and using knowledge wherever it resides, in order to meet existing and emerging needs, to identify and exploit existing and acquired assets and to develop new opportunities. For example, Nonaka and Takeuchi (1995) classified knowledge into two main types, and have adopted this distinction and applied it to the organisation. Explicit knowledge describes the type of knowledge that is documented and public, structured, fixed-content, externalised and conscious. Tacit knowledge can be generally understood as the form of knowledge that exists within an individual, and is intuitive and unarticulated. Tacit knowledge has been conceptualised by a myriad of academics from differing perspectives. Collins (1995) sees three types of tacit knowledge that present challenges to epistemological concerns of management. Embodied knowledge describes a type of knowledge that is a function of the physical environment. It cannot be easily transferred from one brain to another, as it is specific to the unique 'hardware' that accompanies an individual's brain; it is an integral part of the unique make-up of the human body. For example, a boxer's knowledge of fighting may be transferred to a professor, but the latter may not be physically able to use that knowledge in practice. Second, 'embrained knowledge' describes a type of knowledge that is specified by the exclusive physicality of an individual brain. Finally, 'encultured knowledge' describes a type of knowledge that is embedded within a social context, and cannot exist apart from it.

A thorough review of literature in the area of knowledge management suggests that it is diverse in perspective, approach and focus, yet there is a common bond that knowledge is a valuable asset in organisations throughout the industrial world. From an economic perspective, Drucker in the 1990s (as cited in McAdam and McCreedy, 1999: 93) noted 'we are entering the knowledge society in which the basic economic resource is no longer capita... but is and will be knowledge.' In a similar vein, Edvinsson (2000) claimed that there has been a shift in focus from local and physical to global and intangible. Knowledge is celebrated as a resource that is as important for organisations to understand and manage as labour and capital was in the 'old economy'. Similarly, O'Regan and O'Donnell (2000) also noted 'we are living in an era where the intangible is rapidly assuming economic, social and psychological supremacy over the tangible.'

10.3   Knowledge Management and Innovations in Project Based Environments

It is now generally accepted that the ability to innovate depends partly on the way in which an organisation uses and exploits the resources available to it, and a vital organisational resource, at the heart of innovation, is knowledge. Innovation can be viewed as the successful introduction and exploitation of an idea (product, service, technology and market). Knowledge management is highly associated with innovation because of its ability to convert tacit knowledge of people into explicit knowledge. This is grounded in the notion that unique tacit knowledge of individuals is of immense value to the organisation as a whole, and is the 'wellspring of innovation'. Given the close connection between knowledge possessed by personnel in an organisation, and the products and services obtainable from the organisation, it is uncontroversial that an organisation's ability to produce new products and other aspects of performance are inextricably linked to how it organises its human resources. It is tacit rather than explicit knowledge, which will typically be of more value to innovation processes. Yet, tacit knowledge is knowledge, which cannot be easily communicated, understood or used without the 'knowing subject'. The implication of the above discourse is that knowledge management that focuses on creating network structures to transfer only explicit knowledge will be severely limited in terms of its contribution to innovation and organisational and project success.

Construction organisations will therefore have to meet the needs and expectations of their clients, by drawing on the knowledge and expertise of project teams. Deploying experts into project teams is a mechanism by which professional service organisations coordinate and apply the diverse expertise and experience embedded within individuals, creating a knowledge network within the team, although sharing knowledge between team members may also create some tensions concerning competitive positions of the team members. The ability to match expertise with client needs and expectations is key to improving the competitive advantage of project-based professional or technical service organisations, where collaborative knowledge gained through successive projects is captured by organisations to continually enhance their market position. From a knowledge-based strategic management perspective, the creation of an optimal mix of a project team (i.e. having expertise in its membership drawn from across organisations) that has requisite skills and competence matched to the client's project requirements, will more often than not lead to client satisfaction and the project objectives being met. In addition, it provides opportunity for innovations through the maximisation of knowledge and expertise within a team.

Many innovation processes in the management and procurement of construction activities are becoming increasingly interactive, requiring simultaneous networking across multiple 'communities of practice', such as professional groups, functional groups and business units. This networking involves communication and negotiation amongst different social communities with distinctive norms, cultural values and interest in the innovation process. This therefore means that knowledge needed for innovation is distributed within organisations and across organisational boundaries through different supply chains. In the UK construction industry, there is a steady increase in collaborative working practices, such as partnering, alliances, Public Private Partnerships and Joint Ventures. In addition, projects are growing in complexity and cost, and clients' demands and expectations are also increasing more than ever before. This presents a situation where organisations have to collaborate and share knowledge, skills and expertise, in order to meet the needs of the clients. However, in sharing knowledge, organisations need to be both mindful of the communicative behaviours and practices associated with knowledge exchange as well as the 'knowledge paradox'. Organisations will have to be open to formal and informal information and knowledge flows from both networks and markets. At the same time, they must protect and preserve their intellectual capital and knowledge base, because it is upon this latter point that survival depends.

10.4   Managing Knowledge in Construction: Challenges Facing Project Based Organisations

Understanding how organisations manage knowledge assets involves due cognisance of a number of factors. It involves understanding the strategies that underpin KM practices within organisations, the structure and culture that sustains KM, the tools and technologies that support KM and how organisations measure the effects of KM. All these raise real challenges to managers in organisations and in projects. The construction industry is a project-based industry, and it remains the case that the UK construction industry contains a small number of relatively large firms and a large number of small firms. About 95% of construction firms employ fewer than eight people. The characteristics of the UK construction industry are well rehearsed and documented; however, it is important to see this from a knowledge management perspective.

The fragmentation of the construction industry reflects the economics of production, encouraging small firms organised by trade or craft. Construction firms typically involve relatively low capital investment. There are also relatively low barriers to entry and exit of firms within the construction industry. There are also characteristics of small firms, which affect their ability to access and transfer knowledge. First, this is related to their perceived technological weakness, such as a specialised range of technological competencies, inability to develop and manage complex systems, inability to fund long-term and risky knowledge management programmes. Similarly, investment in formal and informal training and education in the acquisition and sharing of requisite knowledge could be said to be more challenging for smaller than for larger organisations. Other perceived disadvantages of small organisations include little management experience, power imbalance when collaborating with large firms, difficulty in coping with complex regulations and associated cost of compliance. However, small size organisations could be said to have organisational strengths, which could stand them in good stead for managing knowledge assets. In the main, small firms often do not need the formal strategies used in large firms to ensure communication and coordination. These less-formal strategies in small firms, it could be argued, ease the communication of knowledge, improve informal networks, increase speed of decision-making, and improve the degree of employee commitment and receptiveness of novelty. Smaller organisations also tend to react faster to changing market requirements.

The construction industry is also characterised by projects, by short-term employment and by temporary coalitions of contractors and subcontractors. It is also perceived to have an adversarial culture. It could be argued that the nature of projects does not lend itself to knowledge management practices. For a start, projects tend to be temporary events; whereas, knowledge management programmes are usually seen as long-term investments. Projects can also be seen as temporary coalitions of individuals and teams who come together for the duration of the project, and are then disbanded after the end of the project. This latter characteristic of a project poses some challenges in terms of knowledge management. These include the difficulty of building trust amongst project team members, and motivating project staff and operatives during the project period. These are important issues for knowledge management.

Project culture is also likely to be different from an organisational culture. The project manager also has the project objectives (e.g. cost, time, quality, safety and environmental issues) as the overriding project concerns, as it is on the fulfilment of these objectives that they are normally judged. There is also the added complication in project environment, which is the fact that there may be members of the wider project team (i.e. those involved in the project supply chains) who, for many reasons, may not have the interest, drive and commitment towards knowledge management. There are also those who are committed to the project for a short period (e.g. a week or two, or even less) and who are bound to ask the question: 'What is there for me (WITFM) in knowledge management anyway?' However, it is important to point out that project knowledge is not just made up of the explicit type that is easily documented and archived. There is also the tacit knowledge. There is a need to be reminded that projects are made up of people, many of whom have requisite skills, knowledge, competence and wisdom. Some will also bring with them knowledge dimensions from previous jobs, which could be of use in solving problems and for providing innovative solutions to their new projects. In essence, knowledge management from a project perspective is about harnessing individual and project knowledge to the benefit of the project. The challenges therefore, for project managers and leaders of projects, are firstly to recognise the particular constraints imposed on knowledge management processes by the project environments, and secondly to find the means of creating, transferring, sharing, implementing and exploiting individual and project knowledge in such a way that they lead to project success and provide benefits to project clients.

Addressing the challenges of knowledge management calls for effective leadership and an understanding of the fundamental roles that people play in knowledge capture and knowledge sharing. If construction leaders are interested in knowledge capture, sharing and exploitation, then it is necessary to consider that knowledge workers (project staff and team members) should be included in a dynamic knowledge management process. This process is one that demands the support of motivation, creativity and the ability to improve an intellectual and comprehensive vision of the relationship between the project and the project team members. In other words, individual and project knowledge should be seen as a project's intellectual capital, and an important factor in project success. Targeted education and training geared towards construction personnel is important to improve cultural awareness in these important areas.

10.5   Knowledge Management Strategy – Issues and Contexts

It is important to stress at the outset, that there is no one KM strategy that best addresses knowledge management issues in all organisations. Organisations are different. Their needs are different and their capacity and capabilities to address issues are also different. Similarly, the wider contexts in which they operate are different, and so these are likely to impact on whatever strategy they decide to adopt and implement. However, whatever knowledge management strategy is being considered, sufficient attention should be given to both informal knowledge processes and formalised knowledge management initiatives. Attention should also be given to the complexities of *knowledge*, rather than any proxies, such as information. It is the case that many accounts of 'Knowledge Management' default to a focus on information management. Such a view underestimates the richness of the subject of knowledge, and the opportunities a knowledge focus offers for re-thinking business processes. It is argued elsewhere (Egbu *et al.*, 2003a) that whereas certain types of knowledge can be codified and treated as information, much knowledge is personal, being based on experience and reflection, and remains tacit.

Knowledge also has a social dimension, being created and shared in social groupings, within which tacit knowledge sharing may often occur (Brown and Duguid, 1991). Related to its social nature, knowledge is also created in specific contexts, and is to varying degrees 'situated' (Lave and Wenger 1991) or context specific, and may be 'sticky' and difficult to transfer or share (von Hippel, 1994). This reduces the potential for the simple and costless transfer of lessons learned between contexts, such as companies or industries. Thus, there are different approaches to knowledge management, and many of these emphasise the capture and processing of knowledge resources or assets (or intellectual capital) that the organisation already possesses. There is also an approach that focuses on knowledge processes and dynamic capabilities, as well as knowledge resources. An emphasis on organisational processes, such as knowledge creation and sharing, provides a greater indication of dynamic capabilities. Another approach is one that emphasises on practitioner knowledge (Gibbons *et al.*, 1994). In *The New Production of Knowledge*, they identify a shift in modes of knowledge creation from Mode 1 to Mode 2. Mode 1 knowledge is produced in institutions, is disciplinary, with a hierarchical control system through peer review, and an emphasis on generating codified knowledge that is transferable. Mode 1 knowledge grows cumulatively, is stored in libraries and forms the content of university syllabuses and professional qualifications. In contrast, Mode 2 knowledge is created in the context of application – it results from practice, is transient and often unrecorded. It is trans-disciplinary, and the capability to produce it is widely diffused. Gibbons *et al.* (1994) argue that Mode 2 has hitherto been undervalued, and also that it is increasing in importance.

A further approach focuses on cross-boundary knowledge processes, and particularly the need for organisations to acquire knowledge from external sources. Arguably, no firm has ever been independent in knowledge terms and today all organisations are likely to be increasingly dependent on external sources of knowledge. The capability to track, make sense of, understand and assimilate externally sourced knowledge is known as *absorptive capacity* (Cohen and Levinthal 1990). The skills required to absorb knowledge include techniques of sourcing, sense-making and learning.

So, it is important for a knowledge management strategy to give due consideration to organisational knowledge capabilities or processes as well as resources, to take account of informal knowledge processes as well as formal knowledge management and values practitioner (or Mode 2) knowledge, and consider context and address the issue of absorptive capacity. It should also acknowledge the richness and complexity of the subject of knowledge, whilst recognising that firms need to address practical issues. Thus, there is need for the following knowledge processes occurring within and between firms:

- creating/generating/producing knowledge;
- communicating, sharing knowledge;
- searching/sourcing knowledge;
- synthesising/transforming/combining knowledge;
- capturing/codifying/storing/classifying knowledge;
- mapping knowledge or knowledge proxies; and
- applying and re-using knowledge.

10.6   Knowledge Management Techniques and Technologies

For knowledge management, the term 'tools' is used loosely. Too often, knowledge management 'tools' is used to mean only IT tools. There is a need for a better understanding of Information Technology (IT) and non-IT tools, their differences and characteristics. A host of technologies (IT-based) and techniques (non-IT based) exist for knowledge management in organisations. In the main, the selection of appropriate technologies appears to follow a more structured approach than the selection of techniques for knowledge management. In industrial settings, there are two main approaches for selecting appropriate knowledge management technologies. The first approach is based on knowledge management 'sub-processes'. The second approach is based on 'technology families'. The former appears more popular, as it allows the 'users' to identify sub-processes that they need to manage and then select the most appropriate technologies geared for the need of the identified sub-processes. In construction organisations, the potential benefits of technologies and techniques for KM are not fully understood. There is a need for some guidance in the approaches employed by organisations for selecting appropriate technologies and techniques for knowledge management.

In this respect, knowledge management techniques do not depend on IT, although it provides support in some cases. Examples include brainstorming, communities of practice, face-to-face interactions, recruitment, training, story-telling, coaching, job-rotation, shadowing and quality circles. KM techniques, arguably, require strategies for learning, more involvement of people, and are more focused on tacit knowledge. They are relatively more affordable to most organisations. This is because no sophisticated infrastructure is required, although some techniques require more resources than others (e.g. training requires more resources than face-to-face interactions). KM techniques are relatively easy to implement and maintain due to their simple and straightforward nature. They focus on retaining and increasing the organisational tacit knowledge, a key asset to organisations. Furthermore, KM techniques are not new, as organisations have been implementing them for a long time, where their use has been under the umbrella of several management approaches (e.g. organisational learning and learning organisations). Using these tools for the management of organisational knowledge requires their use to be enhanced so that benefits from them, in terms of knowledge gain/increase, are properly managed.

Communities of Practice (CoPs) consist of a group of people of different skill sets, development histories and experience backgrounds, who work together to achieve commonly shared goals. These groups are different from teams and task forces. People in CoPs can perform the same job, collaborate on a shared task (software developers), or work together on a product. They are peers in the execution of 'real work.' What holds them together is a common sense of purpose and a real need to know what each other knows. Usually, there are many communities of practice within a single company and most people normally belong to more than one.

Post-Project Reviews are an important KM technique. They are debriefing sessions used to highlight lessons learnt during the course of a project. These reviews are important to capture knowledge about, causes of failures, how they were addressed, and the best practices identified in a project. This increases the effectiveness of learning, as knowledge can be transferred to subsequent projects. However, if this technique is to be effectively utilised, adequate time should be allocated for those who were involved in a project to participate. It is also crucial for post-project review meetings to take place immediately after a project is completed, as project participants may move or be transferred to other projects or organisations. Whilst project reviews are important, their limitations need to be acknowledged. There is also increasing importance for live-capture of knowledge as projects progress.

Apprenticeship as an important KM technique is a form of training in a particular trade carried out mainly by practical experience or learning by doing (not through formal instruction). Apprentices often work with their masters and learn craftsmanship through observation, imitation and practice. The masters focus on improving the skills of the individuals so that they can later perform tasks on their own. This process of skill building requires continuous practice by the apprentices until they reach the required level. For example, mentoring is a process whereby a trainee or a junior staff member is attached or assigned to a senior member of an organisation for advice relating to career development. The mentor provides a coaching role to facilitate the development of the trainee (mentee) by identifying training needs and other development aspirations. This type of training usually consists of career objectives given to the trainee, whereby the mentor checks if the objectives are achieved and provides feedback.

There are many KM tools and techniques used in construction (Egbu *et al.*, 2003b), to capture knowledge through knowledge mapping tools, knowledge bases, Case-based Reasoning, etc. KM technologies consist of a combination of hardware and software technologies. In this respect, hardware technologies and components are important for a KM system as they form the platform for the software technologies to perform and the medium for the storage and transfer of knowledge. Some of the hardware requirements for a KM system are:

- the personal computer or workstation to facilitate access to the required knowledge;
- highly powerful servers to allow the organisation to be networked;
- open architecture to ensure inter-operability in distributed environments;
- media-rich applications requiring Integrated Services Digital Network (ISDN) and fibre optics to provide high speed;
- Asynchronous Transfer Mode (ATM) as a multi-media switching technology for handling the combination of voice, video and data traffic simultaneously; and
- use of the public network (e.g. Internet) and private networks (e.g. Intranet, Extranet) to facilitate access to and sharing of knowledge.

Software technologies also play an important part in facilitating the implementation of KM. The number of software applications has increased considerably in the last few years. Solutions provided by software vendors take many forms and perform different tasks. The large number of vendors that provide KM solutions makes it extremely difficult to identify the most appropriate solutions. This has resulted in organisations adopting different models for establishing KM systems. KM software technologies have seen many improvements in the last decade due to many alliances, and mergers and acquisitions between KM and Portal tool vendors. Data and text mining is a technology for extracting knowledge from masses of data or text. The process of data/text mining enables meaningful patterns and associations of data (words and phrases) to be identified from one or more large databases or 'knowledge-bases'.

The more commonly used technologies for knowledge capture and sharing in the construction industry include groupware, Intranet, Internet, Extranet and knowledge bases. Groupware, as a software product, helps people to communicate, share knowledge and information, perform their work efficiently and effectively, and helps in decision-making. It supports distributed and virtual project teams, where team members are from multiple organisations and in geographically dispersed locations. Groupware tools usually contain email communications, instant messaging, discussion areas, file area or document repository, information management tools (e.g. calendar, contact lists, meeting agendas and minutes) and search facilities. An Intranet is an internal organisational Internet that is guarded against outside access by special security tools called firewalls, and an Extranet is an Intranet with limited access to outsiders – often making it possible for them to collect and deliver certain knowledge over the Intranet. This technology is useful for making organisational knowledge available to geographically dispersed staff members, and is therefore used by many organisations. Knowledge bases are repositories that store knowledge about a topic in a concise and organised manner. They present facts that can be found in a book, a collection of books, websites or even human knowledge. This is different from the knowledge bases of expert systems, which incorporate rules as part of the inference engine that searches the knowledge base to make decisions.

10.7   Effective Knowledge Management Practices in Turbulent Economic and Market Conditions

Effective knowledge management practices depend on a host of factors. There is no one strategy or theory that explains successful knowledge management in all organisational or project contexts. It depends on an appropriate balance of such issues a:

- organisational strategy, choices and path dependency;
- people in organisations, organisational culture and structure;
- technological and financial resources; organisational processes and routines; and
- a thorough consideration of internal and external stimuli – the context in which knowledge this is to be managed.

However, leaders of organisations and projects need to see knowledge management as important, and an integral part of strategic decision making, which can influence profitability and competitiveness of the organisation, as well as project success. Leaders of organisations should therefore endeavour to establish at all levels of the organisation a strategic intent of knowledge acquisition, creation, accumulation, protection and exploitation of knowledge. In the same vein, the linkages between strategic management and human values need to be examined carefully, along with the role of a KM orientation in an effort to support adequately successful strategies. Leaders of organisations should not pay 'lip service' or just be interested in knowledge development, but should proactively support it.

As part of effective knowledge management, it is important that leaders consider employees as a fundamental part of the dynamic knowledge process; this will not only help support motivation and enhance creativity, but also help provide a comprehensive vision of the relationship between the organisation, people, the project and its environment. Leaders should also endeavour to determine and put in place an effective communication infrastructure for the effective capture, transfer and leveraging of knowledge. Above all, leaders would have to create an appropriate culture for effective knowledge management. Such culture should encourage workers' autonomy, so that they may express and share the knowledge they possess in a 'free environment'. An effective knowledge management should also be able to determine methods for mapping knowledge, determining who has what knowledge, needs what knowledge, and where this knowledge resides, and measuring the extent of KM effectiveness. Auditing the knowledge present at, or accessible to, the organisation, and managing adequately the inventory of 'knowledge repositories', is vital for continuing improvement and organisational renewal.

The role that leaders play in effective knowledge management is therefore vital, and should provide opportunities for:

- sharing culture where there is openness and willingness to share information, experience and knowledge across organisation and project teams. This should allow 'flexibility' in the lines of communications allowing top-down, bottom-up and lateral communications within project and organisational structures;
- creating a risk tolerant climate, where it is accepted that lessons could be learned through mistakes;
- developing knowledge teams, i.e. staff from all disciplines to develop or improve methods and processes;
- introducing knowledge webs (networks of experts/communities of practice who collaborate across project teams) and the provisions of collaborative technologies, such as Intranets or GroupWare for rapid information access;
- defining and communicating knowledge performance behaviours, and identifying key knowledge workers and knowledge performance positions;
- make knowledge performance part and parcel of organisation and project performance;
- rewarding knowledge-sharing behaviours and incentivise key knowledge management actions.

As part of effective knowledge management practices, networking, CoPs, storytelling, coaching, mentoring and quality circles are important mechanisms for sharing and transferring tacit knowledge in organisation and project environments. These should be considered, encouraged and promoted more by project leaders. Communities of practice are needed to encourage individuals to think of themselves as members of 'professional families' with a strong sense of reciprocity.

Human networking processes can encourage sharing and the use of knowledge for project and organisational innovations are important. Leaders of projects should also espouse 'the law of increasing returns of knowledge' as a positive way of encouraging knowledge sharing – shared knowledge stays with the giver, while enriching the receiver. Intuitive knowledge is managed by individuals being valued and not by being heavy-handed through project 'controlled processes'. It is folly to believe that any project or organisational environment can make people have ideas and force them to reveal intuitive messages or share their knowledge in any sustained manner. An individual's intuitive knowledge cannot be manipulated in any meaningful way nor controlled without the individual being willing and privy to it. The process of trying to manipulate or control intuitive knowledge in fact creates their destruction. The issues of trust, respect and reciprocity are vital elements of a conducive environment for managing tacit knowledge. It is through these that individual members of the project or organisation can be motivated to share their experiences and exploit their creativity. Leaders would need to recognise, provide incentives and reward knowledge performance and sharing behaviour patterns. They should also take action on poor knowledge performance. The regular communication of the benefits of knowledge management is important in sustaining the co-operation of project team members. A variety of ways exist for doing this, including regular meetings, project summaries, project memos and through project GroupWare/Intranet facilities where they exist. Every organisation and project strategy for KM should consider training, recruitment and selection of project team members (e.g. subcontractors and suppliers). It should also pay due cognisance to the team members' competencies, requisite knowledge and their willingness and effectiveness in sharing knowledge for the benefit of the project and the organisation.

10.8   Conclusion

Knowledge management continues to have a significant role to play in organisational competitiveness and innovation. People are at the heart of knowledge management and organisations need to view their people as their greatest asset. It is understandable that for many organisations the current economic climate provides added challenges for managing organisational assets. However, through effective leadership and a clear strategic vision, organisations are able to manage their knowledge assets effectively in the current turbulent economic environments. In this respect, hardware technologies and components are important for a KM system, as they form the platform for the software technologies to perform, and the medium for the storage and transfer of knowledge. Software technologies also play an important part in facilitating the implementation of KM. The number of software applications has increased considerably in the last few years. Solutions provided by software vendors take many forms and perform different tasks. The more commonly used technologies for knowledge capture and sharing in the construction industry include groupware, Intranet, Internet, extranet and knowledge bases.

Many factors confront organisations in the management of knowledge assets. These include organisational culture, motivation and incentivising their workforce. Their strategic choices, knowledge assets and capabilities, and financial and technological capabilities, and also the effect of internal and environment stimuli are all inextricably linked.

References

Alvesson, M. (1993) Organisation as rhetoric: knowledge intensive firms and the struggle for ambiguity. *Journal of Management Studies,* **30**(6), 997–1020.

Anumba, C., Egbu, C., & Carrillo, P. (2005) *Knowledge Management in Construction.* Blackwell Publishing, Oxford.

Barney, J.B. (1991) Firm resources and sustained competitive advantage. *Journal of Management,* **17**(1), 99–120.

Brown, J.S. & Duguid, P. (1991) Organizational learning and communities of practice: towards a unified view of working, learning and organization. *Organization Science,* **2**(1), 40–57.

Cohen, W.M. & Levinthal, D.A. (1990) Absorptive capacity: a new perspective on learning and innovation. *Administrative Science Quarterly,* **35**, 128–152.

Collins, H.M. (1995) Humans, machines, and the structure of knowledge. In: ed. Ruggles, R.L., *Knowledge Management Tools.* Butterworth-Heinnemann, Newton, MA. 145–165.

Davenport, T.H. & Prusak, L. (1998) *Working Knowledge.* Harvard Business School Press, Boston, MA.

Despres, C. & Chauvel, D. (1999) Knowledge management(s). *Journal of Knowledge Management,* **3**(2), 110–120.

Edvinsson, L. (2000) Some perspectives on intangibles and intellectual capital 2000. *Journal of Intellectual Capital,* **1**(1), 12–16.

Egbu, C.O. (2005) Knowledge Management as a Driver for Innovation. In: Anumba, C., Egbu, C.O. & Camillo, P. *Knowledge Management in Construction.* Blackwell Publishing, Oxford. 121–131.

Egbu, C.O., Kurul, E., Quintas, P., Hutchinson, V., Anumba, C., Al-Ghassani, A. & Kuitar, R. (2003a) *A Systematic Analysis of Knowledge Practices in Other Sectors: Lessons for Construction.* September, Report: Work Package 2; Department of Trade and Industry (DTI) – Partners in Innovation project: CI 39/3/709.

Egbu, C.O., Kurul, E., Quintas, P., Hutchinson, V., Anumba, C., Al-Ghassani, A. & Kuitar, R. (2003b) *Techniques and Technologies for Knowledge Management.* September, Report: Work Package 3; Department of Trade and Industry (DTI) – Partners in Innovation project: CI 39/3/709.

Egbu, C.O. & Botterill, K. (2001) Knowledge management and intellectual capital: Benefits for project based industries. *Proceedings of the RICS Foundation – Construction and Building Research Conference (COBRA),* vol. 2. Glasgow Caledonian University, 3–5 September, ed. Kelly, J. & Hunter, K. 414–422.

Gibbons, M., Limoges, C., Nowotny, H., Schartzman, H., Scott, P. & Trow, M. (1994) *The New Production of Knowledge: the Dynamics of Science and Research.* Sage, London.

Habermas, J. (1984) *The Theory of Communicative Action: The Rationality of Action and the Rationalization of Society,* vol. 1. Polity Press, Cambridge.

Hippel, E. von (1994) 'Sticky information' and the locus of problem solving: implications for innovation. *Management Science*, **40**(4), 429–439.

Lave, J. & Wenger, E. (1991) *Situated Learning: Legitimate Peripheral Participation.* Cambridge University Press, Cambridge.

McAdam, R. & McCreedy, S. (1999) A critical review of knowledge management. *The Learning Organisation*, **6**(3), 91–100.

Nonaka, I. & Takeuchi, H. (1995) *The Knowledge-Creating Company.* Oxford University Press, Oxford.

O'Regan, P. & O'Donnell, D. (2000) Mapping intellectual resources: Insights from critical modernism. *Journal of European Industrial Training*, **24**(2/3/4), 118–127.

Innovation through Collaborative Procurement Strategy and Practices

Akintola Akintoye and Jamie Main

In recent times, the way in which construction activities are managed and procured to achieve clients' requirements is receiving greater attention. There is an increasing concern over the 'ineffectiveness' of existing procurement systems. In some ways, this has been attributable to the fragmented nature of the construction industry, mainly because of a large number of relatively small firms, combined with a large number of relatively small construction projects and low barriers to entry, particularly in the (small) contracting sub-sector. In this respect, the 'industry is fragmented because of the many disciplines involved – designers, constructors, professional consultants and engineers, and specialist contractors. It is fragmented because of long and complex supply chains, bringing together the different specialists. Low profit margins combined with traditional procurement in construction led to adversarial relationships and poor service to clients (Fairclough, 2002). These issues have not been limited to a particular system of procurement, although the management forms appear to have taken the main criticism. This is not to say that any one system is better or worse than its contemporaries, merely that the industry is concerned that, at present, there is no universal solution for solving the problems associated with procurement systems.

Procurement on a project-by-project basis has been regarded as stifling innovation, ostensibly through the lack of security, continuity and critical mass of projects to encourage longer-term work programmes that help to develop in-depth research skills and development (Fairclough, 2002). Another contention is that the problem is not about procurement *per se*, but rather the process of tendering that is stifling innovation. Sidwell and

Construction Innovation and Process Improvement, First Edition.
Edited by Akintola Akintoye, Jack S. Goulding and Girma Zawdie.
© 2012 Blackwell Publishing Ltd. Published 2012 by Blackwell Publishing Ltd.

Budiawan (2001) highlight problems with the competitive tendering process in relation to contractor-led innovation, and explore ways in which owners can develop procurement procedures that allow and encourage innovation from contractors. In addition, Khalfan and McDermont (2006) demonstrated that innovation procurements with integrated supply chain participants within the construction industry often requires innovative thinking, supported by trust and transparency.

This chapter presents procurement in relation to the criteria needed for the choice of procurement methods available, the problems associated with conventional procurement approaches, and how innovative procurement methods can be used to address some of the inherent problems facing the industry.

11.2   Construction Procurement and the Procurement Cycle

Procurement can be described in several ways: as 'the act of obtaining by care or effort, acquiring or bring about' (*Oxford English Dictionary*). Mohsini and Davidson (1989) describe this as 'the acquisition of new buildings, or space within buildings, either by directly buying, renting or leasing from the open market, or by designing and building the facility to meet a specific need.' Lenard and Moshsini, (1998) define this as a strategy to satisfy client's development and/or operational needs with respect to the provision of constructed facilities for a discrete life cycle. Notwithstanding these differences, typically, the procurement involves:

1. identification of requirements;
2. market sourcing;
3. selection of tenders;
4. evaluation of tenders;
5. contract award; and
6. managing and evaluation of delivery.

The CIB W92 Procurement Systems Working Group (CIB, 2011) identified that construction procurement as a framework within which construction is brought about, acquired or obtained. In essence, building or construction procurement is the organisational structure adopted by the client for the management of the design and construction of a building project (Masterman, 1992). In this respect, Rowlinson and McDermott (1999) noted that elements such as contract strategy and the client are functional parts of the procurement system, and that the effectiveness of the client organisation or the contract strategy is modified by other procurement system variables such as culture, sustainability, economic and political environment and more practical concepts such as partnering. Similarly, the Byatt Report (2001) regards procurement as 'the whole process of acquisition of goods, services and construction projects spanning the whole life cycle from the initial concept and definition of business needs through to the end of the useful life of an asset, services contract or need for the activity.'

NEDO (1985) 'Think about Building' lists seven steps to successful construction procurement, specifically:

1. selection of an in-house project executive;
2. appointment of a 'Principal Advisor' if required;
3. careful definition of the requirements of the project;
4. realistic determination of project timing;
5. selection of an appropriate procurement path;
6. considered choice of organisation to be employed; and
7. commitment to a building or site held back until it has been professionally appraised.

In addition, they identified eight criteria for the selection of procurement path as follows:

1. timing
2. controllable variation;
3. complexity;
4. quality level;
5. price certainty;
6. completion;
7. division of responsibility, and
8. risk.

Notwithstanding these nuances, the procurement process often leads to a contract between the parties in order to deliver goods and services. One of the fundamental reasons for entering into a contract is to allocate the risk inherent in a project. In this respect, a construction contract, for example, typically specifies the roles of the client and the contractor, along with the dates by which tasks must be completed, and the formal mechanisms for payment. These issues are particularly important, as all too often contract documents contain uncertainties, and organisations then enter into contracts with conflicting assumptions. This is compounded where the complexity of the activities involved in the procurement cycle involves many professions and practitioners that are expected to work together to deliver a project. Consequently, OGC (2010) views procurement as a cyclical activity rather than a linear project, which is expected to engage with informing and supporting strategic management of organisational deliverables.

11.3   Procurement Strategies

It is possible to divide procurement methods into two broad categories: conventional procurement methods, and innovative or collaborative procurement methods – the details of which can be seen in Table 11.1. The procedure for the selection of the contract and the building team is predominantly influenced by the procurement method. However, fundamental to the categorisation of procurement methods and contract

Table 11.1   Procurement: categories, types and strategies.

Procurement	Type	Strategies
Conventional	Designer-led competitive tendering Designer-led construction managed for a fee Package Deal	Traditional Lump Sum approach Management contracting Construction Management Turnkey Design and Build (Construct) British Property Federation
Innovative/ Collaborative	Package Deal Framework agreements Partnering	Public Private Partnership Design, Build, Finance and Operate Prime contracting Partnering: strategic and project Joint Venture Supply Chain Management

selection is the apportionment of risk between the parties to a construction contract, which naturally include the client and contractor. In this respect, risk apportionment can be influenced by various factors, including:

1. the complexity and uniqueness of the project;
2. the employer's involvement with the design process;
3. the client's involvement with the construction process;
4. the required speed from inception to completion;
5. the required degree of price certainty; and
6. the amount of risk to be transferred to the construction team, etc.

Recently, clients (particularly public sector clients) have moved from traditional or conventional methods to more innovative procurement methods in order to tap into private sector finance, transfer more risk to contractors, and/or achieve better efficiency in the management of construction development to garner value for money for the taxpayers.

11.4   Conventional Procurement Methods

The last 30 years have seen the development of three main procurement systems in the construction industry. The systems that generally reflect the position of the design team in relation to the contractor can be seen as:

- *Traditional*: Designer Lead;
- *Design and Build*: Contractor Lead;
- *Management*: Design Team (including Contractor) Lead.

There are further subdivisions of each of the procurement systems, each system having its own unique characteristics, advantages and drawbacks.

11.4.1   Traditional Procurement System

The traditional procurement system is often referred to as being designer-lead, and has the unique characteristic of the separation of the responsibility for the design of the project from that of its construction. Most architects regard this approach as the 'normal' way to build, and therefore continue to support it (Pain, 1993). This system was widely used on construction projects, although the popularity of this method is now decreasing. The main features of this approach include:

- the contractor not being involved at all during the pre-construction stage;
- long gestation period until commencement on site;
- construction period agreed at tender stage;
- full contractual and commercial risks/opportunities accepted by contractor;
- complete competition on price;
- lump sum contract and agreed construction period;
- adversarial roles between contractor and design team likely;
- changes and variations lead to claims and extensions of time; and
- lack of contractor advice available to client/design team.

11.4.2   Design and Build Procurement System

The Design and Build (D&B) procurement system tends to bring together the responsibilities for design and construction within one organisation, that is, a single point of responsibility for design and construction with one party, the contractor. This direct set up often improves lines of communication. Thus, the level of integration of design and construction in D&B depends on the level of organisation of the contractor. It is recognised that the nature of this integration improves 'buildability', which has a significant effect on cost, time and quality. The major attractions of this method to the client are a 'one-stop-shop' technique of building production, along with the cost security associated with the guaranteed maximum price and the transfer of risk to the contractor. The term design and build now encompasses a multitude of hybrids, including develop and construct, innovation design and build, traditional, turnkey and package deal (Akintoye, 1994; Akintoye and Fitzgerald 1995). In general, some of these hybrids have been introduced by clients to bring more competition into the process, and by contractors to relieve them of some of the project risks.

11.4.3   Management Procurement System

The management procurement system is industry's attempt to reduce the 'them and us' feeling, often embedded in procurement systems. This is an entwined design and construction processes. Management forms include

Table 11.2 Employer's objectives in relation to procurement methods.

Employer's Objective	Procurement methods			
			Management	
	Traditional	Design and Build	Management Contract	Construction Management
Complex and unique project			X	X
Employer's involvement with the design	X		X	X
Employer's involvement with the construction				X
Employer's requirement for early completion		X	X	X
Price certainty prior to construction start	X	X		
Requirement to transfer majority of risk		X		

management contracting, construction management, fee management, fast track construction, and design and manage. The most popular of the management forms is management contracting. A management contractor is appointed as a professional to the design team to offer expertise, which is usually reimbursed on a fee basis. The actual construction work is carried out solely by works contractors appointed by the management contractor. The management contractor is therefore responsible for setting out, managing, organising and supervising the project. This system ensures early appointment of the management contractor, allows buildability to be increased, and because design often runs in parallel to construction, it provides a high degree of flexibility for changes to building costs. However, because of this flexibility there tends to be a greater number of variations in the management contracting process. This coordinated approach, and potential flexibility results in greater operational speed and efficiency, but this idea is only apparent on larger, more complex and innovative projects. This system apportions large amounts of risk to the client and a low, but not a no-risk option to the contractor.

11.4.4   Comparison of Conventional Procurement Methods

Table 11.2 presents a comparison of employer's objectives and apportionment of risk between the employer (client) and contractor for the four main conventional contract procurement methods. This identifies that Design and Build apportions the least risk to the employers, while construction management allocates the greatest risk to the employer.

This is also demonstrated through a case study relating to a Heath Service project in the UK. The project had two main significant requirements, which were:

1. the critical programming requirements dictated by the occupation by a specified date; and
2. the absolute requirement for effective cost planning, tender documentation and cost control/monitoring activities to be carried out in accordance with the procedures laid down by the Health Board.

Table 11.3 shows the various features of contract routes used in conventional procurement methods. This compares the routes, and allows an evaluation to be made of each method relative to the requirements of the particular project. It was considered based on the evaluation that the most appropriate contract route in terms of the prime objectives of the project (i.e. time scale and control) was management contracting and two stage tendering (fast track). For this reason, the management contract was selected for the procurement of this project.

11.4.5   Problems with Conventional Procurement Methods

Typical problems often involved in conventional procurement methods have been articulated in Black (1998). Traditional approaches to procurement usually involve a chain of separate firms who add value to items purchased from other organisations via arms-length one-off contracts. In this respect, according to Lewis (1995), buyers source the market for the best price trading suppliers off against each other. Typically, buyers focus on transaction exchange relationships with suppliers, and source requirements from two or more organisations to avoid becoming dependent on any one source. Buying organisations usually stipulate their requirements via the specification, and the supplier is required to meet the specification at the most competitive price and within an acceptable delivery period to secure the contract. This focus on short-term advantage therefore results in a significant degree of uncertainty and volatility for suppliers (Chadwick and Rajagopal, 1995). MacBeth and Ferguson (1994) provide a framework to describe traditional procurement relationships as follows:

11.4.5.1 Time Span

The interaction between client and supplier is regarded as a discrete occurrence, although negotiating parties make use of their experience of each other. It is relatively inexpensive to change suppliers, therefore contracts only cover a few months at most, and suppliers must recover all their costs on each transaction. In addition, competitive tendering exercises are commonly used to keep prices low and play suppliers off against each other. This forces suppliers to tender for more work than their resources can meet, in anticipation that only a proportion of tenders will be successful.

Table 11.3 Comparison of conventional procurement methods for a health project.

Type of contract	Early involvement of contractor	Time scale	Level of competition/ Accountability	Degree of cost control	Early agreement of final account	Degree of Risk	Total	Comments
Traditional – Single stage	0	0	10	10	6	10	36	Unsuitable due to pre-contract time scale required
Traditional – Two stage	8	0	6	8	6	8	36	Unsuitable due to pre-contract time scale required
Fast track: Two stage	8	8	5	6	8	7	42	Considered appropriate for this project
Management Contract	9	8	8	8	10	8	51	Considered appropriate for this project
Prime cost	8	6	0	0	10	0	24	Unsuitable due to lack of accountability and cost control
Negotiated contract	7	6	2	6	6	6	33	Unsuitable due to lack of accountability and cost control
Design and Build	10	0	3	2	6	6	27	Not considered appropriate for this project
Target cost	8	8	2	2	8	4	32	Unsuitable due to lack of accountability and cost control

* Individual features are rated on a scale of 1 to 10.

11.4.5.2 *Personal Attitudes*

Organisations are fiercely protective of their expertise, and will only usually part with this in return for financial reward. Therefore, negotiators are constantly wary of giving too much away, and try to reveal as little information as possible. There is thus a tendency for buyers to use their purchasing power as a threat against poor pricing or service, while suppliers rely on threatening to withdraw their services. The buying organisation's specification is rigid, and suppliers learn not to make suggestions (as these are usually ignored), and often the head office design team is blamed for the inflexibility of the specification. In addition, contact between the buyer and supplier tends to occur at the contract negotiation stage, and it is typically thereafter when problems are encountered.

11.4.5.3 *Behaviour*

Parties tend to concentrate on personal gain, even at the expense of the other party. This results in a short-term view, whereby individuals are apathetic about problems that will be faced by those that follow them. Individuals can therefore be perceived to be aggressive in their interactions, often with both colleagues and with external organisations.

11.4.5.4 *Organisational Processes*

Personal interactions are often avoided where possible, and individuals are not kept informed of the big picture. Procedures are neither questioned nor changed over time.

11.4.5.5 *Measurements*

Clients measure only price, quality and delivery, while suppliers only consider the ability of the customer to pay, and the likelihood of winning a tender as a basis for selecting tender lists or deciding to bid respectively. Feedback is rarely provided and is used only as 'ammunition' when conducting future negotiations. Organisations concentrate on efficiency as the best way in which to make savings, and often carry out multiple inspections, for example, regarding the quality of goods received, without seeking to identify and eliminate root causes of problems encountered.

11.5   Collaborative Procurement or Innovation Procurement Methods

Innovation procurement methods are often devised as a result of the shortcomings of conventional procurement methods, and therefore attest to achieve best value for the parties involved in project development. In support of collaborative procurement approaches and innovation, Egan (1998) encouraged long-term procurement relations to improve construction development quality and efficiency. This was followed by Egan (2003), who argued for integrated supply chains for construction procurement. Khafan and McDermott (2006) noted how innovative procurement methods could bring about improvements in existing processes, resulting in the development of innovative solutions by different supply chain partners to different problems through an integrated approach.

Collaborative relationships or procurements are now used in many industries, including manufacturing, retailing, construction and service sectors. In the construction industry, this has been encouraged by two major reports produced by Latham (1994) and Egan (1998). These reports have a recurring theme in that they suggest the industry would be improved through greater teamwork, not only at site level and organisational level, but also with clients and suppliers. Recommendations within these reports have led to some construction clients and companies using collaborative arrangements, such as long-term/strategic arrangements, project and strategic partnering, joint venture, partnership, prime contracting and SCM in order to improve the construction development process. This usage has brought many advantages to companies, where balanced collaborative relationships are achieved. Despite these benefits, the intensity of these relationships and the central philosophy of commitment embedded in such relationships can lead to a high level of pressure to perform, whereby partners under pressure may be encouraged to take unnecessary risks to prove their worth. Lorange and Roos (1991) commented that it was an overstatement to say that all collaborative relationships were successful.

Typically, the conventional procurement process is comprised of six stages, specifically:

1. identify the requirement
2. source the market
3. seek tenders
4. evaluation of tenders
5. contract award, and
6. managing and evaluation of delivery.

This is mainly limited to pre-contract and contract award stage activities and less on what happens at the post-contract stages. This is a particular shortcoming that collaborative procurement has attempted to address.

11.5.1   An Overview of Collaborative Procurement Methods

Collaboration procurement relies on cooperation and teamwork, openness and honesty, trust, equity and equality, if it is to succeed (Bennett and Jayes, 1998). Collaboration can provide a framework for the establishment of mutual objectives amongst the building team as well as encouraging the principle of continuous improvement. This framework encourages trust, co-operation and teamwork into a fragmented process, which enables the combined effort of the participants of the industry to focus upon project objectives (Naoum, 2003). Some authors advocate that collaboration is a long way from returning tangible benefits to the contractor, mainly because clients still have a deep-rooted cost-driven agenda (Green, 1999; Taylor, 1999). Consequently, they expect to reduce costs, or to pass costs and risks down the supply chain, and thereby do not genuinely adopt a win-win attitude (Wood and Ellis, 2005).

One of the key elements and common feature of collaborative procurement is a high level of commitment between parties at management level. This was found in a study by Black *et al.* (2000), where organisations with experience of partnering rated management commitment more highly than those without. Communication between stakeholders is essential whenever an organisation is dealing with change, and it is equally true when introducing or managing a partnership, as communication between parties is vital for understanding each party's expectations, attitudes and limitations. This was considered crucial for success (Black *et al.*, 2000).

Continuous evaluation of collaborative procurement is needed in order to ensure that it is developed according to the expectations of the parties involved, as Bennett and Jayes (1995) highlighted that continual performance improvement was necessary in order to deliver the benefits of collaboration. Furthermore, as advocated by Egan (1998), the use of integrated teams is a common feature of collaborative arrangements, and by involving the team at the earliest stage in a project, improvement can be made in quality, productivity, health and safety and cash flow, along with reduced project durations and risks (Egan, 2003).

Additional exemplars include a study by Burnes and New (1996), which revealed many examples of the ways in which different industries and organisations have sought to use collaborative relationships. Examples of the benefits realised include the minimisation of waste, improvements in operational efficiency and productivity, and improved supply chain coordination (Hamza and Hibberd, 1999). On this theme, collaboration through innovative procurement methods can encourage openness and communication as 'neither side benefits from exploitation of the other, innovation is also encouraged and each partner is aware of the others needs, concerns and objectives and is interested in helping their partner achieve such' (Cook and Hancher, 1990). This can therefore lead to better mutual understanding of each other needs. Therefore, the working process becomes more efficient, which in turn can reduce wastage (McGeorge and Palmer, 1997) and promote organisational flexibility, which is beginning to be seen as a means for developing an environment supportive of innovation and learning (Bennett and Jayes, 1995).

Litigation is also a major problem in many construction projects, which does not help realise potential saving. However, in collaborative arrangements, dispute problems, claims and litigation can be greatly reduced through open communication and improved working relationships (Cook and Hancher, 1990). Similarly, Bennett and Jayes (1995) noted the financial benefits of collaboration, highlighting that collaborative workshops and other related collaborative efforts could achieve savings of around 10% of total costs. Furthermore, collaboration has the potential to improve cost performance, as it can reduce the risk of budget overruns through improved cost control by alleviating rework and reducing schedule time through improved communication and clearer project goals (Albanese, 1994). Thus, by improving communication on projects, parties are less likely to be 'surprised' by schedule delays and additional costs, which often lead to disputes and litigation (Moore *et al.*, 1992). This was reinforced by Arntzen *et al.* (1995), noting that collaboration could improve project quality by building an atmosphere that fosters a team approach along with improved communication. This enables potential problems and quality issues to be recognised earlier (Albenese, 1994), and can also enhance customer satisfaction as the customer is closer to the construction process and better informed (Nielsen, 1996). There is therefore general consensus that collaboration has the potential to bring consistently better results than the more traditional approach. Typical benefits from partnering would be:

- reduced exposure to litigation;
- improved project outcomes in terms of cost, time and quality;
- lower administrative and legal costs;
- increased opportunity for innovation and value engineering; and
- increased chances of financial success (CIIA, 1996).

In summary, collaborative relationships have brought many advantages to companies where a balanced collaborative relationship is achieved, which include:

- ability to leverage internal investments;
- ability to focus on core competencies;
- leverage core competencies of other organisations;
- reduce capital needs, broaden products offerings;
- gain access or faster entry to new markets;
- share scarce resources;
- spread risk and opportunity;
- improve quality and productivity;
- having access to alternative technologies;
- provide competition to in-house developers;
- use a larger talent pool and satisfy the customer (Crouse, 1991).

However, Lamming (1993) noted that despite these benefits, the intensity of the relationship and the central philosophy of commitment embedded in such relationships could lead to a high level of pressure to perform, especially taking unnecessary risks. Notwithstanding this, it is also important to acknowledge that several success factors can be espoused, especially through benchmarking and continuous improvement studies. These collaboration issues (risk, success factors, etc) are well acknowledged (Akintoye and Main 2006b).

11.5.2   Comparison of Collaborative Procurement Methods

There are various types of collaborative procurement approaches in use in the construction industry (McDermott *et al.* 2004), the main issues of which can be seen as follows:

11.5.2.1 Joint Ventures

The IPPR (2002) stated that 'Joint ventures benefit from not relying on arms length relationships between the public and private sectors – the organisational form of joint ventures gives tangible expression to the commitment to work in partnership. However, they can prove unstable in the face of changing circumstances and also raise difficult issues concerning accountability and risk transfer.' In addition, Cheatham (2004) noted that more often than not, joint ventures form because of a mutual interest by two contracting parties in sharing and spreading the risk associated with large, complex or long-term contracts, which could have dire consequences if not all goes as planned. Other reasons for entering into joint ventures can be as simple (or as varied) as:

- to leverage unique skills and/or assets (equipment, personnel, etc);
- to gain complimentary skills (or pool resources);
- to take advantage of local geographic and/or sub-trade knowledge (or expand market penetration);
- for political connections and/or owner relationships;
- to maximise surety capacity and/or influence surety underwriting;
- to meet set aside requirements;
- to obtain pricing security (i.e. second opinion);
- to strengthen financial structure and/or pre-qualification validation; or
- to meet pre-qualification requirements (experience, capacity, licensing, etc).

11.5.2.2 Partnering

Partnering has been defined as 'a long-term commitment between two or more organisations for the purpose of achieving specific business objectives by maximising the effectiveness of each participant's resources' (NEDO, 1991: 5). Therefore, partnering can be seen as a structured management approach for facilitating team working across contractual boundaries. The fundamental components of this are formalised mutual objectives, agreed problem resolution methods, and an active search for continuous measurable improvements (CBP, 2003). Cognisant of this, in recent years, influential reports (Latham Report, 1994; Egan Report, 1998) have garnered an increased awareness amongst clients, that the lowest priced bids do not always present the best value for money. These reports have also highlighted the importance of partnering arrangements to improve team working across contractual boundaries. Thus, 'partnering and integration strategies attempt to address a fundamental characteristic of the industry that it is fragmented, as individuals from different organisations which are geographically and temporally dispersed are involved in the construction process' (Luck, 1996:.1). Therefore, particular attention has focused upon improving relations between the project team and client, as well as encouraging feedback and adjustment between design and construction processes (Banwell, 1964; Higgin and Jessop, 1965; Harvey and Ashworth, 1993). In this respect, the partnering concept represents the latest in this series of initiatives (NEDO, 1991; Latham, 1994), such that 'the confrontational culture which is endemic in the sector has resulted in the development of inefficient business processes, which feed through, as overheads, to total project costs' (ACTIVE, 1996: 7). Partnering as a concept is intended to reduce the adversarial nature of the construction industry, and central to the ethos of partnering, is a move away from adversarialism and litigation, towards ways of resolving problems jointly and informally through more effective forms of collaboration. In this respect, Khalfan and McDermott (2006) highlighted two partnering cases studies, with 'the potential benefits at the start of the framework approach which motivated the client to introduce the innovative procurement route, and also the other parties involved especially the main contractors.' These benefits included savings on tendering/procurement costs, time savings on programme, lesson learned and rolled forward within the delivery team, benefits of performance management systems, fewer delays and added value.

11.5.2.3 Alliances

In the manufacturing industry, there are several methods of long-term buyer-supplier relationships. However, unlike the construction industry, these methods are used in controlled production environments where the supply of goods is a repeating process. This is generally not applied to the construction industry, as repetition is rare, and works are procured typically on a one-off project-by-project basis. Furthermore, the design, construction and supply chain parties tend to work together in constantly changing coalitions on different building projects, and these temporary coalitions disappear on the completion of these projects (O'Brien et al., 1995). In the traditional method of construction, the contractor wins a building project by tendering, and the client often chooses the contractor who offers the lowest price. However, due to the cost-driven nature of the industry, the contractor who wins the project in turn tries to get the lowest price from its suppliers and subcontractors. The project therefore consists of several contracts for which each firm involved allocates resources accordingly (Winch, 1989). Thus, members of the project team are only responsible for their specific project input – the corollary of which creates professional and organisational boundaries that are rarely crossed. Therefore, problems tend to occur, as the design, construction and supply firms often differ from each other in size and culture. This can be exacerbated, where firms have their own methods of construction and organisational approach to construction, which consequently give rise to conflicts between parties during the building process (Pries and Janszen, 1995).

In the construction industry, there are few clients that are able to offer repeat orders for work over a long time periods, as works are typically procured on a one-off project-by-project basis, therefore suppliers tend not to work with each other on a regular basis. In addition, projects tend to attract unique uncertainties, which require new design and production solutions. This, combined with time pressures and project complexity, can often result in insufficient information supply and poor planning, which in turn can lead to conflicts between project parties (Winch, 1989). Cognisant of this, partnering/alliances have become prevalent in today's construction industry. Altruistically, the core essence of alliances is the quest for mutual benefit, that is, working together to address a market need – acknowledging that the combined alliance would be more valuable/successful than the contributors could deliver by themselves. Thus, alliances are a specific type of partnering arrangement, where organisations come together to form a new joint organisation to target a particular market or category of end customer. Consequently, by their nature, they can often be more complex to achieve than normal partnering arrangements. This increased complexity arises from greatly enlarging the scope and scale of projects; conversely, however, this can also lead to greater rewards than more conventional contractual arrangements. Therefore, a key component of an alliance is to agree up front all expectations of each partner in that alliance. This agreement is important, so that any gaps uncovered at a later stage do not hold up progress. Thus, by forming an alliance, organisations are able to change the nature and the scope of what they do; and in some instances, upsizing organisations can enter a more intensive and competitive market where the competition is more intense. Therefore, all aspects of business need to be taken into consideration, as it often influences the choice of organisation a company chooses to go into partnership with.

In summary therefore, the risks and rewards of an alliance/partnering arrangement can be shared and allocated purposefully and appropriately, in accordance with the key business drivers of each particular instance, such that all parties share fully in the planning and organisation of the operation, along with an appreciation of the risks and responsibilities are allocated therein. This requires trust as an essential ingredient (Bresnen and Marshall, 2000), particularly to resolve areas of possible contention. Therefore, organisations in the alliance need to rely implicitly on each other to act in full accord with the aims and objectives of the alliance arrangement, and not for individual or vested interests.

11.5.2.4 Strategic Alliances

The migration away from the traditional approach towards collaborative long-term relationships often involves the formation of partnerships or strategic alliance, particularly with manufacturer-supplier relationships (Kalwani and Narayandas, 1995). However, Macneil (1981) recognised that this departure from the traditional approach required the establishment of relationships with an orientation focus and 'closeness' aligned to relation-specific goals. This is particularly important, as expectations of continuity (in relationships) require understanding the unique behaviours of firms (Noordewier et al., 1990).

Strategic alliances are inter-firm co-operative arrangements aimed at achieving the strategic objectives of the partners (Das and Teng, 2001). They therefore provide a way for organisations to pool their resources to create value that each partner could not achieve if they acted alone (Inkpen and Ross, 2001). In this respect, Agapiou *et al.* (1998) emphasised the importance of active participation of senior management in the formation of strategic alliances, as this is a continuous development process, which requires inter-relationships and confidence between parties to be established. These voluntary organisational relationships therefore often involve the exchange, sharing or co-development of new knowledge, products, services or technologies (de Rond, 2003: 90). This is especially true, as strategic alliances tend to come in many forms, including horizontal alliances between competitors, vertical alliances between buyers and suppliers, and diagonal alliances between firms in different industries (Nooteboom, 1999: 1). They can also take the form of outsourcing, franchises, joint ventures, joint product development, joint R&D and joint marketing arrangements. Strategic alliances can therefore be said to be a source of competitive advantage (Das and Teng, 2000; Ireland *et al.*, 2002). However, there is also a growing body of evidence intimating a high failure rate exists in such arrangements (de Rond, 2003; Gerwin, 2004), one cause of which relates to the high level of risk associated with alliances, compared to 'in-house' activities (Das and Teng, 2001). This risk may be caused by the difficulties inherent in gaining co-operation with partners who have different objectives and orientations, and the potential for partners to opportunistically exploit the dependence relationship (Dekker, 2003).

11.5.2.5 *Prime Contracting*

Holti *et al.* (1999) noted that 'prime contracting replaces short-term, contractually driven single project adversarial inter-company relationships with long-term, multiple project relationships based on trust and cooperation.' The 'prime contracting' arrangement to which government procurement guidance refers is essentially similar to management contracting, an arrangement with which construction professionals are largely familiar. In this respect, the prime contractor needs to be an organisation with an ability to bring together all of the parties (consultants, contractors and suppliers) necessary to meet the client's requirements effectively (Scottish Executive, 2004).

A Prime Contract incorporates certain fundamental principles, such as whole service procurement, economies of scale and collaborative working. All parties to the contract must therefore have a common interest and willingness to co-operate in order to achieve mutual goals to the highest degree. Thus, under Prime Contracting (PC), one organisation acts as a single point of responsibility (the Prime Contractor) between the client and the supply chain. The prime contractor is normally an organisation with an ability to bring together all of the parties (the supply chain) necessary to meet the clients requirements effectively, such that there is nothing to prevent a designer, facilities manager, financier or any other organisation from acting as the prime contractor. A key part of the PC route is the development of a whole life cost model before construction commences. Clients therefore need prime contractors to provide details of all the parties in their supply chain when tendering. However, it may not be possible to appropriately assess the technical capacity of a prime contractor under the EC procurement rules unless a significant number of the other organisations that make up the supply chain are known and taken into account during the assessment (Scottish Executive, 2004). Notwithstanding this, PC is traditionally best suited to large, complex projects, particularly where there are many variables, but where an early start to work is required, thus allowing work to start whilst other sections are still being designed. Furthermore, as PC does not have price certainty, this by default only makes it financially viable on larger projects.

Target pricing (i.e. set a target cost and design the product to achieve that cost) is employed in PC. This 'design to cost' approach enhances innovation in the achievement of clients' (owners and users) expected outputs. The essence of PC is to achieve superior underlying values rather than lower margins. Therefore, the prime contractor engages 'lead' specialist-contractors (cluster leaders) to design and construct various aspects of the project, such as building structure, building envelope, mechanical services, electrical services, etc. These various subcontracts are then often let on a competitive basis. Cluster leaders thus prepare, submit and unite their method statements in accordance with the structured activities. The prime contractor must therefore have project management capabilities that can coordinate the activities of all the cluster leaders to derive value. However, value gains in PC can only be sustained by a structured programme of continuous innovation, where 'tried and tested' ways of doing things are continually challenged (and if possible) changed, to develop better alternatives.

11.5.2.6 Public Private Partnerships

Public Private Partnerships (PPP), particularly Private Finance Initiative (PFI) projects, were created for the provision of services and not specifically for the exclusive provision of capital assets such as buildings. However, PPPs are not a single model, they need to be tailored to individual circumstances, but the breadth of these include a full range of partnerships from PFI, joint ventures and concessions, to the sale of equity stakes in state-owned businesses (HM Treasury, 2000). Under PPP arrangements, private sector contractors become long-term providers of services, rather than simply up-front asset builders, combining the responsibilities of designing, building and operating (and possibly financing) assets, in order to deliver the services needed by the public sector. The overall aim of PPP is to increase the flow of capital projects against a background of restraint on public expenditure, and with a particular remit of transferring risk from the public to the private sector. Thus, through PPP, the contractor has an incentive to design a facility that will have a low operating cost, or provide an incentive to generate the lowest operational cost over the long term. This may result in the delivery of a higher specification facility than might otherwise have been expected, which is particularly effective where there are limitations on capital spend under traditional procurement routes. Under this form of procurement, authorities can also make significant savings in operational costs, which may also involve the transfer of existing employees to the contractor. One of the main advantages of PPP is that the entire construction risk rests with the contractor. Therefore, in the event of a cost or time overrun, this will (depending upon the terms of the principal agreement) be the contractor's responsibility (HM Treasury, 2003).

From a PFI perspective, this normally involves a complex collection of agreements, including a principal agreement (between authority and contractor), a construction contract (between contractor and subcontractors) and a Facilities Management agreement (for maintenance of the completed facilities), together with loan agreements with banks and a range of collateral agreements and warranties. Due to the long-term commitment involved, the contracts are often drafted on a bespoke basis, which tends to involve many months of negotiation with contractors, bankers, lawyers, insurers, operators and their respective advisors. However, as far as the transfer of financial risk is concerned, Akintoye et al. (1999) noted that many bodies involved in PFI viewed financial risk (e.g. debt risk, banker's risks) as less important than many other risks (i.e. design risk, construction cost risk, performance risk, etc).

Another advantage of PPPs is the provision of an improved quality of service, which is often better than that achieved by traditional procurement (EC, 2003). This may reflect the better integration of services with supporting assets, improved economies of scale, the introduction of innovation in service delivery, or the performance incentives and penalties typically included within a PPP contract. However, one disadvantage of PPP is the long lead-time for completion of the project compared to capital project work. For example, it normally take many months to negotiate a completed contract with a contractor, which often involves complex drafting in order to incorporate specifications both for the facility and its maintenance, revenue and payment arrangements and compatibility of construction contract, facilities management agreement, Project Agreement and credit agreements. However, although there are financial and construction risks associated with the construction of the PPP project, risks remain with the management of the project throughout the concession period. This requires careful monitoring to ensure continuity of desired service, and often requires the use of risk management skills to identify any potential threats. Good risk management is also an effective way of developing innovative solutions to service delivery challenges, and coping with adverse consequences in order to maximise value for money (National Audit Office, 2002). It is generally recognised that this form of public sector contracting is already changing the climate within which decisions about procurement are made and managed, the operationalisation of which encourages a longer-term perspective to be taken throughout the whole life of a built asset.

11.5.2.7 Framework Agreements

Framework Agreements (FA) are agreements undertaken with suppliers, to establish the terms governing contracts to be awarded during a given period, with particular regard to price and quantity. In other words, an FA can be seen as a general term for agreements with suppliers, which set out terms and conditions under which specific purchases (call-offs) can be made throughout the term of the agreement. Therefore, FAs (including call-off contracts) with a single supplier or a limited number of suppliers can result in significant savings to both parties. The resource implications for the client of managing more than one FA for each type of work should however be borne in mind when deciding whether to award more than one FA (Scottish Executive, 2004).

FAs are generally inappropriate for clients that only occasionally procure buildings. In this respect, these agreements often cover procurement routes such as prime contracting and design and construct. They can be particularly appropriate for maintenance projects, or where a stream of similar projects are planned. These agreements have several advantages, as this approach tends to provide contractors with a continuous workflow; the familiarity and continuity of which can help improvements to be secured by transferring the learning from one project to another. In addition, FAs enable the contractor to make potential savings in respect of bidding costs, as there is no requirement for re-bidding for each individual project. Cost savings from FAs tend to come from:

- no requirements for re-bidding of each individual project;
- continuous improvement through the transference of learning from one project to another;
- reduced confrontation (conflicts and disputes); and
- workflow continuity.

FAs do not normally have a standard form of contract, therefore considerable time and effort may have to be invested to secure the FA required. In addition, the time taken to enter into a framework can be considerable, particularly regarding EC procurement compliance and time taken to pre-qualify contractors. Details of a typical FA can be seen in Table 11.4.

11.5.2.8 Supply Chain Management

SCM is not a procurement route in itself; however, it is used in many forms of construction procurement such as PPP and prime contracting. SCM deals with the management of materials and information resources across a network of organisations that are involved in the design and production process. It therefore recognises the inter-connectivity between materials and information resources within (and across) organisational boundaries, and seeks systematic improvements in the way these resources are structured and controlled. Applications of the SCM techniques in

Table 11.4   Example framework agreement. Reproduced by permission of Place North West (www.placenorthwest.co.uk).

PLACE North West (2010) http://www.placenorthwest.co.uk/homepage.html [accessed 3 June, 2011]

(PLACE North West - For property and regeneration professionals

Contractors share £1bn public sector framework

27 Jul 2010, 17:29

Fourteen construction companies have been appointed to the North West Construction Hub Medium Value Framework established by councils across the region.

The panel covers 13 lots broken down by sub-region and type of work. The contracts that will be handed out according to the contract range from £500,000 to £10 m, for duration of four years. The four-year framework will be used by local authorities and other public sector organisations in the North West.

Manchester City Council, supported by the Centre for Construction Innovation, led the framework development with financial support from the North West Improvement and Efficiency Partnership.

The contractors appointed are: Bramall Construction; Cruden Group Limited; Eric Wright Construction; FMP Construction; GB Building Solutions; Herbert T. Forrest; ISG Regions; Kier Regional; Laing O'Rourke; Mansell Construction Services; Morgan Ashurst; Seddon Group; Wates Group; Willmott Dixon Construction.

This is the second of three regional frameworks to be delivered by the North West Construction Hub aimed at creating a better deal for public sector authorities in the region. The larger panel for contracts over £10 m was decided in April. The smallest of the three frameworks, for works under £500,000, will be published later this year.

The framework will enable public sector organisations and the construction companies to work in partnership to deliver projects more efficiently, on time and cost, with an enhanced focus on quality and added value / sustainability outcomes.

John Lorimer, NWCH Lead said: "I am pleased with the outcome of the tendering process and look forward to working with our new partners over the coming years. The Medium Value Framework will eliminate costly procurement exercise for individual projects whilst providing benefits including cost savings for authorities, reductions in contract time, improved quality of work and the ability to deliver added value objectives that are not obtainable through traditional procurement methods."

Andrew Thomas, chief executive of the Centre for Construction Innovation North West, said of the framework: "We would like to commend the local authorities' efforts in its creation. The basis of integrated supply chain working and adoption of fair payment practices are more important than ever in these challenging times. Innovative and collaborative initiatives can provide solutions to some of the challenges, especially when delivered in a stable framework environment. CCI will continue to provide ongoing support of the Hub's work and deliver the benefits this framework will bring in providing a stable workload for local contractors and better value solutions to clients in the North West."

---

manufacturing environments have saved hundreds of millions of pounds while improving customer service (Arntzen *et. al.*, 1995); where, as subcontractors and the supplier production comprise the largest value of project cost, supply-chain approaches may have similar benefits. However, in contrast to manufacturing, construction projects are usually unique and temporary. This implies a temporary organisation of production for each project, again characterised by a short-term coalition of participants with frequent changes of membership, often called 'temporary multi-organisation' (Cherns and Bryant, 1984; Burbridge *et al.*, 1993). This fragmentation has been cited as a contributor to low productivity, cost and time overruns, conflicts and disputes, additional claims, and time-consuming litigation. Other issues include:

- inadequate capture, structuring, prioritisation and implementation of client needs;
- fragmentation of design, fabrication and construction data, with data being generated in isolation not being readily re-used downstream;
- development of pseudo-optimal design solutions;
- lack of integration, coordination and collaboration between the various functional disciplines involved in the life-cycle aspects of the project;
- poor communication of design intent and rationale, which leads to unwarranted design changes, inadequate design specifications, unnecessary liability claims, and increase in project time and cost.

There is therefore a need to integrate the supply chain to achieve mutual benefits, develop organisational learning, promote efficiency, improve quality, and leverage full value and profitability opportunities (Akintoye and Main, 2007). The key to SCM is the information flows associated with inter-organisational communications. As a result, a core issue is the effective management of information, in the form of information flows that permit rapid inter-organisational transactions between supply chain partners.

In summary, the relationship between construction project participants is often multifarious, complex and intertwined. Thus, attention and focus must be given to the intensive collaboration amongst project participants to synchronise both the input and output of the supply chain. A key enabler to successful collaboration is the ability to communicate, and share and exchange project information in a timely and accurate manner.

11.6   Conclusion

The need for innovation through procurement strategies is now more important than ever before. The industry is faced with the need to deliver enhanced value for money, with increasingly complex projects, enhanced competition, and additional pressures to comply with legislative demands and requirements, for example, sustainable development. Innovative construction procurement methods have developed out of these new demands, which have helped to improve risk management, value for money, etc. These new methods are now transforming the industry.

This chapter presented an overview of new procurement methods and how these can be used to leverage innovation, from the construction supply chain, through to collaborative relationship between the demand and supply sides of the industry. Acknowledging that some collaborative relationships might involve a limited number of construction industry players working with a single client group in an FA to garner benefits, innovation procurement can provide avenues for integrated teams to explore in order to secure similar benefits. This is in line with the four main areas identified in 2002 through the Strategic Forum's 'Accelerating Change' (cited in BERR, 2008); where:

1. *Client leadership*: clients procure projects in a way that allows all in the integrated team to maximise the added value their expertise can deliver;
2. *Integrated teams and supply side integration*: created at the optimal time in the process to release fully the contribution each can make and share risk and reward in a non-adversarial way;
3. *Culture change in people issues*: a positive image, an emphasis on education and training and behaviour based on mutual respect) and
4. *a focus on the end product*.

Integrated procurement in the form of construction joint ventures, partnering, alliances (including project and strategic), prime contracting, public private partnerships, and all the associated FAs and SCM opportunities have been identified as being able to deliver construction innovation and process improvement (as opposed to the conventional adversary procurement method types). These integrated methods provide a framework for establishing mutual objectives amongst the building team, as well as encouraging and engaging the principles of continuous improvement. These methods also provide an enabling environment to engender trust, co-operation and teamwork (within these fragmented processes), which thereby enables the combined efforts of the project participants to focus upon the actual project objectives. Having said that, the principal success factors for innovative procurement methods would be commitment of adequate resources from the partners, equity of relationship, recognition of the importance of non-financial benefits, and clarity of objectives, while the principal failure factors are lack of trust and consolation, and lack of experience and business fit (Akintoye and Main, 2007). However, if augmented and managed appropriately, these relationships can deliver real (tangible) innovation and process improvements, but such collaboration methods need to be carefully considered to ensure that they fit into the business plans of all contributory organisations.

References

ACTIVE (1996) ACTIVE Engineering Construction Initiative, The Action Plan, ACTIVE, London.

Agapiou, A., Flanagan, R., Norman, G. & Notman, D. (1998) The changing role of builders merchants in the construction supply chain. *Construction Management and Economics*, **16**, 351–361.

Akintoye, A. (1994) Design and build: A survey of construction contractors' views. *Construction Management and Economics*, **12**, 155–163.

Akintoye, A. & Fitzgerald, E. (1995) Design and build: A survey of UK architects. *Engineering, Construction and Architectural Management*, **2**(1), 27–44.

Akintoye, A. & Main, J. (2006a) Collaborative advantages in construction development. In: ed. Baldwin, A., Hui, E. & Francis Wong, F., *Construction Sustainability
and Innovation*. Proceedings of CIB W89 International Conference on Building
Education and Research (BEAR), 10–13 April, Hong Kong Polytechnic University,
Hong Kong.

Akintoye, A. & Main, J. (2006b) Perception on success and failure factors for construction collaborative relationships. In: ed. Dulami, M., *Sustainable Development
through Culture and Innovation*. Proceedings of the Joint International Conference
of Construction Culture, Innovation and Management, 26–29 November, The
British University of Dubai, Dubai. 143.

Akintoye, A. & Main, J. (2007) Collaborative relationships in construction: The UK
contractors perception. *Engineering, Construction and Architectural Management*,
14(6), 597–617.

Albanese, R. (1994) Team-building process: key to better project results. *Journal of
Management in Engineering, ASCE*, **10**(6), 36–44.

Arntzen, B.C., Brown, G.G., Harrison, T.P. & Trafton, L.L. (1995) Global supply
chain management at digital equipment corporation. *Interfaces*, **25**(1), 69–93.

Banwell, H. (1964) *Report of the Committee on the Placing and Management of
Contracts for Building and Civil Engineering Works*. HMSO, London.

Bennett, J. & Jayes, S. (1995) *Trusting the Team: the Best Practice Guide to Partnering
in Construction*. Centre for Strategic Studies in Construction/Reading Construction
Forum, Reading.

Bennett, J. & Jayes, S. (1998) *The Seven Pillars of Partnering*. Reading Construction
Forum, Reading.

BERR (Department of Business Enterprise and Regulatory Reform (2008) *Innovation
in Construction Services*. BERR.

Black, C., Akintoye, A. & Fitzgerald, E. (2000) An analysis of the success factors and
benefits of partnering in construction. *International Journal Project Management*,
18(6), 423–434.

Black, C. (1998) *Partnering in Construction: Opportunities and Reward*. MSc thesis,
Construction   Management   dissertation,   Glasgow   Caledonian   University,
Glasgow.

Bresnen, M. & Marshall, N. (2000) Partnering in construction: a critical review of issues,
problems and dilemmas. *Construction Management and Economics*, **18**, 229–237.

Burbridge, J. L & Falster, P. (1993) Reducing delivery times for OKP products.
Production Planning and Control, **4**(1), 77–83.

Byatt, I. (2001) *Delivering Better Services for Citizens, A review of local government
procurement in England*. Department for Transport, Local Government and the
Regions, June.

Burnes, B. & New, S. (1996) *Strategic Advantage and Supply Chain Collaboration*.
Manchester School of Management, UMIST, Manchester.

Chadwick, T. & Rajagopal, S. (1995) *Strategic Supply Management*. Butterworth-
Heinemann Ltd, London. 92–117.

Cheatham, W. (2004) *Surety Bonding Joint Ventures*. Construction Executive Surety
Bonding, November.

Cherns, A.B. & Bryant, D. (1984) Studying the client's role in construction management. *Construction Management and Economics*, **2**(2), 177–184.

CIB (2011) International Council for Research and Innovation in Building and
Construction.   Available   from   http://www.cibworld.nl/site/home/index.html
(accessed June 2011).

Construction Industry Institute Australia (CIIA) (1996) *Partnering: Models for
Success*. Research Report No. 8., Sydney.

Construction Best Practice (2003) *Partnering Fact Sheet*. Available from http://www.constructingexcellence.org.uk/ (accessed June 2011).

Cook, E.L. & Hancher, D.E. (1990) Partnering: contracting for the future. *Journal of Management in Engineering, ASCE*, 6(4), 431–447.

Crouse, H.J. (1991) The power of partnerships. *The Journal of Business Strategy*, **November/December**, 4–8.

Das, T. & Teng, B. (2000) A resource-based theory of strategic alliances. *Journal of Management*, **26**(1), 31–61.

Das, T. & Teng, B. (2001) Trust, control and risk in strategic alliances: an integrated framework. *Organisation Studies*, **22**(2), 251–283.

de Rond, M. (2003) *Strategic Alliances as Social Fact*. Cambridge University Press, Cambridge.

Dekker, H.C. (2003) Value chain analysis in inter-firm relationships: a field study. *Management Accounting Research*, **14**, 1–23.

Egan, J. (1998) *Rethinking Construction*. Department of the Environment, Transport and the Regions, London.

Egan, J. (2003) *Accelerating Change*. A report of Strategic Forum for Construction for Rethinking Construction, London.

Fairclough, J. (2002) *Rethinking Construction Innovation and Research*. A Review of Government R&D Policies and Practices, UK Department of Trade and Industry/February.

Gerwin, D. (2004) Coordinating new product development in strategic alliances. *Academy of Management Review*, **29**(2), 241–257.

Green, S.D. (1999) Partnering: the propaganda of corporatism? In: ed. Ogunlana, S.O., *Profitable Partnering in Construction Procurement, CIB W92 and CIB TG23 Joint Symposium*. E&FN Spon, London, 3–14.

Hamza, A., Djebarnu, R. & Hibberd, P. (1999) The implications of partnership success within the UK construction industry supply chain. In: ed. Ogunlana, S.O., *Profitable Partnering in Construction Procurement, CIB W92 and CIB TG23 Joint Symposium*, 39–46.

Harvey, R.C. & Ashworth, A.A. (1993) *The Construction Industry of Great Britain*. Butterworth-Heinemann Ltd, London.

Higgin, G. & Jessop, N. (1965) Communications in the Building Industry. Tavistock Publications, Tavistock Institute of Human Relations, London.

HM Treasury (2003) *PFI: Meeting the investment challenge*. HMSO, London.

HM Treasury (2000) *Public Private Partnerships: The Government's Approach*. HMSO, London.

Holti, R., Nicolini, D. & Smalley, M. (1999) *Prime Contractor Handbook of Supply Chain Management*. Sections 1 & 2. Crown, London.

Inkpen, A.C. & Ross, J. (2001) Why do some strategic alliances persist beyond their useful life? *California Management Review*, **44**(1), 132–148.

Institute of Public Policy Research (IPPR) (2002) *Building Better Partnerships*, IPPR, London. 41.

Ireland, R.D., Hitt, M.A., & Vaidyanath, D., 2002. Alliance management as a source of competitive advantage. *Journal of Management*, 28, 413–446.

Kalwani, M.U. & Narayandas, N. (1995) Long-term manufacturer-supplier relationships: Do they pay off for supplier firms? *Journal of Marketing*, 59(1), 1–16.

Khalfan, M.M.A. & McDermott, P. (2006) Innovating for supply chain integration within construction. *Construction Innovation*, **6**, 143–157.

Lamming (1993) *Beyond Partnership: Strategies for Innovation and Lean Supply*. Prentice Hall International (UK) Ltd, London. 168–175.

Lenard, D. & Mohsini, R. (1998) Recommendations from the organisational workshop. *Procurement – The Way Forward, Proceedings of CIB W92 Conference*, University of Montreal, CIB Publication 203.

Lewis, J.D. (1995) *The Connected Corporation*. Free Press, New York. 1–13.

Latham, M. (1994) *Constructing the Team: Joint Review of Procurement and Contractual Arrangements in the United Kingdom Construction Industry*. HMSO, London.

Lorange, P. & Roos, J. (1991) Why some strategic alliances succeed and others fail. *The Journal of Business Strategy*. **January/February**, 25–30.

Luck, R.A.C. & Newcombe, R. (1996) The case for the integration of the project participants' activities within a construction project environment. In: ed. Langford, D.A. & Retik, A., *The Organisation and Management of Construction: Shaping Theory and Practice*, vol. 2. E & FN Spon, London. 458–470.

MacBeth, D.K. & Ferguson, N. (1994) *Partnership Sourcing*. Pitman Publishing, London. 96–140.

Macneil (1981) Economic analysis of contractual relationship: Its shortfall and the need for a rich classificatory apparatus. *Northwestern University Law Review*, 77, 1022.

Masterman, J.W.E. (1992) *An Introduction to Building Procurement Systems*, 1st edn. E & FN SPON, London.

McDermott, P., Khalfan, M.M.A. & Swan, W. (2004) An exploration of the relationship between trust and collaborative working in the construction sector. *Construction Information Quarterly*, 6, 140–146.

McGeorge, D. & Palmer, A. (1997) *Construction Management: New Directions*. Blackwell Science Ltd, Oxford.

Mohsini, R. & Davidson, C.H. (1989) Building procurement – key to improved performance. In: ed. Cheetham, D., Carter, D., Lewis, T. & Jagger, D.M., *Contractual procedures for building: proceedings of the international workshop*, 6–7 April, University of Liverpool, Liverpool.

Moore, C., Mosley, D. & Slagle, M. (1992) Partnering guidelines for win–win project management. *Project Management Journal*, **22(1)**, 18–21.

Naoum, S. (2003) An overview into the concept of partnering. *International Journal of Project Management*, **21**, 71–76.

National Audit Office (2002) Managing the relationship to secure a successful partnership in PFI projects. Report by the Controller and Auditor General, HC 375 Session 2001–2002.

NEDO (1985) *Thinking about Building – A Successful Business Consumer's Guide to Using the Construction Industry*. Building Economic Development Committee, London.

NEDO (1991) *Partnering: Contracting without Conflict*. HMSO, London.

Nielson, C. C. (1996). An empirical examination of switching cost investments in business-to-business marketing relationships. *The Journal of Business and Industrial Marketing*, **11**, 38–60.

Noordewier, G.N., John, G. & Nevin, J.R. (1990) Performance Outcomes of Purchasing Arrangements in Industrial Buyer-vendor Relationships. *Journal of Marketing*, 54(October), 80–93.

Nooteboom, B. (1999) *Inter-firm Alliances – Analysis and Design*. Routledge, London.

OGC (2010) *Faster Procurement, Procurement Process Study Report*. Available from http://www.ogc.gov.uk/documents/Faster_Procurement_Process_Study_Report.pdf (accessed 15 August 2010).

O'Brien, M.J., Fischer, M.A. & Jucker, J.V. (1995) An economic view on project. *Construction Management and Economics*, **13**(5), 93–400.

Pain. J. (1993) Design and Build compared with traditional contracts. *Architects' Journal*, **November**, 34–35.

Pries, F. & Janszen, F. (1995) Innovation in the construction industry: the dominant role of construction. *Construction Management and Economics*, **13**(1), 43–51.

Rowlinson, S. & McDermott, P. (1999) *Procurement Systems: A Guide to Best Practice in Construction*. E&FN Spon, London.

Scottish Executive (2004) *Construction Works Procurement Guidance*. Client Pack.

Sidwell, A.C. & Budiawan, D. (2001) The significance of the tendering contract on the opportunities for clients to encourage contractor-led innovation. *Construction Innovation*, **1**, 107–116.

Taylor, S. (1999) Can partnering work for you? *Contract Journal*, **January**, 18–19.

Winch, G. (1989) The construction firm and the construction project: a transaction cost. *Construction Management and Economics*, 7, 331–345.

Wood, G. & Ellis, R. (2005) Main contractor experiences of partnering relationships on UK construction projects. *Construction Management and Economics*, **23**, 317–325.

Concurrent Engineering in Construction

Chimay J. Anumba and John M. Kamara

12.1 Introduction

This chapter focuses on the concept of Concurrent Engineering (CE) and its contribution to construction innovation and process improvement. CE has been widely implemented in the manufacturing industry with notable benefits in reducing product development times, and improvements in product quality through the early consideration of life-cycle issues and the systematic incorporation of customer requirements in the product development process. The benefits arising from the implementation of CE in manufacturing, and the realisation that construction can be considered as a manufacturing process, led to calls for the adoption of CE in construction. The goal was to exploit the benefits of CE to bring about innovation and process improvements in the construction industry, the problems of which relate ostensibly with fragmentation in the design and construction process.

The main aim of this chapter is to provide an overview of the key issues and technologies needed for the adoption of CE in construction, and the implications for construction innovation and process improvement in the industry. It begins with an overview of the concept of CE, highlighting the benefits garnered in the manufacturing sector, and describes the applicability and implementation considerations of this along with organisational and technological enablers for the adoption of CE in Construction (CEC). The chapter concludes with a discussion on the contribution of CEC to construction innovation and process improvement.

Construction Innovation and Process Improvement, First Edition.
Edited by Akintola Akintoye, Jack S. Goulding and Girma Zawdie.
© 2012 Blackwell Publishing Ltd. Published 2012 by Blackwell Publishing Ltd.

12.2   The Concept of Concurrent Engineering

Concurrent Engineering was coined in the late 1980s to explain the systematic method of concurrently designing both the product and its downstream production and support processes (Evbuomwan and Anumba, 1995; Huovila *et al.*, 1997). CE was proposed as a means to minimise product development time (Prasad, 1996). This was necessitated by changes in: manufacturing techniques and methods, management of quality, market structure, increasing complexity of products, and demands for high quality and accelerated deliveries at reduced costs. These changes resulted in a shift in corporate emphasis with the result that the ability to rapidly react to changing market needs and time-to-market became critical measures of business performance (Constable, 1994: Thamhain, 1994). The earliest definition of CE by Winner *et al.* (1988) refers to 'integrated, concurrent design of products and their related processes, including manufacture and support'; with the ultimate goal of customer satisfaction through the reduction of cost and time-to-market, and the improvement of product quality. CE embodies two key principles: integration and concurrency. Integration here relates to the process and content of information and knowledge, between and within project stages, and of all technologies and tools used in the product development process. Integrated concurrent design also involves up-front requirements analysis by multidisciplinary teams and the early consideration of all life-cycle issues affecting a product. Concurrency relates to the way that tasks are scheduled, and the interactions between different actors (people and tools) in the product development process. Table 12.1 identifies a concurrency matrix that can be used to assess the level of 'concurrency' within a project team (Prasad *et al.*, 1993).

The rows represent modes of operation and the columns represent the possible work-group configurations. A co-operating user is '*a person* who completes the work left unfinished by previous users' (Prasad *et al.*, 1993).

Table 12.1   Matrix of concurrency (Prasad *et al.* 1993).

No.	Modes of interactions	Single user	Work-Group configurations		
				Simultaneous users	
			Cooperating users	Different versions	Same version
1	Access own products' interaction tools or applications (PITA)	Sequential Engineering (SE)	SE	SE	SE
2	Run against their own data	SE	SE	SE	SE/CE
3	Access PITA belonging to other work-groups	SE/CE	CE	CE	CE
4	Access data belonging to other work-groups	CE	CE	CE	CE
5	Access both PITA and data from other work-groups	CE	CE	CE	CE

Simultaneous users refer to other members of the project team who may access 'the same design, tool or application concurrently, or …different versions of product information tools or applications (PITA) at the same time' (Prasad *et al.*, 1993). The level of concurrency depends on the type of interactions, and this increases as one moves from top to bottom and from left to right. It is observed that some situations are described as both sequential and concurrent, when simultaneous users run their own data, and when a single user accesses the Product Interaction Tools and Applications (PITA) belonging to other work groups (Table 12.1). The interaction is sequential if two or more users cannot edit and save changes to a document until another user has finished with it, even though they may be working in parallel.

The key features of CE can therefore be summarised to include:

- concurrent and parallel scheduling of all activities and tasks as much as possible;
- integration of product, process and commercial information over the life cycle of a project;
- integration of life-cycle issues during project definition (design);
- integration of the supply chain involved in delivering the project through effective collaboration, communication and coordination;
- integration of all technologies and tools utilised in the project development process (e.g. through inter-operability).

12.3   Implementation of CE

CE is a philosophy that contains (or is implemented by) several methodologies. The attainment of 'integrated, concurrent design' requires a variety of enablers, which include tools (software applications), techniques, technologies and support structures. These enablers can be generic and can be used to support other concepts. The extent to which these principles are implemented determines the level to which the objectives of CE (e.g. shorter lead times) are realised. Figure 12.1 shows a framework for the implementation of CE with respect to the inter-relationships between the goals, objectives, strategies and tactics (tools and technologies) for CE.

The goal of fully satisfying the customer and operating a competitive business is made possible by shorter lead times (time-to-market), lower costs and high-quality products. These in turn, are achieved through rigorous requirements analysis, early consideration of all life-cycle issues affecting a product, integrated and concurrent product development, and the use of multi-disciplinary teams and other strategies. The overall CE framework is facilitated by various tools and techniques, which include quality function deployment (QFD), agent and knowledge based tools, computer aided design (CAD) and computer aided manufacturing (CAM) tools, and other relevant tools (Evbuomwan and Sivaloganathan, 1994; Prasad, 1998). These CE enablers (tools and techniques) can be grouped into two broad categories, which are inter-related: organisational and technological. Organisational enablers provide the framework for people and machines to work

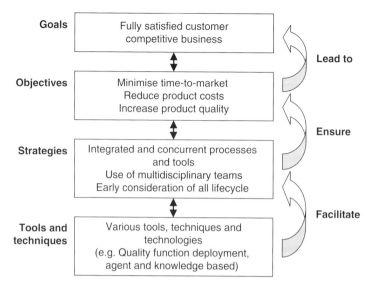

Figure 12.1   Framework for understanding CE (Kamara *et al.*, 2000).

Table 12.2   Support requirement matrix for concurrent engineering (Harding and Popplewell, 1996).

	Levels		
Dimensions	Organizational	Team	Individual
Distribution	Move information between multiple sites.	Reduce remoteness and promote exchange of information between team members at different physical locations.	Make information available to individuals.
Heterogeneity	Support organizations to achieve different missions.	Support Project Teams to achieve different goals.	Support Individuals to perform different jobs.
Autonomy	Discourage multiple individual stores of information.	Support team members to work as individuals, or as a group, and transitions between these two types of working.	Support individual's preferred manner of working.

'concurrently'. This includes facilitating the work of multi-disciplinary teams, involving all relevant parties in the product development process, and managerial/technological support for organisational, team and individual levels of working. Technological enablers facilitate concurrent working within organisations. They include all the information and communications technology (ICT) and computer-based applications required for integration, concurrent working, communication and collaboration. Table 12.2 summarises the kind of support required for CE at the organisational, team and individual levels with respect to distribution, heterogeneity and autonomy.

Table 12.3   Savings in product development time using concurrent engineering (Prasad, 1996).

Company	Product	Best development time (months)			Major contributing factors*
		Before CE	After CE	Reduction	
ABB	Switching Systems	48	10	79%	CMS
AT & T	Phones	24	12	50%	ACM
British Aerospace	Aeroplanes	36	18	50%	MS
Digital Equipment	Personal Computers	30	12	60%	ACMS
Ford	Cars	60	42	30%	—
General Motors	Engines	84	48	43%	MS
GM/Buick	Cars	60	41	32%	MS
Goldstar	Telephone Systems	18	9	50%	CM
Honeywell	Thermostats	48	12	75%	MS
Honda	Cars	60	36	40%	—
Hewlett-Packard	Printers	54	22	59%	ACMS
IBM	–	48-50	12-15	70-75%	ACM
Motorola	Mobile Phones	36	7	81%	ACM
Navister	Trucks	60	30	50%	MS
Warner Electric	Clutch Brakes	36	9	75%	M
Xerox	Copiers	60	24	60%	ACM
Xerox	–	53	36	32%	ACM

Legend:
*M: Multi-functional Teams
*A: Analytical Methods and Tools'
*C: Computer Integration
*S: Suppliers in the Project Team

12.4   Benefits of Concurrent Engineering

The benefits of CE derive from the fact that it is predominantly focused on the design phase (Koskela, 1992), which determines and largely influences the overall cost of a product; as much as 80% of the production cost of a product can be committed at the design stage (Dowlatshasi, 1994). Therefore, addressing all life-cycle issues up front at the design stage to ensure that the design is 'right-first-time' should lead to cost savings, products that precisely match customers' needs, and those which are of a high quality. The adoption of CE can also result in reductions in product development time of up to 70% (Madan, 1993; Carter, 1994; Constable, 1994; Dowlatshahi, 1994; Evbuomwan *et al.*, 1994; Frank, 1994; Nicholas, 1994; Thamhain, 1994; Smith *et al.*, 1995; Prasad, 1996). Table 12.3 provides a summary of documented savings in product development time through the adoption of CE by particular companies in the manufacturing sector.

12.5   CE in Construction

The construction industry has largely depended on collaborative working between a number of professional teams brought together, often in an *ad hoc* manner, for the translation of its clients' requirements into physical

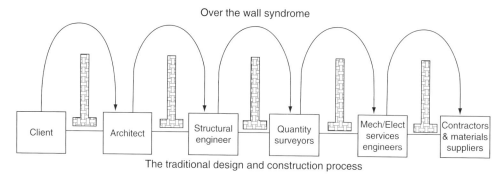

Figure 12.2   The Over the Wall Approach (Evbuomwan and Anumba, 1998).

constructed facilities. Whilst this has entrenched the practice of collaborative working, it has also reinforced traditional disciplines to the extent that, on many projects, an adversarial environment prevails and the fundamental ethos of collaboration is not fully evident. This has resulted in numerous problems for the construction industry, with the result that the industry is now highly inefficient compared to other sectors (Anumba *et al.*, 1995) and is characterised by serial processes, as illustrated in the Figure 12.2, which depicts the 'over the wall' syndrome (Evbuomwan and Anumba, 1998). Competitive pressures from within the industry as well as external political, economic and other considerations are now forcing the industry to re-examine and improve its *modus operandi*. Therefore, CE within the construction sector seeks to optimise the design of a facility and its construction process in order to achieve reduced lead times, improve quality and cost by the integration of design, fabrication, construction and erection activities, and maximise concurrency and collaboration in working practices (Evbuomwan and Anumba, 1995, 1996).

The success of CE in manufacturing is one of the main motivators for adopting this in construction (de la Garza *et al.*, 1994; Anumba and Evbuomwan, 1995; Huovila and Serén, 1995; Evbuomwan and Anumba, 1995, 1996; Hannus *et al.*, 1997; Kamara *et al.*, 1997; Love and Gunasekaran, 1997; Anumba *et al.*, 1999). It is based on the assumption that because construction can be considered as a manufacturing process, concepts which have been successful in the manufacturing industry could bring about similar improvements in the construction industry. Furthermore, the goals and objectives of CE directly address the challenges that currently face the construction industry.

12.5.1   Construction as a Manufacturing Process

The interest in modelling construction as a manufacturing process is primarily based on the similarities between the two industries, and the assumption that aligning business processes of the construction industry to those of the manufacturing industry will significantly improve its competitiveness (Sanvido and Medeiros, 1990; Anumba *et al.*, 1995; Anumba and

Table 12.4   Rationale for adopting concurrent engineering in construction (Anumba et al, 2007).

Need for change in construction	Goals and principles of CE
The need for change in construction is brought about by the uncompetitive nature of the industry, and the inability to fully satisfy its clients with respect to costs, time and value.	The goals and objectives of CE (Figure 12.1) include: customer satisfaction, competitive business, reduction of product development time and cost, improvement of quality and value.
Integration of the construction process is seen as one of most important strategies to improve the notoriously fragmented construction industry.	The use of CE facilitates the integration of the members of the product development team, and the manufacturing process, thereby improving the product development process
Emerging strategies for improving the construction process are inadequate; they only address one aspect of the problem, resulting in 'islands of automation' as in the case of computer-integrated construction strategies.	As an amalgam of other methodologies, tools and techniques, CE provides a framework for not only integrating the construction process, but also the various tools and technologies that are used in the process.

Evbuomwan, 1995; Crowley, 1996; Egan, 1998). Both the manufacturing and construction industries:

- produce engineered products that provide a service to the user;
- are involved in the processing of raw materials and the assembly of many diverse pre-manufactured components in the final products;
- utilise repeated processes in the design and production of their products;
- experience similar problems, e.g. the high cost of correcting design errors due to late changes, poor resource utilisation and inadequate information management.

The differences between manufacturing and construction with regard to the location of production activities, and the production of 'one-off' facilities in construction, as opposed to mass production in manufacturing, have led to suggestions that the two industries are profoundly different (Sanvido and Medeiros, 1990; Crowley, 1996; Egan, 1998). However, the parallel between construction and manufacturing is not with respect to repeated (or mass-produced) *products*, but rather to the repeated *processes* that are involved in the design and production of products in both industries. The implication of this is that developments in manufacturing, such as CE which have led to improvements in productivity (as a result of *process* reengineering), can be used in construction.

12.5.2   Relevance of CE Principles to Construction

Another justification for the adoption of CEC is because the goals and strategies (principles) of this directly address the problems facing the construction industry. Table 12.4 provides a summary of how the needs in construction (previously discussed) could be addressed by CE. This pairing of needs versus capabilities in support of CEC is further reinforced by the

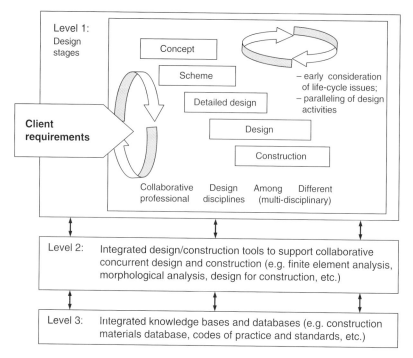

Collaborative   Design   Among   Different
professional   disciplines   (multi-disciplinary)

Level 2:   Integrated design/construction tools to support collaborative
concurrent design and construction (e.g. finite element analysis,
morphological analysis, design for construction, etc.)

Level 3:   Integrated knowledge bases and databases (e.g. construction
materials database, codes of practice and standards, etc.)

Figure 12.3   Simplified CLDC framework (Kamara *et al.*, 2000).

fact that existing practices in the construction industry, which are similar to CE, could readily facilitate its successful implementation in construction.

Given these espoused benefits, CE has considerable potential in construction. In this respect, its capacity to provide an effective framework for integrating and improving the construction process is now also widely acknowledged in the industry (Anumba and Evbuomwan, 1997; Egan, 1998). From both the context in which it evolved (manufacturing) and its inherent features, CE can be matched to the construction process. Evbuomwan and Anumba (1998) developed a Concurrent Lifecycle Design and Construction (CLDC) framework as a basis for the implementation of CE in the construction industry. A simplified version of the CLDC framework is presented in Figure 12.3. However, CE implementation in construction needs to be tailored to the particular needs of the construction industry; this is discussed in the next section.

12.5.3   Implementation of CEC

The construction industry, otherwise referred to as the Architecture, Engineering and Construction (AEC) industry, is typically organised around projects that are paid for by clients who are technically not part of the industry. Construction projects are also delivered by many firms, unlike the manufacturing industry, where a greater proportion of the skills required are often held within one organisation. Achieving 'true concurrency' in AEC (Table 12.2) for example, might require users from one firm (e.g. structural engineering consultants) to access both PITA (product interaction tools and applications – e.g. CAD workstations) and data from other work groups that might be located in other firms.

Because of the project-based nature of the AEC industry, CE implementation in construction should be considered at both the project and organisational levels (i.e. individual consulting/contracting firm). At the organisational level, it is relatively easy to devise strategies that reflect the requirements set out in Tables 12.1 and 12.2, which are ostensibly based on a single-organisation model. At the project level, 'concurrency' and 'integration' should focus only on issues pertaining to the project. The matrix of concurrency in Table 12.1 is also applicable at this level, but rather more difficult to implement; some aspects of Table 12.2 (e.g. organisational, team and individual support for heterogeneity) may not be applicable, since a specific project can be considered as a homogenous entity.

Other challenges for CE in AEC include the linkages between organisational (i.e. firm level) support structures and project-level support requirements. For example, someone operating at the organisational level may store data on different projects in their PITA; therefore, access to information relating to a specific project by someone outside the organisation could become problematic. Another challenge relates to the role of clients who govern the nature and form of the project organisation, through procurement and contractual strategies adopted, and in some cases, even the range of technologies that can be used. The fact that the project and organisational levels are influenced by different (and sometimes) opposing forces (i.e. client and industry), poses challenges for the linkages between the two.

Construction organisations intending to adopt CE therefore need to address a number of key issues in order to ensure that they maximise the potential benefits. It is particularly important that organisations undertake a readiness assessment exercise, to ensure that CE implementation is tailored to its specific objectives and business strategy. Some of the main considerations in CE implementation include the following (Anumba *et al.*, 2007):

- the availability of a robust project development process, which is documented, adaptable, periodically evaluated, and facilitates concurrency;
- the existence of an organisational framework and policies that support both individuals and teams, and enables the project development process to be controlled;
- the need for a clear business strategy that outlines an organisation's objectives with regard to interaction with clients and other project team members;
- the agility of an organisation and its capacity to respond quickly to changes in its operating environment;
- the appropriateness of strategies for team formation and operation, including the need to ensure that team members understand their roles and work towards a common purpose;
- appropriate selection and delegation of authority to team leaders;

- the need for appropriate guidelines for maintaining team discipline;
- the provision of training to enable team members to fulfil their roles and the institution of reward structures that recognise both individual and team achievements;
- maintaining focus on the client's requirements and having the capacity to respond to any changes that might occur;
- the institution of appropriate procedures and policies for quality assurance;
- the development of designs that are flexible, robust and informed by the client's requirements;
- the availability of appropriate technologies to facilitate information exchange and knowledge sharing;
- the use of an integrated project model and systems that facilitate integration between members of a project team;
- use of common hardware and software platforms to ensure the seamless exchange of information on projects;
- use of standard and proven information and communications technologies.

The above list is not exhaustive, but includes the majority of issues that need to be considered.

12.6   Critical Enablers of CE Adoption in the Construction Industry

There are several developments in the construction industry that could enable the adoption of CE. For simplicity, these can be broadly categorised into organisational enablers and technological enablers.

12.6.1   Organisational Enablers

The term 'organisational enablers' is used here to describe those elements that relate either to an individual organisation or project organisation, and which have an influence on CE with respect to processes, procurement, client and organisational readiness for CE.

In this respect, Khalfan *et al.* (2007) identified readiness assessment approaches for CE implementation within the construction industry. Following a detailed study of existing readiness assessment tools, they developed a new tool – BEACON (Benchmarking and Readiness Assessment for CEC) – specifically for the construction industry. Based on a synthesis of existing tools, the model enables construction sector organisations to assess themselves on a range of critical success factors grouped under four key elements, Process, People, Project and Technology. The BEACON model was subsequently used to assess different sectors of the construction supply chain (Khalfan *et al.*, 2007) and demonstrated its utility in facilitating a targeted and effective implementation of CE at the level of individual organisations as well as within project organisations.

Another important enabler for CE implementation in the construction industry is the availability of tools for capturing and incorporating the 'voice of the client' in construction projects. Such tools can facilitate both a CE approach and a focus on the needs of the client, which is critical to delivering and exceeding customer/client expectations – a key aspect of CE. Kamara and Anumba (2007) describe a model (the Client Requirements Processing Model) and an associated software tool, ClientPro, which facilitates the capture and translation of client requirements into actionable design attributes. The growth in the number of so-called 'expert clients' who procure construction services on a routine basis also augurs well for the implementation of CE, as these clients are well placed to understand and implement CE, as well as reap the attendant benefits. In addition, the effective implementation of CE in the construction industry is dependent on a move away from adversarial procurement and contractual arrangements that often reinforce the fragmentation within the industry. Thus, the move towards more collaborative and integrated procurement and contractual methods, such as Design and Build and Integrated Project Delivery (AIA, 2007), is a positive one from a CE perspective. Bowron (2002) went further and developed a tailor-made procurement process for CEC. Walker (2007) also sees opportunities for the adoption of CE in practices such as partnering and framework agreements (FA's), which enable both clients and industry practitioners to adopt a more long-term collaborative perspective in projects.

The increasing focus within the industry on process modelling, process mapping, and the standardisation of processes is seen as a key enabler for CE adoption. Kagioglou *et al.* (2007) view process management as crucial to the adoption of CEC. They outline the key principles of an improved holistic process for CEC that allows for continuous learning through feedback mechanisms. This approach is encapsulated in Process Protocol (Kagioglou *et al.*, 2000) and offers an appropriate industrial platform for the adoption of CEC projects. Other organisational enablers include the Integration Toolkit, which was developed by the Strategic Forum for Construction in the UK. The toolkit provides guidance in setting up integrated project teams and supply chains. This is based on the principles of committed leadership, a change in culture and values and the adoption of appropriate processes, tools and commercial arrangements (SFC, 2003).

12.6.2   Technological Enablers

Numerous technological developments have the capacity to facilitate CE implementation in construction. These include ontologies and standards-based approaches to inter-operability, integrated product and process modelling, document management systems, 4D CAD, telepresence and end-user support tools. In this respect, ontologies provide a means of describing common concepts within a given domain. They are now increasingly being developed for use in the construction industry and, along with emerging inter-operability standards such as IFCs, XML, BIM standards, etc, provide a solid basis for the integration of tools and technologies within a CE framework. Pouchard and Cutting-Decelle (2007) provide a discussion on the role of interoperability standards in supporting the exchange of information and the collaborative teamwork that are integral components of CE. Given this, many modelling systems now enable the integration of product and process information. This is also an important enabler for CEC, as the co-evolution of the process information in line with the product representation is vital for the integration of design and construction. Kimmance (2002) developed an integrated product and process model that provided dynamic linkages between product representations and the associated production process; while Anumba *et al.* (2007a) presented a practical example within the context of the ProMICE project.

The primary medium for communication and exchange of project information in the construction industry remains predominantly through formal documentation. As such, advanced electronic document management systems have a key role to play in the implementation of CEC. In addition to geometric information, which is now increasingly being exchanged or communicated through 3D/4D models, construction projects involve many other documents. Amor and Clift (2007) show that the proper management of documents not only provides information about all aspects of a project, but also facilitates the effective coordination of project activities and processes within a CE framework. Furthermore, developments in CAD and BIM also serve as important enablers for CE implementation in construction. For example, BIM facilitates the digital generation of a computer model of a building, which contains, 'precise geometry and relevant data needed to support the construction, fabrication and procurement activities needed to realise the building' (Eastman *et al.*, 2008). The ability to visualise designs in 3D, 4D and nD provides project teams with the capacity to purposefully evaluate the potential downstream implications of early decisions. This makes it easier to 'get it right first time', thereby reducing expensive late changes, minimising rework and improving the quality of the end product. Staub-French and Fischer (2007) discussed the use of 4D models for effective coordination and project planning during the pre-construction phase of a project, and illustrated this with a case study. Integrated computing environments have been demonstrated in, for example, the Heathrow Terminal 5 project, where a Single Model Environment was created to facilitate collaborative working on the project (Lion, 2004).

An important tenet of CE is collaborative teamwork. While this can be easily achieved when team members are collocated, it becomes problematic when they are geographically dispersed – sometimes across time zones, continents and cultures. The emergence of distributed collaboration tools and telepresence systems has made this aspect of CE implementation more readily implementable. Telepresence provides the means whereby collaborating team members can be virtually collocated within a given 3D environment in which they are able to interact with one another or with virtual objects that are also present in the environment (Anumba and Duke, 1997). Specific areas in which Telepresence could support CE implementation in construction include facilitating multi-disciplinary teams, integrating communication facilities with design tools and supporting project team communications with the use of collaborative virtual environments (Anumba and Duke, 2007). Scherer and Turk (2007) see support for teamwork as a vital enabler for CEC. They developed an open, human-centred web-based collaboration environment (otherwise known as 'project extranets'), maintained by a service provider, which supports CE while working on multiple projects simultaneously and offers easy access to specialised engineering services distributed over the web on a rental basis.

12.6.3   Other Core Enablers

In addition to organisational and technological enablers, there are also several other developments that make CE implementation far more feasible in many construction project settings. However, only a few examples will be provided here. The recognition that 'Best Value' is a more appropriate basis for tender decisions than 'Least Cost' is one specific example. This has enabled government agencies and other public bodies to dispense with the outmoded compulsory competitive tendering practices that usually insisted on least cost bidders. The move towards public-private partnerships (PPPs) and novel project financing models all support the formation of the entire project team early in the project life cycle, and often require collaborative working at those early stages to ensure the best possible outcome for the team and the client. The most prevalent form of PPP is the Private Finance Initiative (PFI), where the private sector funds, builds and owns public assets and the public sector purchases the flow of services from the assets through a long-term commitment (Group, 1997; Li *et al.*, 2005). While the key motivation behind PFI projects is to minimise the Public Sector Borrowing Requirements of governments (Dixon *et al.*, 2005), the consideration of life-cycle issues early on in the project development process reflects a key principle of CE.

12.7   Overcoming Barriers to CE Adoption

There are still many barriers to the uptake of CEC and consideration needs to be given to overcome these to ensure successful CE implementation (Anumba *et al.*, 2007b). In this regard, some of the main barriers that need to be overcome include:

- fragmentation and traditional adversarial relationships between team members;
- lack of trust between team members;
- lack of a recognised stakeholder for overall process improvements;
- traditional adherence (usually by government bodies) to a 'lowest bidder' model of tendering rather than best value;
- conservative nature of the construction industry;
- low levels of awareness and understanding of the principles and benefits of CE.

These barriers can be addressed in a variety of ways, but by far the most promising approaches include the following (Anumba *et al.*, 2007b):

- adoption of collaborative procurement and contractual arrangements that promote collaborative working and knowledge sharing;
- improvements in education and training for both new entrants and established practitioners in the construction industry;
- provision of incentives for collaborative working;
- use of demonstration projects with innovative clients to showcase the benefits of the CE approach;
- changes in government regulations, particularly with regard to compulsory competitive bidding (which is no longer required by many public bodies);
- adoption of established information and communications technologies (e.g. groupware, 3D modelling, Web-based project collaboration systems, etc) that facilitate collaborative working; and
- establishment of strategic alliances and partnerships.

Current industry interest in the adoption of integrated project delivery methods and the enabling technologies (such as BIM) is strong evidence that the basic principles of collaborative and CE principles are being surreptitiously embedded in industry practice. This bodes well for the future of the AEC sector, but more needs to be done to ensure the long-term sustainability of CEC.

12.8   Benefits of CE to the Construction Industry

The benefits of CE to the AEC sector derive from similar benefits achieved in other industry sectors, while others are based on the anecdotal evidence from construction organisations and project teams that have implemented aspects of CE. Nevertheless, it is useful here to reiterate these benefits (Anumba *et al.*, 2007b):

- improved quality of facilities relative to cost;
- reduced duration of capital projects;
- enhanced efficiency and productivity due to reduction in rework;
- better coordination and management of the construction process;
- better informed decision making and coordination, with decisions taken at the right time and by the right person(s);
- improved competitiveness of the construction industry relative to other industry sectors;
- better project definition due to more time provision at the early project stages;
- improved integration of life-cycle considerations;
- enhanced collaboration and teamwork between members of the project team;
- more robust information exchange between team members and across the stages in the project delivery process;
- improved quality of the end product – the constructed facility;
- greater client satisfaction, given the improved focus on the client's requirements and the delivery of greater value;

- waste reduction;
- reduced scope for conflicts and litigation;
- greater profits for construction companies due to the ability to control more aspects of the project, reducing overall construction time, and improved interaction with designers and other team members;
- improved safety and 'uptime' for existing operations.

Realisation of the above potential benefits therefore depends largely on the effectiveness of CE implementation within the whole construction supply chain, rather than in individual firms. However, it is important to note that there is scope for all participants in the construction process to benefit from this.

12.9   Conclusion

This chapter presented the key principles of CE and its applicability to the construction industry. It argues that, while the fundamental principles have much to offer the sector, there is a need for the construction industry, which is primarily project-based, to adopt these in a way that maximises their potential benefits. The main considerations for the implementation of CEC were outlined, along with the importance of adopting a multi-faceted approach, in order to leverage enablers and mitigate barriers to CE adoption. The implementation of CEC represents a new way of working, as it promotes a change in culture and practice with respect to integration (of tools, processes, teams, etc) and the upfront consideration of life-cycle issues in the project development process. This in essence represents innovation in the construction process with respect to bringing about change through the introduction of new methods and techniques in the development and implementation of construction projects. Innovation is not only restricted to the introduction of new methods and working practices. It is also '…aimed at improving efficiency in some way – whether the goal is shortened construction time, improved quality of product, reduced rework or rectification work during construction or any of a number of other possible outcomes' (Marosszeky, 1999). The goal and benefits of CE are in keeping with the general aims of innovation in construction. Indeed, the key motivation for the adoption of CEC was fuelled by the need for change in the industry due in part to its uncompetitive nature and its inability to satisfy its clients with respect to costs, time and value (Table 12.4).

Finally, it should be pointed out that while there are several enablers for the adoption of CEC, these tools and techniques may not necessarily be marketed (or explicitly described) as 'Concurrent Engineering' tools. However, these enablers can facilitate the realisation of the core CE strategies (e.g. integrated and concurrent processes). Notwithstanding this, there is currently no single tool that supports every aspect of CE at every level, and a precise measurement (or assessment) of its level of uptake in the industry is therefore unrealistic. However, the concept of CE embodies (in a comprehensive and holistic way) the aspirations for an integrated and efficient construction process – which are key ingredients of the quest for innovation and process improvement in the industry.

12.10   Acknowledgement

Much of the content of this chapter is reproduced, with permission, from Chapter 1 – 'Introduction to Concurrent Engineering' (and other sections) *Concurrent Engineering in Construction Projects*, edited by Anumba, C.J., Kamara, J.M. and Cutting-Decelle, A-F., © 2007 Taylor and Francis, Abingdon (ISBN: 978-0-415-39488-8).

References

AIA (2007) *Integrated Project Delivery: a Guide*. American Institute of Architects, California.

Amor, R. & Clift, M. (2007) Document management in concurrent life cycle design and construction. In: ed. Anumba C.J., Kamara J.M. and Cutting-Decelle A.F., *Concurrent Engineering in Construction Projects*. Taylor and Francis Books Ltd, London. 183–200.

Anumba, C.J., Evbuomwan, N.F.O. & Sarkodie-Gyan, T. (1995) an approach to modelling construction as a competitive manufacturing process. *12th Conference of the Irish Manufacturing Committee*, Cork. 1069–1076.

Anumba, C.J. & Evbuomwan, N.F.O. (1995) The Manufacture of Constructed Facilities. *Proceedings, ICAM '95*, A.9.1–A.9.8.

Anumba, C.J. & Duke, A.K. (1997) Telepresence in Virtual Project Team Communications. *Proceedings ECCE Symposium on Computers in the Practice of Building and Civil Engineering*, 3–5 September, Lahti, Finland. 80–84.

Anumba, C.J. & Evbuomwan, N.F.O. (1997) Concurrent Engineering in Design-Build Projects. *Construction Management and Economics*, 15(3), 271–281.

Anumba, C.J., Baldwin, A.N., Bouchlaghem, N.M., Prasad, B., Cutting-Decelle, A.F., Dufau, J. & Mommessin, M. (1999) Methodology for integrating concurrent engineering concepts in a steelwork construction project. *Advances in Concurrent Engineering, Proceedings of the CE99 Conference*, Bath, September, Technomic Publishing Company.

Anumba, C.J. & Duke, A.K. (2007) Telepresence environment for concurrent lifecycle design and construction. In: ed. Anumba C.J., Kamara J.M. & Cutting-Decelle A.F., *Concurrent Engineering in Construction Projects*. Taylor and Francis Books Ltd, London, 218–249.

Anumba, C.J., Kamara, J.M. & Cutting-Decelle, A.F. (2007b) *Concurrent Engineering in Construction Projects*. Taylor and Francis Books Ltd, London. 290 pp.

Anumba, C.J., Bouchlaghem, N.M., Baldwin, A.N. & Cutting-Decelle, A.F. (2007a) Integrated product and process modelling for concurrent engineering. In: Anumba C.J., Kamara J.M. & Cutting-Decelle A.F., *Concurrent Engineering in Construction Projects*, Taylor and Francis Books Ltd, London. 161–182.

Bowron J. (2002) *Reengineering the Project Procurement Process through Concurrent Engineering*, PhD Thesis, Loughborough University, Loughborough.

Carter, D.E. (1994) Concurrent engineering. *Proceedings of the 2nd. International Conference on Concurrent Engineering and Electronic Design Automation (CEEDA '94)*, 7–8 April. 5–7.

Constable, G. (1994) Concurrent engineering – Its procedures and pitfalls. *Measurement and Control*, **27**(8), 245–247.

Crowley, A. (1996) Construction as a manufacturing process. In: ed. Kumar, B. & Retik, A., *Information Representation and Delivery in Civil and Structural Engineering Design*. Civil-Comp Press, Edinburgh. 85–91.

de la Garza, J.M., Alcantara, P., Kapoor, M. & Ramesh, P.S. (1994) Value of concurrent engineering for A/E/C Industry. *Journal of Management in Engineering*, **10**(3), 46–55.

Dixon, T., Pottinger, G. & Jordan, A. (2005) Lessons from the private finance initiative in the UK: Benefits, problems and critical success factors. *Journal of Property Investment and Finance*, **3**(5), 412–423.

Dowlatshahi, S. (1994) A comparison of approaches to concurrent engineering. *International Journal of Advanced Manufacturing Technology*, **9**, 106–113.

Eastman, C., Teicholz, P., Sacks, R. & Liston, K. (2008) *BIM handbook – A Guide to Building Information Modelling for Owners, Managers, Designers, Engineers and Contractors*. John Wiley and Sons, Inc, Chichester, UK.

Egan, J. (1998) Rethinking construction. *Report of the Construction Task Force on the Scope for Improving the Quality and Efficiency of UK Construction*. Department of the Environment, Transport and the Regions, London.

Evbuomwan, N.F.O., Sivaloganathan, S. & Jebb, A. (1994) A state of the art report on concurrent engineering. In: ed. Paul, A.J. and Sobolewski, M., *Proceedings of: Concurrent Engineering: Research and Applications, 1994 Conference*, August 29–31, Pittsburg, Pennsylvania. 35–44.

Evbuomwan, N.F.O. & Sivaloganathan, S. (1994) The nature, classification and management of tools and resources for concurrent engineering. In: ed. Paul, A.J. & Sobolewski, M., *Proceedings of: Concurrent Engineering: Research and Applications, 1994 Conference*, August 29–31, Pittsburgh, Pennsylvania. 119–128.

Evbuomwan, N.F.O. & Anumba, C.J. (1995) Concurrent life-cycle design and construction. In: ed. Topping, B.H.V., *Developments in Computer Aided Design and Modelling for Civil Engineering*, Civil-Comp Press, Edinburgh. 93–102.

Evbuomwan, N.F.O. & Anumba, C.J. (1996) Towards a concurrent engineering model for design-and-build projects. *The Structural Engineer*, **74**(5), 73–78.

Evbuomwan, N.F.O. & Anumba, C.J. (1998) An integrated framework for concurrent lifecycle design and construction. *Advances in Engineering Software*, **5**(7–9), 587–597.

Frank, D.N. (1994) Concurrent engineering: A building block for TQM. *Annual International Conference Proceedings – American Production and Inventory Control Society (APICS)*, Falls Church, VA, 132–134.

Hannus, M., Huovila, P., Lahdenpera, P., Laurikka, P. & Serén, K-J. (1997) Methodologies for systematic improving of construction processes. In: ed. Anumba, C.J. & Evbuomwan, N.F.O., *Concurrent Engineering in Construction: Proceedings of the First International Conference*, Institution of Structural Engineers, London. 55–64.

Group, P.A. (1997) The economics of the Private Finance Initiative. *Oxford Review of Economic Policy*, **13**(4), 53–66.

Harding, J.A. & Popplewell, K. (1996) Driving concurrency in a distributed concurrent engineering project team: a specification for an engineering moderator. *International Journal of Production Research*, **36**(3), 841–861.

Huovila, P. & Serén, K-J. (1995) customer-oriented design methods for construction projects. *International Conference on Engineering Design, ICED 95*, Praha, August 22–24, 444–449.

Huovila, P., Koskela, L. & Lautanala, M. (1997) Fast or concurrent: The art of getting construction improved. In: ed. Alarcon, L., *Lean Construction*. A. A. Balkema, Rotterdam. 143–159.

Kagioglou M., Cooper R., Aouad G. & Sexton M. (2000) Rethinking construction: The generic design and construction process protocol. *Engineering, Construction and Architectural Management*, **7**(2), 141–154.

Kagioglou, M., Aouad, G., Wu, S., Lee A., Fleming, A. & Cooper, R. (2007) Process management for concurrent life cycle design and construction. In: ed. Anumba C.J., Kamara J.M. & Cutting-Decelle A.F., *Concurrent Engineering in Construction Projects*. Taylor and Francis Books Ltd, London, 98–117.

Kamara, J.M., Anumba, C.J & Evbuomwan, N.F.O. (1997) Considerations for the Effective Implementation of Concurrent Engineering in Construction. In: Anumba, C.J. & Evbuomwan, N.F.O., *Concurrent Engineering in Construction: Proceedings of the First International Conference*. Institution of Structural Engineers, London. 33–44.

Kamara, J.M., Anumba, C.J. & Evbuomwan, N.F.O. (2000) Developments in the implementation of concurrent engineering in construction. *International Journal of Computer-Integrated Design and Construction*, **2**(1): 68–78.

Kamara, J.M. & Anumba, C.J. (2007) The 'Voice of the Client' within a concurrent engineering design context. In: *Concurrent Engineering in Construction Projects*. Taylor and Francis Books Ltd, London. 57–79.

Khalfan, M.M.A., Anumba, C.J. & Carrillo, P.M. (2007) Readiness Assessment for Concurrent Engineering in Construction. In: Anumba C.J., Kamara J.M. & Cutting-Decelle A.F., *Concurrent Engineering in Construction Projects*. Taylor and Francis Books Ltd, London. 30–56.

Kimmance, A. (2002) An integrated product and process information modelling system for construction, PhD Thesis, Loughborough University, Loughborough.

Koskela, L. (1992) Application of the new production philosophy to construction. *Technical Report No. 72*, Centre for Integrated Facility Engineering (CIFE), Stanford University.

Li, B., Akintoye, A., Edwards, P.J. & Hardcastle, C. (2005) Critical success factors for PPP/PFI projects in the UK construction industry. *Construction Management and Economics*, **23**(5), 459–471.

Lion, R. (2004) Terminal 5 single model environment – vision, reality and results. *Design Productivity Journal*, **3**(3), 53–58. Available from: http://www.excitech.co.uk/DPJ/library.asp

Love, P.E.D. & Gunasekaran, A. (1997) Concurrent engineering in the construction industry. *Concurrent Engineering: Research and Applications*, **5**(2), 155–162.

Madan, P. (1993) Concurrent engineering and its application in turnkey projects management. *IEEE International Management Conference*. 7–17.

Marosszeky, M. (1999) Technology and innovation. In: ed. Best, R. & de Valence, G., *Building In Value: Pre-Design Issues*. Arnold, London. 342–355.

Nicholas, J.M. (1994) Concurrent engineering: overcoming obstacles to teamwork. *Production and Inventory Management*, **35**(3), 18–22.

Prasad, B., Morenc, R.S. & Rangan, R.M. (1993) Information management for concurrent engineering: research issues. *Concurrent Engineering: Research and Applications*, **1**(1), 3–20.

Prasad, B. (1996) *Concurrent Engineering Fundamentals*, vol. 1: *Integrated Products and Process Organization*. Prentice Hall PTR, NJ.

Prasad, B. (1998) Editorial: How tools and techniques in concurrent engineering contribute towards easing co-operation, creativity and uncertainty. *Concurrent Engineering: Research and Applications*, **6**(1), 2–6.

Pouchard, L. & Cutting-Decelle, A.F. (2007) Ontologies and standards based approaches to inter-operability for concurrent engineering. In: Anumba, C.J., Kamara, J.M. & Cutting-Decelle, A.F., *Concurrent Engineering in Construction Projects*. Taylor and Francis Books Ltd, London. 118–160.

Sanvido, V.E. & Medeiros, D.J. (1990) Applying computer-integrated manufacturing concepts to construction. *Journal of Construction Engineering and Management*, **116**(2), 365–379.

Scherer, R.J. & Turk, Z. (2007) Support for users within a concurrent engineering environment. In: Anumba, C.J., Kamara, J.M. & Cutting-Decelle, A.F., *Concurrent Engineering in Construction Projects*. Taylor and Francis Books Ltd, London. 250–277.

Strategic Forum for Construction (SFC) (2003). Integration Toolkit. Available from http://www.constructingexcellence.org.uk/tools/sfctoolkit/ipt_workbooks/00.html (accessed June 2011).

Smith, J., Tomasek, R., Jin, M.G. & Wang, P.K.U. (1995) Integrated system simulation. *Printed Circuit Design*, **12**(2), 22,63.

Staub-French, S. & Fischer, M. (2007) Enabling concurrent engineering through 4D CAD. In: Anumba C.J., Kamara J.M. & Cutting-Decelle A.F., *Concurrent Engineering in Construction Projects*. Taylor and Francis Books Ltd, London. 201–217.

Thamhain, H.J. (1994) Concurrent engineering: Criteria for effective implementation. *Industrial Management*, **36**(6), 29–32.

Walker, P. (2007) Procurement, contracts and conditions of engagement within a concurrent engineering context. In: Anumba C.J., Kamara J.M. & Cutting-Decelle A.F., *Concurrent Engineering in Construction Projects*. Taylor and Francis Books Ltd, London. 80–97.

Winner, R.I., Pennell, J.P., Bertrend, H.E. & Slusarczuk, M.M.G. (1988) The role of concurrent engineering in weapons system acquisition. *IDA Report R-338*, Institute for Defence Analyses, Alexandria, VA.

13

Complexity Theory: Implications for the Built Environment

Mark D. Sharp

13.1 Introduction

This chapter introduces 'entangled complexity' and its relationship with the various aspects of the built environment. Particular emphasis is placed on its interaction in process, as this may be used to leverage innovation. The first section of this chapter presents the formal foundations of complexity, its evolution, and relevance to current thinking. This is followed by an analysis of the extant literature and key determinants of complex procedures, and how entangled complexity interacts with the built environment. The final section presents the transition of complexity to innovation and performance improvement, using worked exemplars as paradigms for exploitation.

Thus, the core aim of this chapter is to present readers with a cogent message of just how entangled and complex the built environment is – from inception through to use, demolition and reuse, across several longitudinal continuums.

13.2 Complexity Overview

Complexity is probably one of the most commonly used terms without being recognised in its truest sense. It is often used to describe something that is either not easily understood, or has many components (variables) that go to make up a larger situation. The difficult part of complexity is how to define this as a tool for analysis. What is clear is that complexity (*per se*) is not particularly easy to define. There are numerous definitions that have appropriated themselves to complexity, with varying degrees of understanding, from vague meaningless labels through to descriptive definitions that also become

Construction Innovation and Process Improvement, First Edition.
Edited by Akintola Akintoye, Jack S. Goulding and Girma Zawdie.
© 2012 Blackwell Publishing Ltd. Published 2012 by Blackwell Publishing Ltd.

meaningless in their description (Edmunds, 1996). However, there are two main descriptions that can be used to define the meaning of complexity; first, the origins of the word followed by the historic beginnings of complexity. Thus, the Latin word '*complexus*' indicates 'entwined', or 'twisted together'; and the appropriateness of using Latin was validated by Aristotle, who wrote about the need for balance in *Metaphysics*, Vol. II (Aristotle, 1045, a10).

The discovery of complexity is drawn from Oscar II (Swedish nineteenth-century king) who offered a prize if someone could solve the problem collectively known as the three celestial bodies. He identified that when two bodies were in motion, Newton's Laws of motion were sufficiently accurate to calculate the path. However, when a third body is added, so one orbits around a central body and the third body orbits around the second body; like the moon, earth and sun, predicting where the path is going to be becomes far more complicated (Sharp, 2007a). Much in the same way that the discovery of the electron being more than 2D model by Sommerfield in 1919, which was the final clue to Bohr's model and led to the splitting of the atom. In the later nineteenth century, Jules Henri Poincaré concluded that the celestial problem could never be solved, and identified that infinitely complicated behaviours could arise in simple non-linear systems. Therefore, if it was not possible to calculate the motions of three celestial bodies, then there was no way to predict the outcome of systems we see about us every day with millions, trillions, or more, intensely interacting parts.

Complexity has often been confused with the misnomer of chaos (Yorke, 1975), which relates to a 'simple lack of order', whereas Poincare's Deterministic chaos (Poincaré, 1957) is founded on mathematical measurement principles. Thus, complexity finds itself building the bridge between unpredictability and determinism (Geyer, 2005). However, complexity cannot exist without its own paradigm criteria, i.e.:

- The system must have *causality*.
- *Reductionism* should be kept to a minimum, although it is accepted that some reductionism must exist.
- The system will probably be *predictive* and *determinism* is likely to be within the system.
- The system is also likely to be *probabilistic*.
- *Interpretation* is likely by the users and *emergence* will most likely be present.

The term used to describe the above is 'entangled complexity'. The built environment holds all of the terms described above in *causality*. The original intention of the built environment from its inception is to be a shelter from the earth's elements (Neolithic man was essentially the last of the cave dwellers). Ancient Greeks (7000 BC) became some of the first known house builders, building 'megaron' houses, which formed the basis of the types of structures recognised in today's residential accommodation. Thus, the determination of the above is that none of these elements can operate in isolation; it is only when they are conjoined can they operate as a process. These relationships therefore entwine to make a process.

Innovation is found in the Innovation (from the Latin *innovatus*, 'to renew or change', from *in* – 'into' + *novus* 'new') of these processes. From this, processes are therefore the fuel that makes other disassociated elements operate (acting as the enabler). The entanglement of these elements is determined by its operational efficiency. For example, inducted elements such as aggregate, cement and water on their own do not make concrete. It is the chemical reaction that occurs when these elements are brought together, when the process becomes more complex, which also relates to the strength of the concrete (by means of prior calculation). However, due to the number of variables in the process (weather conditions, slump of concrete, type of cement, etc), each batch of concrete is measured to determine the mix and quality of mix, which is particularly important with its consideration of load bearing components of a building. Thus, the use of reductionism in the built environment is evidenced by the means of desire to only build what is needed (Seeley, 1976). On this theme, although they did not know it at the time, early construction methods utilised a range of inter-related processes, the primary purpose of which was to withstand the elements (e.g. to ensure that the roof kept the water out). The use of innovation in these processes has now evolved to ensure raw materials are optimised, along with time, cost and efficiency measures.

Predictiveness within the built environment is clear; it is known that when buildings are built, they are somewhat predictable; for example, they have a predictability of being able to withstand certain loads on the foundations, the floors, roof, etc. It is also predictable on how buildings will look, how an estate will look (through plans and now visualisation software), or how an urban environment will behave. Other aspects of the built environment can be *determined*, for example, the amount of water and electricity consumed, as well as the amount of carbon to be produced from the built environment. The same applies to the performance of the internal activities within the building (Clements-Croome and Baizhan, 2000). For validity and reliability of statistics and validation of scientific reasoning, see Büttner (1997) and Kendall (2003). However, in terms of causality on process and innovation, predictiveness is not always present, especially where untried and untested methods are being used for the first time. An example is when Terminal 5 at Heathrow was being planned; a new type of scaffold was to be used to aid the optimisation and specification of construction as well as to improve safety. This new type of scaffold was also used in the construction process (note: there was a specific requirement to construct offsite for joining elements together); therefore, replica scaffolding was constructed on a test site to ensure that the new scaffolding system would work (in order that the construction process would continue at optimum loads of efficiency). Thus, predictiveness can be seen as a viable option, especially in complex systems where untried and untested methods have not been used before.

Probability also exists in the built environment. For example, it is *probable* (in some countries) that in the winter, some form of heating will be required (but *probably* not in the summer). In an urban environment, it is also probable that more consumables will be needed, such as energy for street lighting, which would probably not be needed in a rural environment. See also Bayesians (Stigler, 1986; Aldrich, 2008; Lucien Le Cam, 1986; Fienberg,

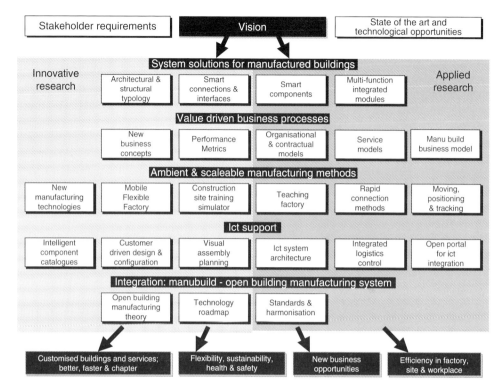

Figure 13.1   ManuBuild vision (ManuBuild, 2009). Reproduced by permission of Samir Boudjabeur, ManuBuild (www.manubuild.org).

2006), on how probability can be interpreted from a mathematical perspective of its use in the built environment. These types of predictable connotations and permutations have both a direct and indirect causality on innovation and process, as there is often uncertainty when calculations are too large to show immediate impact, and an example of this would be climate change. Thus, complex calculations can be used, but the question is, are the outcomes predictable?

Interpretation by users is likely to be present in this context by taking the form of the users interpreting different styles (fashion) to personalise both the external and internal environments. This is best exemplified by the innovative interpretation of quantum mechanics that plays such a vital role in the built environment (Carnap, 1939; Reichenbach, 1944; Popper, 1963; Bub and Clifton, 1996; Omnès, 1999). For example, even Modern Methods of Construction can be used to keep the external façade of a building aesthetically near to (mock) Georgian or Victorian, in order to entertain a certain sense of establishment and security (which is deemed important by building purchasers, and indeed mortgage vendors). However, the end user is often given little consideration of individual interpretation, as the designer has already interpreted this in the design (load variation and material limitations accepted).

This leads to the final part of determinations of entangled complexity, which is emergence. Goldstein (1999) noted that emergence was 'the arising of novel and coherent structures, patterns and properties during the process of self-organisation in complex systems' (Corning, 2002). This can also be perceived as a definition of innovation. Thus, in the context of the innovative built environment, there have been numerous attempts to use emergence to improve processes. This is exemplified by the recent hive of activity directed towards the use of production methodologies used by manufacturing industries in the built environment. A recent European initiative – ManuBuild (Figure 13.1) was the first of these that attempted to join the numerous variables of the building process in a form of open building manufacture to using the numerous components available at the time (Sharp, 2007b; ManuBuild, 2009).

This can be seen as an example of emergence, as it presents a good case study of how the built environment uses innovation to move the body of knowledge forward.

13.3   Complexity in the Built Environment

This section identifies the boundaries of complexity in the built environment, especially where entangled complexity interacts with construction processes in general. However, due to the diversity and range of variables offered, it may be difficult to ascribe sufficient value to this as an holistic tool kit of understanding. The key here is in the scalability of entangled complexity. For example, Havel (1995) identified 'scale-thin', where if its distinguishable structure extends only over one or a few scales, for example, a building seen from the outside may have a distinguishable structure on two or three scales: the building as a whole, the windows and doors, and perhaps the individual bricks. Reverse scalability of reverse differentiation (linearity), therefore extending outwards from the brick to the house, to the estate, to the county, to the world etc., shows how scalability can be used throughout the built environment and so forms part of the entangled complexity body of knowledge. This relates to only the physical components of the building, but does not take into consideration the least known variable, human beings.

Complexity has been traditionally associated with the physical sciences and human biology (Heylighen, 1990; Solè and Goodwin, 2000; Mazzocchi, 2008). Furthermore, both Kraugman (1996) and Stacey (1996) report on the macro and local simulations of behaviour, from the flocking migration of birds to the trail laying of ants. This is further complicated by complexity within the social domain, as Mitteleton-Kelly (1997) recognised that behaviour in the former can be 'assumed' to be governed by laws, whereas the latter claimed law may itself generate changed behaviour. In this respect, social systems (including organisations and their management) are fundamentally different from all other complex systems. This is important, as complex relations have an effect on both innovation and process, not in one specific area but throughout the whole process. For example, the UK construction industry accounts for approximately 1.5 million people (DTI,

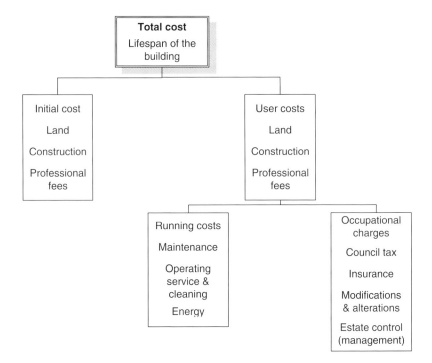

Figure 13.2   Building life total costs (Sharp, 2007a). Reproduced by permission of the CIB World Congress.

2010), with an incredibly complex infrastructure that embraces socio–political and economic subsystems. It is therefore advocated that in order to be able to predict future needs, then complex models could be a viable option. For example, predictions are still used predominantly for new build construction, which consider refurbishment as a type of exogenous constant (Kohler, 2002); and several existing models are available in extant literature (Briggs, 1992; Bezelga, 1991; Johnstone, 1995; Schwaiger 1997). This naturally leads to the market orientated cost dimension, combined with life-cycle models (Figure 13.2).

13.4   Complexity in Organisations

Having looked at the entangled complexity issues that impinge upon macro factors of residential markets, the following section looks at the way that organisations in the built environment approach the market.

Generally, organisational management has focused on how to maintain and promote a common culture within an organisation. The need for 'more of the same, but with larger profit margins', does not promote divergence (Cohendet and Llerena, 1997). However, the need to think about 'how' to promote a divergent 'thriving' business is a prerequisite for the promotion of staff within organisational culture (Darvas et al., 2005); and political stamina and general risk adversity of individuals to business continuity is also a prerequisite – which leads to lack of ideas and innovation within business culture (Coombs and Tomlinson, 1998). Generally, a construction organisation is structured on a cohesive management structure reliant on hierarchical precedence, based on strategic planning and supported by common culture. The organisation likes to stick to the things that the business was built on (core business), with its focus firmly on the bottom line (Yin, 2008; Hui et al., 2009). The emphasis for the organisation is not on survival but growth. Why? Because that is the way business, given the constraints of the 'market', exists; except in times of recession where there is a complete revision of this core business value, and where there is a knee jerk reaction that organisations prefer to adopt in the contracting of the business (Druker, 1954). Thus, in dynamic markets, the role of creative disorder shows that organisations need to be aware of market change and diversification. Stacey (1993) summarises these issues, noting:

- Analysis loses its primacy.
- Contingency (cause and effect) loses its meaning.
- Long-term planning becomes impossible.
- Visions become illusions.
- Consensus and strong cultures become dangerous.
- Statistical relationships become dubious.

Organisations therefore need to look and learn from other perspectives. If the future is unknowable, you cannot predict a long-term future, thus making any forward planning irrelevant. Whilst it is not easy for an organisation to adjust to these models of divergence, they will surely not survive if they do not adapt to the complex markets in which they operate. Successful organisations will be those who do not just set organisational objectives, but mobilise the organisation around these objectives for the sake of conformity. In a similar guise, sustainability in the market will be attained by those who recognise that the organisation has to adapt to complex and continued interactions, both from within the organisation and those outside it. Thus, innovation must show adaptability and service offering to end-users, as emergence often evolves from diversity which, as Wheatley (1992) notes, requires an 'act of faith'. Thus, the difficulty for entangled complexity is not in the evidence within specific fields, as the acceptance that these findings can be substantiated outside their fields; if they are substantiated, can they then be 'generalised' to all systems outside that field? The following section looks at the methods available to substantiate this, by relating this to theory and practice examples.

13.5   Toolkits

It is clear from the above that old approaches to complex issues in the built environment will no longer be valid once the full realisation of entangled complexity is explored. Whilst research into complexity continues to advance, it is in the application of the tacit knowledge of applied theory to the built environment, especially where entangled complexity links to innovation. In this respect, innovation has been acknowledged as being a means of successfully applying ideas into a meaningful application in practice (Schumpeter, 1934). Innovation is also frequently associated with creativity (Amabile, 1996; Keegan, 1996; Luecke and Katz, 2003) but this is more than just creativity, as it requires the means of being applied to a situation. This section looks at complexity, not just from a value to innovation practice *per se*, but using a whole life value approach applied to the built environment. This in turn highlights the benefits to organisations and industry, in order to identify which complex hierarchical factors to use (along with their priorities).

There have been various attempts to model the whole life cycle of a building, ranging from transactions that take place over the construction project life cycle (Winch, 2001), through to the incorporation of maintenance and refurbishment (Clements-Croome *et al.*, 2003). In this respect, the factors that affect the prioritisation have been identified both diagrammatically and mathematically. For example, Spedding *et al.* (1995) identified these factors diagrammatically, as shown in Figure 13.3.

Whilst these factors are important to the prioritisation, Holmes and Shen (1995) also identified five criteria used to set up maintenance, which can be used in complex systems such as the built environment:

1. *Building Status* – the relative importance compared to others in terms of function;
2. *Physical Condition* : physical condition of the defective element, and possibility of its breakdown;
3. *Importance of Usage* : the effects of breakdown to the occupants and users of the building;
4. *Effects on Fabrics* : the cost implication of defective element; and
5. *Effects on Service Provision* : the cost implication of the breakdown or failure.

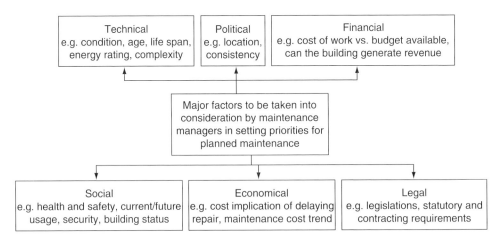

Figure 13.3   Factors of prioritisation for planned maintenance (Spedding *et al.*, 1995).

These factors can then be delineated into levels, depending on the number of situations.

Thus, probabilistic models have concentrated on Whole Life-Cycle Costing (WLCC) issues of how to design out problems, with general emphasis being placed on the designer to design out complex processes. This gives a true cost value over the buildings life, rather than simply building as cheaply as possible. Furthermore, Aarseth and Hovde (1999) and Moser (1999) proposed a methodology for dealing with uncertainty in the factor method, by treating the individual factors as stochastic inputs; thus, providing a probability distribution of the service life prediction, with an associated confidence interval. In this respect, the EuroLifeForm Model (ELF) (Kirkham et al., 2004) is an integrative attempt to respond to some of the problems identified in normal WLCC methods. The ELF model consists of three distinct elements, a probabilistic life-cycle cost model, a probabilistic deterioration model and a decision support application (The Logbook). In addition, a simple environmental impact analysis is supported that can be accessed through the decision support application and maintained throughout the design process. These elements work in synergy to enable the user to produce stochastic WLCCP results, based on accurate predictions of life-cycle replacement and maintenance costs, in a logical and iterative manner, as represented in the ELF model.

The logbook acts as the gateway to the WLCC and performance modelling process, and is designed to work as a repository of information for the project, from design through to eventual disposal. It not only stores this information, but also provides a framework in which to carry out the logical and iterative application of WLCC techniques to a particular design scenario. The logbook is therefore not just confined to the design process and construction phase of the project. The logbook also facilitates Post Occupancy Analysis, the idea being that the building owner utilises this logbook to record up-to-date operational cost data, thereby providing the necessary information to update the WLCC model forecasts if required.

In the ELF application, a deterioration model is provided that uses a combination of deterministic methods, for example, the Factorial Approach (Bird, 1987) and stochastic models including the Weibull Distribution (Weibull, 1951) to provide this information at the detailed design stage of the WLCC process. It also includes a semi-quantitative risk analysis, which can be used to assign a likelihood and consequence score to the failure of any element under consideration, to allow users to prioritise complex actions. One highlighted problem with this method is the reliance on deterministic values, and then applying cost once the process is identified; this, combined with the need to derive data from the two models, clearly makes this the best of the probabilistic models.

There have been numerous attempts at quantifying complexity process issues through numerical methods (Ostwald and Tucker, 2007). Numerical methods are often divided into elementary ones, such as finding the root of an equation, integrating a function, or solving a linear system of equations to intensive ones like the finite element method. Intensive methods are often needed for the solution of practical problems, which often require the systematic application of a range of elementary methods, often thousands or millions of times over. In the development of numerical methods, simplifications need to be made to progress towards a solution; as 'numerical methods do not usually give the exact answer' to a given problem, they can only tend towards a solution, getting closer and closer with each iteration, which lends itself to be a bridge towards determinism (Salinaros, 1999).

Maintenance policies are often based on a perfect monitoring assumption, and the status of each unit (running, failed, level of degradation, etc) is supposed to be continuously and perfectly known (Cho and Parlar, 1991). However, in many realistic situations, it is impossible or too expensive to verify this assumption, as the state of a unit can be totally unknown or submitted to many diagnosis errors: measurement errors, false alarm, non-detection, detection delay, etc (Basseville and Nikifirov, 1993). On this theme, developed by Thomas Saaty (1977), the Analytical Hierarchical Approach (AHP) provides a proven, effective means to deal with complex decision-making. It can also assist with identifying and weighting selection criteria, analysing the data collected and expediting the decision-making process. AHP can help capture both subjective and objective evaluation measures, thereby providing a useful mechanism for checking the consistency of the evaluation measures, whilst also using alternatives to reduce bias in decision-making.

The first step is to decompose the goal into its constituent parts, progressing from the general to the specific. In its simplest form, this structure comprises a goal, criteria and alternative levels. Each set of alternatives are then further divided into an appropriate level of detail, recognising that the more criteria included, the less important each individual criterion may become. Next, a relative weight is assigned to each, where criterion has a local (immediate) and global priority. The sum of all the criteria beneath a given parent criterion in each tier of the model must therefore equal one; and the global priority shows its relative importance within the overall model. Finally, after the criteria are weighted and the information is collected, the information is uploaded to the model. Scoring is made on a relative basis, not an absolute basis, comparing one choice to another. Relative scores for each choice are then computed within each leaf of the hierarchy, and synthesised to yield a composite score for each choice at every tier, as well as an overall score. A worked example of AHP can be seen in Figure 13.4.

Thus, assume that a set of objectives have been established along with a normalised set of weighting for comparing alternatives. For simplicity, assume four objectives exist: O1, O2, O3 and O4. From a pair-wise comparison matrix A, where the number in the i_{th} row and j_{th} column gives the relative importance of O_i as compared with O_j.

Using a Likert Scale (1–9), where:

a_{ij} = 1 if the two objectives are equal in importance;
a_{ij} = 3 if O_i is weakly more important than O_j;
a_{ij} = 5 if O_i is strongly more important than O_j;
a_{ij} = 7 if O_i is strongly more important than O_j;
a_{ij} = 9 if Oi is absolutely more important than O_j;
a_{ij} = 1/3 if Oj is weakly more important than O_j.

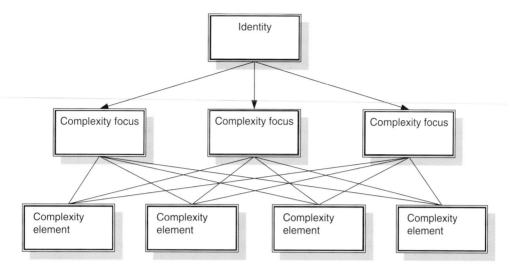

Figure 13.4   Worked example of Saaty's hierarchy applied to complexity.

the following matrix can then be formulated:

$$A = \begin{bmatrix} 1 & 1/5 & 1/3 & 1/7 \\ 5 & 1 & 3 & 5 \\ 3 & 1/3 & 1 & 3 \\ 7 & 1/5 & 1/3 & 1 \end{bmatrix} = \begin{bmatrix} 1.000 & 0.200 & 0.333 & 0.143 \\ 5.000 & 1.000 & 3.000 & 5.000 \\ 3.000 & 0.333 & 1.000 & 3.000 \\ 7.000 & 0.200 & 0.333 & 1.000 \end{bmatrix}$$

To normalise the weightings, the sum of each column are computed, then each column is divided by the corresponding sum. Using an over-bar to denote normalisation:

$$\overline{A} = \begin{bmatrix} 0.063 & 0.115 & 0.071 & 0.016 \\ 0.313 & 0.577 & 0.643 & 0.547 \\ 0.188 & 0.192 & 0.214 & 0.328 \\ 0.438 & 0.115 & 0.071 & 0.109 \end{bmatrix}$$

Because the numbers in the second row are generally larger than the rest of the matrix (except for those in column 1), this would indicate there was some inconsistency in comparisons used in the original matrix. Ideally, the four normalised columns would all be identical if the pair-wise comparisons were consistent. In practice, the consistency measure would normally be measured using Eigenvalues of the normalised comparison matrix. Eigenvalues are a special set of scalars associated with a linear system of equations (i.e. a matrix equation) that are sometimes also known as characteristic roots, characteristic values (Hoffman and Kunze, 1971), proper values or latent roots (Marcus and Minc, 1988: 144). The determination of the Eigenvalues and Eigenvectors of a system is extremely important, where it is equivalent to matrix diagonalisation. Each Eigenvalue is paired with a corresponding so-called Eigenvector (or, in general, a corresponding right Eigenvector and a corresponding left Eigenvector; there is no analogous distinction between left and right for Eigenvalues).

The decomposition of a square matrix λ into Eigenvalues and Eigenvectors is known as Eigen decomposition, and the fact that this decomposition is always possible as long as the matrix consisting of the Eigenvectors of λ is square, is known as the Eigen decomposition theorem. It is on this theory that Saaty (1980) based the principles of AHP and like many theories, came under scrutiny. For example, Spedding *et al.* (1995) reviewed the AHP methodology and identified that the system was inherently flawed for use in manufacturing systems. The reason for this was because the AHP approach is based on deriving ratio scale measurements. However, it is a valuable tool to identify entangled complexity evaluation, as it has the flexibility and characteristics to evaluate either individual buildings, projects or multiple connotations of projects, or indeed project versus project evaluation.

13.6   Complex Innovation in Organisations

Processes within organisations must constantly be dynamic to improve the quality of both the company's internal and external services. This calls for 'appropriate' measurement of the process, thereby requiring diligence in monitoring, controlling and assessment in both the technical and non-technical (or socio-cultural) approaches (Stewart, 1995); and technical and instrumental tools, rather than socio-cultural (or non-technical) aspects (Low, 1993). In addition, Spekknink (1995) argued that to maintain a Quality Management System effectively in construction, a balance of the technical requirements and theoretical approach was necessary.

Spekknink's (1995) concept can be seen in Figure 13.5, highlighting a theoretical approach for effective maintenance of a complex system, using

Figure 13.5 Technical and non-technical approaches for improving service quality (Spekknink, 1995).

the application of an integrative approach in an organisation rather than a segmentalist approach. While the integrative approach has a structure and culture that advances towards an egalitarian and meritocratic ideal, the segmentalist approach, on the other hand, has a structure that insists on traditional bureaucracy, isolating labour from management, and focusing on uncrossable boundaries between functions for effective management. It appears that segmentalism is currently prevalent in many construction companies because of their conservative attitude towards change and innovation. The management of change and innovation should therefore be an integral part of the building process (Tan and Low, 1991).

Rather than analyse these reasons from a historical perspective, it is important to understand organisational excellence, which potentially leads to the success of businesses (Kanji, 2001), especially in terms of entangled complexity, where it is not always obvious what these success factors may be. Again, the AHP approach can be applied to organisational needs. Other indicators of business performance (i.e. quality, customer satisfaction, innovation and market share) also reflect an organisation's economic condition and growth prospects better than reported earnings (Eccles and Pyburn, 1992). Therefore, performance measures must go beyond the presentation of financial figures to serve as the driver for fostering performance, not purely in financial terms, but also in non-financial aspects such as quality, customer satisfaction, innovation and market share. These performance measures must also be prioritised so that appropriate decisions can be made. Thus, the assessment of process performance using complexity can be used, as, according to Baird (1989), this refers to the relative liking on the part of an evaluator for particular outcomes. Furthermore, assuming a predefined set of performance measures exist, a mathematical function between all possible outcomes of each individual measure and their corresponding relative importance may be developed. Such function is commonly referred to in the literature as the 'utility function', where a multiple attribute utility function integrates these individual utility functions into a single platform, thus providing a collective assessment of engineering performance on a project.

The integration of various measures of process performance in the form of a multiple attribute utility function requires identifying a preference structure that depicts the relative importance of each measure to the others. Clemen (1991) reported on various techniques for developing the preference structure, including pricing out technique, the lottery weight technique, etc. Amongst all of the possibilities, this study uses the Eigenvector prioritisation method commonly employed in the AHP developed by Saaty (1980). This method is a popular alternative for deriving the preference structure in various practical applications of Multi Criteria Decision Making (Zeleny 1982; Mollaghasemi and Pet-Edwards 1997). The major strengths this method brings are its systematic procedure, and its ability to examine the consistency of the evaluator's judgments. However, the evaluation of process performance during the operation and maintenance phase requires several years of full production of the industrial facility (Tucker and Scarlett, 1986). Given this, it is often difficult to collect data measures of process performance pertinent to the operation (within a reasonable timeframe). This is an

Table 13.1 Level of complexity in building processes.

Type of Complexity	Building Phases									
	Identifying need	Score	Design	Score	Contract	Score	Construction	Score	Payment	Score
Causality	✓	0	✓	2	✓	27	✓	24	✓	26
	✓✓	2	✓✓	26	✓✓	2	✓✓	5	✓✓	4
	✓✓✓	28	✓✓✓	2	✓✓✓	1	✓✓✓	1	✓✓✓	0
Reductionism	✓	0	✓	2	✓	18	✓	1	✓	3
	✓✓	1	✓✓	0	✓✓	0	✓✓	1	✓✓	0
	✓✓✓	29	✓✓✓	0	✓✓✓	0	✓✓✓	0	✓✓✓	0
Predictive	✓	6	✓	9	✓	3	✓	2	✓	27
	✓✓	20	✓✓	21	✓✓	1	✓✓	0	✓✓	2
	✓✓✓	4	✓✓✓	0	✓✓✓	27	✓✓✓	27	✓✓✓	1
Determinative	✓	6	✓	5	✓	1	✓	3	✓	1
	✓✓	22	✓✓	19	✓✓	2	✓✓	5	✓✓	26
	✓✓✓	2	✓✓✓	6	✓✓✓	27	✓✓✓	23	✓✓✓	3
Probabilistic	✓	19	✓	0	✓	3	✓	1	✓	2
	✓✓	7	✓✓	3	✓✓	1	✓✓	0	✓✓	0
	✓✓✓	4	✓✓✓	27	✓✓✓	0	✓✓✓	29	✓✓✓	0
Interpretive	✓	8	✓	25	✓	23	✓	21	✓	4
	✓✓	19	✓✓	5	✓✓	7	✓✓	9	✓✓	0
	✓✓✓	3	✓✓✓	0	✓✓✓	0	✓✓✓	0	✓✓✓	0
Emergence	✓	0	✓	2	✓	5	✓	27	✓	3
	✓✓	30	✓✓	21	✓✓	25	✓✓	3	✓✓	1
	✓✓✓	0	✓✓✓	7	✓✓✓	0	✓✓✓	0	✓✓✓	0

important part in the utilisation of Building Information Modelling and IT in order to identify, collect, analyse and use decisions. These issues were used in a case study of a commercial international lending organisation to determine the complex nature of the organisations built asset relationship with its business performance. This included undertaking pre- and post-occupational evaluation, as well determinative criteria specific to this organisation. In order to determine whether the same complex issues could be correlated with data outside of the specific requirements of the organisation, a further electronic survey was undertaken to identify how complex the built environment was in terms of processes and innovation. Specific groups involved in the various stages of the construction process were identified, in order to give a holistic approach to the issues of complexity, and how integrated complexity arose in the built environment. Five areas were identified, taking a broad spectrum of activities across the construction processes.

13.6.1   Case Study Findings

The following matrices (Table 13.1 and Table 13.2) were developed as a means of identifying where respondents felt that complexity existed in built environment processes. Scores were derived from the survey regarding whether respondents either did not register an opinion, or where the data was too low to register a score, so this was shown as a nil score, so not considered as wasteful. These results represent an aggregated mix from both the case study organisation and survey respondents. The following discussion presents the results of opinions and views expressed in the telephone survey follow-up.

13.6.2   Discussion

An *ad-hoc* sample of 30 respondents (~10%) was undertaken using telephone follow-up interviews to substantiate the survey in order to both ground and validate findings. In this respect, there are a number of issues that are worthy of comment. Causality in the process identified that there were complex issues in construction. This was not particularly surprising, given the socio-economic and cultural issues already identified at the start of this chapter. The impact of the design process is significant, in that most people thought that it was a moderately complex issue. The steps of the process would impact on the next stage of the process, although this was not

Table 13.2 Aggregated scoring matrix.

Score	<16	17 – 32	33 – 48	>49
Type of Waste	Not complex	Low amount of complexity	Medium amount of complexity	High amount of complexity

the same as the construction phase; most people considered that this followed a logical procedure. The same can be said for payment, as the cause and effect is if you do the work, you are paid, and this was not perceived as a particularly complex issue. The reductionism issues can be explained in the context as epiphenomena (a secondary phenomenon occurring alongside or in parallel to the primary phenomenon), as respondents considered that there was a high level of complexity in identifying need, but not in any other part of this process. Upon further investigation, it was perceived that reductionism was taken as a metonymy, and thus an unsuspecting aspect of construction's physicality.

Most respondents felt that there was a moderate amount of determinative complexity in the first two phases of identifying need and design. However, most respondents determined that the next two phases were highly determinative in complexity of process. Further indication of this was that of uncertainty for bidding for contract, and once on site that issues arose; that meant that what they thought would happen generally did. Some respondents thought that generally you had to determine to expect the unexpected. In terms of probabilistic complexity in the building phases, the majority of respondents felt that it was not a particularly complex process, as the probability of identifying need was either there or not. However, in the design, almost all respondents felt that the design was probabilistic. Upon further investigation, most respondents thought that architects only knew how to design buildings in a certain way. Some respondents also felt that architects were too far removed from the actual building process to see the probability of design flaws and the effect they had on the rest of the project. In addition, the majority of respondents determined that the construction phase was highly complex, and further investigation identified that there was a complete lack of understanding of what goes on in the construction phase. Thus, time, resource management and a huge array of products and services were under a continuous flux of change, which made the construction phase even more complicated and prone to miscommunication.

Interpretive complexity in the process is quite complex for identifying need and it is likely that this result is because of the degree of interpretation needed for identifying demand. It was acknowledged that there was little interpretive complexity in the design stage, as the architect had a brief to work from, and it was only one stage that an insignificant (architects) part of the process interprets the brief to a design. The rest of the process appears not to be too interpretive due to the logical nature of the process. Interestingly, the emergence of processes appears to be at the front end of the process rather than the construction and payment phase. Emergence was determined to mean the constant state of flux felt in identifying needs as a companion in trying to determine how designs were changing to incorporate these changes. This was identified as being partly down to the perception of public opinion, what was acceptable or what was needed, for example, design needed to incorporate possible changes in dwellings to take advantage of technological developments for WiFi or 4G technology. Furthermore, whilst there has been little change in contracts, respondents considered that there was significant amount of complexity in the contract with the amount and different clauses used for the different types of contracts. The rest of the process was deemed determinative rather than emergent.

Complexity in innovation was also identified at this stage due to two factors:

1. how complexity is a way of doing something in a different way; and
2. how to measure it, is also different.

Thus, the traditional 'S' curve of identifying innovation can be used to identify two things for complexity:

1. where you currently are in terms of processes; and
2. where you want to be (even if this is two complete 'S' cycles away).

Therefore, this study identified that this was a way of identifying complexity in process, and also responsive to perception of what complexity is and how this can be measured. The key here is to be able to identify where you are on the initial 'S' curve to start with. It is this identification of the joining of complexity to both known processes, and understanding how this interacts with innovation to determine new entangled complex processes, such that, it is now time for entangled complexity to be able to make a substantial impact on the built environment processes and innovation.

To go beyond this concept, a case study was undertaken to either prove or disprove the results and discussion thus far. The scenario for this was a major commercial organisation that had followed the progress of the primary study. In this respect, a building due for refurbishment was allocated for testing as it had already pre-qualified for the complex nature of the project to be undertaken. The processes for a traditional approach identified that certain processes would mean that the property would not be in use for 32 months. However, by using an innovative approach to the design, and through using different dimensions of resources, it was ensured that this project could be completed in 27 months. This was because a different emphasis was placed on both the use of the building and the change in design to complement its end use. This generality aligns with the discussion to date, and innovative processes were then sought to determine whether the double 'S' curve would enable the building to become more productive (Figure 13.6).

Using a model to decompose the goal into its constituent parts, progressing from the general processes into their elements, each set of alternatives were divided into an appropriate level elemental detail. The resulting condition and performance based model was then developed. New innovative complex entangled processes were found to be at least one and a half 'S' curves (life cycles) in front of the traditional processes. The result of this case study was that not only were new processes used, but also the building presented additional benefits (not just an increased bottom line). The success of this project also enabled the organisation to review other built asset portfolios.

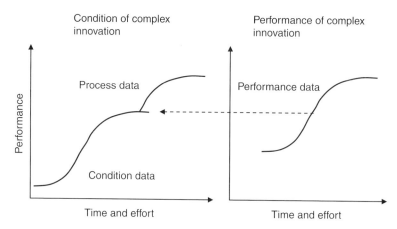

Figure 13.6   Enhanced performance of entangled complexity.

13.7   Conclusion

This chapter presented an holistic approach to 'entangled complexity' and its relationship with the various aspects of the built environment in an attempt to present a cogent picture of conjoined events. It reviewed the theory of complexity, and how contextual information can describe how entangled the process of building our living environment can be. It presented how complexity plays an important part in both innovation and its impact on processes. Thus, through theory and worked examples, entangled complexity in the built environment provides several fruitful avenues to explore. Whilst the issues themselves are often complex, it is also important to acknowledge that complexity can also be used to solve both theoretical and actual problems. This unique quality gives it distinct advantages over other approaches.

Finally, whilst complexity in the built environment is still in its infancy (as far as deterministic and applied research is concerned), it is clear that more research is needed in order to understand the systems, procedures, processes and interactions of variables. Entangled complexity is a pivotal tool that can help identify all of the above (and provide holistic meaning), so that lessons from the past can help shape and generate solutions for the future.

References

Aarseth, L.I. & Hovde, P.J. (1999) A stochastic approach to the factor method for estimating service life. *Proceedings of the 8th International Conference on the Durability of Building Materials and Components*. National Research Council Canada, Vancouver. BC, Canada.

Aldrich, A. (2008) R A. Fisher on Bayes and Bayes' Theorem. *Bayesian Analysis*, 3(1), 161–170.

Amabile, T. (1996) *Creativity in Context*. Westview Press, Boulder, CO.

Aristotle (1045a) Metaphysics. In: Ross, W. (1924) *Aristotle's Metaphysics*, 2 vols. Clarendon Press, Oxford.

Baird, B.F. (1989) *Managerial Decisions Under Uncertainty: An Introduction to the Analysis of Decision Making*. John Wiley & Co. In c, New York.

Basseville, M. & Nikifirov, I. (1993) *Detection Of Abrupt Changes: Theory And Application*. PTR Prentice Hall, Englewood Cliffs, NJ.

Bezelga, A. (1991) *Management Quality & Economics in Building*: CIB Congress in Lisboa. E&FN Spon, London.

Bird, B. (1987) Costs-in-use: principles in the context of building procurement. *Construction Management and Economics*, 5, Special Issue.

Bub, J. & Clifton, R. (1996) A uniqueness theorem for interpretations of quantum mechanics. *Studies in History and Philosophy of Modern Physics*, **27B**, 181–219.

Büttner, J. (1997) Diagnostic validity as a theoretical concept and as a measurable quantity. *Clinical Chim. Acta*, **260**(2),131–143.

Carnap, R. (1939) The interpretation of physics. In: *Foundations of Logic and Mathematics of the International Encyclopedia of Unified Science*. University of Chicago Press, Chicago, IL.

Cho, D. & Parlar, M. (1991) A survey of maintenance model for multi-unit systems. *European Journal of Operation Research*, **51**, 1–23.

Clemen, R. (1991) *Making Hard Decisions: An Introduction to Decision Analysis*. Duxbury Press, Belmont, CA.

Clements-Croome, D. & Baizhan, L (2000) *Productivity and Indoor Environment: Proceedings of Healthy Buildings*, vol. 1. Helsinki.

Clements-Croome, D., Godfaurd, J., Loy, H. & Jones, K. (2003) Through-life environmental business modelling for sustainable architecture. *CIBSE/ASHRAE International Conference on Building Sustainability Value and Profit*, 24–26 September, ASHREA, Edinburgh.

Cohendet, P. & Llerena, P. (1997) Learning, technical change and public policy: How to create and exploit diversity. In: ed. Edquist, C., *Systems of Innovation: Technologies, Institutions and Organizations*. Pinter Publishers, London.

Coombs, R. & Tomlinson, M. (1998) Patterns of UK company innovation styles: new evidence from the CBI innovation trends survey. *Technological Analysis and Strategic Management*, **10**, 3.

Corning, P. (2002) The re-emergence of 'eergence': a venerable concept in search of a theory. *Complexity* 7(6), 18–30. Available from http://www.complexsystems.org/publications/pdf/emergence3.pdf

Darvas, Z., Rose, A. & Szapary, G. (2005) *Fiscal Divergence and Business Cycle Synchronisation: Irresponsibility is Idiosyncratic*. Berkely edu.

Drucker, P. (1954) The Practice of Management: Harper-Collins.

DTI (2010) *Department of Trade and Industry, Constructing the Future*. HMSO, London.

Eccles, R. & Pyburn, P.J. (1992) Creating a comprehensive system to measure performance. *Management Accounting*, **October**, 41–44.

Edmonds, B. (1996) What is Complexity? In: ed. Heylighen. F. & Aerts, D., *The Evolution of Complexity*. Kluwer, Dordrecht.

Fienberg, S. (2006) When did Bayesian inference become 'Bayesian'? *Bayesian Analysis*, **1**(1), 1–40.

Geyer, R. (2005) *Keynote: Centre for Complexity Research*, 13 September. University of Liverpool, Liverpool.

Goldstein, J. (1999) Emergence as a construct: history and issues. *Emergence: Complexity and Organization*, **1**(1), 49–72.

Havel, I. (1995) Scale dimensions in nature. *International Journal of General Systems*, **23**(2), 303–332.

Heylighen, F. (1990) Relational closure: a mathematical concept for distinction-making and complexity analysis. In: ed. Trappl, R., *Cybernetics and Systems '90*. World Science Publishers, 335–342.

Hicks (1999) *Social Democracy and Welfare Capitalism*. Cornell Press, Ithaca, NY.

Holmes, A. & Shen, Q. (1995) A comparative study of priority setting methods for planning maintenance of public buildings. *Facilities* **15**(12/13), 331–339.

Hui, K., Matsunaga, S. & Morse, D. (2009) The impact of conservatism on management forecasts. *Journal of Accounting and Economics*, **47**, 3.

Johnstone, I. (1995) The mortality of New Zealand housing stock. *Arch Science Review*, **37**, 181–188.

Kanji, G.K. (2001). *An Integrated Approach of Organisational Excellence*. Available from http://www.gopal-kanji.com (accessed December 2001).

Keegan, R.T. (1996) Creativity from childhood to adulthood: a difference of degree and not of kind. In: ed. Runco, M.A., *New Directions for Child Development*, No. 72. Jossey-Bass, San Francisco, CA. 57–66.

Kendell R., Jablensky A. (2003) Distinguishing Between the Validity and Utility of Psychiatric Diagnoses: Am J Psychiatry. January; 160(1):4–12.

Kirkham, R.J., Alisa, M., Pimenta da Silva, A., Grindley, T. & Brondsted, J. (2004) Eurolifeform: an integrated probabilistic whole life cycle cost and performance model for building and civil engineering. *Proceedings of COBRA 2004*.

Kohler, N. (1999), The relevance of Green Building Challenge: an observer's perspective, Building Research & Information , Volume, 27, Issue: 4-5, pp 309–320

Krugman, P. (1996) The self organising economy: Blackwell, USA.

Low, S.P. (1993) The rationalisation of quality in the construction industry: some empirical findings. *Construction Management and Economics*, **11**(4), 247–259.

Lucien Le Cam (1986) *Asymptotic Methods in Statistical Decision Theory*, Springer Series in Statistics, Springer-Verlag.

Luecke, R. & Katz, R. (2003) *Managing Creativity and Innovation*. Harvard Business School Press, Boston, MA.

ManuBuild. (2009) *Open Building Manufacturing*. Available from http://www. manubuild.net (accessed June 2011).

Marcus, M. & Minc, H. (1998) Introduction to Linear Algebra. Dover Publications Inc, New York.

Mazzocchi, F. (2008) Complexity in biology. Exceeding the limits of reductionism and determinism using complexity theory EMBO reports. *EMBO*, **9**(1), 10–14.

Mitteleton-Kelly, E. (1997), 'Organisation as Co-Evolving Complex Adaptive Systems', British Academy of Management Conference, London 8–10 September 1997.

Mollaghasemi, M., Pet-Edwards, J (1997) Making Multi-Objective Decisions: IEEE Computer Society Press. Los Alamitos 1997.

Moser, K. (1999) Towards the practical evaluation of service life – illustrative application of the probabilistic approach, *Durability of Building Materials and Components 8: Service Life and Asset Management*, vol. 2, *Service Life Prediction and Sustainable Materials*. National Research Council Canada Press, Ottawa.

Omnès, R. (1999) *Understanding Quantum Mechanics*. Princeton University Press, Princeton, NJ.

Ostwald, M. & Tucker, C. (2007) Calculating characteristic visual complexity in the built environment: an analysis of Bovill's Method. *Symposium: Building Across*

Borders Built Environment Procurement CIB WO92 Procurement Systems, 23–26 September, Hunter Valley, NSW. 305–314.

Poincaré, H. (1957) *Les Méthodes Nouvelles de la Mécanique Céleste*, vol. III. Dover Publications Inc, New York.

Popper, K. (1963) *Conjectures and Refutations*. Routledge, London.

Reichenbach, H. (1944) *Philosophic Foundations of Quantum Mechanics*. University of California Press.

Saaty, T. (1977) A scaling method for priorities in hierarchical structures. *Journal of Mathematical Psychology*, **15**, 234–281.

Saaty, T. (1980) *The Analytic Hierarchy Process*. McGraw Hill, New York.

Salinaros, N. (1999) Architecture, patterns and mathematics. *US Network Journal*, **1**, 75–85.

Schumpeter, J. (1934) *The Theory of Economic Development*. Harvard University Press, Cambridge, MA.

Seeley, I. (1976) *Building Maintenance*: Macmillan, New York.

Sharp, M. (2007a) *Complexity Theory and the Maintenance Paradigm* (Becon, 2007), CIB World Congress, Cape Town, SA.

Sharp, M. (2007b) *Proceedings of the 1st International Conference, The Transformation of the Industry: Open Building Manufacturing*. The Netherlands.

Sommerfield, A. (1919) *Atombau und Spektrallinie*. Friedrich Vieweg und Sohn, Braunschweig.

Spedding, A., Holmes, R. & Shen, Q. (1995) Prioritising planned maintenance in county authorities. In: *International Conference on Planned Maintenance, Reliability and Quality Assurance*. Cambridge University, Cambridge. 72–78.

Spekknink, D. (1995) Architect's and consultant's quality management system. *Building Research and Information*, **23**(2), 97–105.

Stacey (1993) *Strategic Management and Organisational Dynamics*. Pitman Publishing, London.

Stewart (1995) Does God Play Dice? The mathematics of chaos: Blackwell.

Stigler, S. (1986) *History of Statistics*, Chapter 3. Harvard University Press, Cambridge, MA.

Solè, R. & Goodwin, B. (2000) *Signs of Life: How Complexity Pervades Biology*: Basic Books.

Stacey, R. (1996) *Complexity Creativity in Organizations*. Berrett-Koehler, USA.

Schwaiger, K. (1997) Mass flow, energy flow and cost of the German building stock. *CIB 2nd International Conference Buildings and the Environment*.

Tan, O.B. & Low, S.P. (1991) Kanter's strategic approach to managing innovation and change in the construction industry. *The Professional Builder*, **6**(1), 26–31.

Tucker, R. & Scarlett, B. (1986) Evaluation of design effectiveness. *Research Rep. No. SD-16*, Construction Industry Institute, Austin, TX.

Weibull, W. (1951) A statistical distribution function of wide applicability. *Journal of Applied. Mech.-Trans. ASME*, **18**(3), 293–297.

Wheatley, M. (1992) *Leadership In The New Science: Learning About Organization From An Orderly Universe*. Berrett-Koehler, USA.

Winch, G. (2001) Governing the project process: a conceptual framework. *Construction Management and Economics*, **19**, 799–808.

Yin, Y. (2008) *Business and Financial Service: New engine of Economic Growth*. Available from www. Herts.ac.uk

Yorke, J. & Li, T. (1975) Period three implies chaos. *American Mathematical Monthly*, **82**, 985.

Zeleny, M. (1982) *Multiple Criteria Decision Making*. McGraw-Hill, New York.

Part III

Future Technologies

Design Innovation: Advanced Visualisation Futures

Farzad Pour Rahimian

The quality of design can often strongly affect the quality of buildings that we live in (Moum, 2006). Design is an exclusive human activity and a critical aspect of many modern industries, where for example Lawson (1997) acknowledged design as being a multi-aspect, iterative and complicated process. In this context, amongst all the design stages, the early phases of design process tend to be drastically affected by the quality of the communications amongst team members. This chapter categorises design communications into two groups, as follows:

1. the 'design reasoning', which is the interaction between a designer and the design situation; and
2. the 'design collaboration', which is the communication amongst different designers.

Over the last two decades, the communication culture within the design process has changed drastically, influenced predominantly by advances in Information and Communications Technology (ICT) along with increased globalisation.

Lawson (1997) posited that communication culture within every society depends on the tools that people use for their transactions. Today, we are witnessing the revolutions made by globalisation in many societies. In this regard, the culture and the way people interact are often affected by their needs and the emerging communication tools. Cera *et al.* (2002) asserted that ICT has revolutionised product design in the Architecture Engineering-Construction (AEC) industries, as well as any other area where geometric computation and visualisation have proven essential. Friedman (2005) also

Construction Innovation and Process Improvement, First Edition.
Edited by Akintola Akintoye, Jack S. Goulding and Girma Zawdie.
© 2012 Blackwell Publishing Ltd. Published 2012 by Blackwell Publishing Ltd.

recognised the ICT as the most effective phenomenon, which affected the world's interaction culture in the twentieth-first century. From a similar perspective, Moum (2006) argued that participants within the AEC design process have recently faced ICT related benefits and challenges at several levels. Therefore, with progressive globalisation and specialisation trends within the AEC industry, collaboration amongst design stakeholders at distant locations became crucial (Wojtowicz, 1994; Seng *et al.*, 2005). Today, Computer Supported Collaborative Works (CSCWs) (Wojtowicz, 1994) are no longer mere facilities, but an integral part of comprehensive AEC firms in developed countries. An example of this is the design of the Boeing 777 (Dietrich *et al.*, 2007), which comprised over 10,000 designers in 238 teams organised across 17 time zones.

Moum (2006) proposed the high-level use of the advanced visualisation media to address the emerging issues within the AEC industry; whereas, Fruchter (1998) ascertained that integration of design and construction process could better support collaboration within the global AEC teams, hence significantly decrease labour and material costs. This was further reinforced by Goulding *et al.* (2007), concerning the increased tendency of using ICT tools within design and construction, with the acknowledgement that design engineers are able to experiment and experience decisions in a 'cyber-safe' environment in order to mitigate or reduce risks prior to construction. This was also aligned with Petric *et al.*'s (2002) idea regarding capability of such applications in predicting the cost and performance of optimal design proposals to search for good solutions, and the implementation of such applications as they enable design engineers to compare the quality of any one tentative solution against the quality of all previous solutions. Notwithstanding these potential benefits, the use of such applications is not quite evident within the early stages of the building design process, due to the inherent characteristics of existing Computer Aided Design (CAD) tools, which did not fit into the conceptual architectural design process (Lawson, 1997; Suwa *et al.*, 1998; Craft and Cairns, 2006; Levet *et al.*, 2006; Kwon *et al.*, 2005). This chapter therefore highlights the potential differences between the latter technical architectural and the engineering documentations versus the intuitive conceptual design ideation during the early stages of the design process. The resultant heterogeneity in the utilised design tools during different phases of the whole building design process therefore hinders the integration of the whole process, hence leading to the potential tacit knowledge loss through the transitions of the interrupted procedure (Fruchter, 1998). It also provides theoretical foundations for facilitating the digitalisation of the conceptual architectural design process. It proposes the use of tangible, immersive and interactive Virtual Reality (VR) interfaces as a viable solution, particularly for promoting the integration of data simulation and communication through the whole design and construction processes as well as improving the designers' cognition and collaboration. It also discusses how the aforementioned interfaces could provide supreme intuitive interactions to support '*free*' artistic expression, and bridge the gap between artistic experimentation and accurate manufacturing-oriented modelling. It is therefore argued that this approach could proactively foster multidisciplinary teamwork synergies and provide enhanced outcomes for respective collaborative participants in the AEC industries.

This chapter firstly provides a comprehensive literature review to justify the background issues and challenges within the design representation field. Afterwards, it reports the theoretical highlights of the conducted sequential mixed method research methodology, including the qualitative case study and the cognitive-based quantitative protocol analysis experiment. It then presents the tangible VR design interfaces as a viable solution, particularly for promoting the integration of data simulation and communication throughout the whole design and construction processes.

14.2   Design Innovation and Existing Visualisation Tools

Moum (2006) asserted that a good design process is a fundamental mainstay for a successful building project. Design is an exclusive human activity and a critical aspect of many modern societies. Like most of the other design processes, building design often starts with emergence of the particular needs and ends when the users start using the final product (building). In this vision, the building design process consists of five major phases: Originate, FocusDesign, Build and Occupy (AIA, 1993). Through all five stages, a good project requires a comprehensive teamwork and a good degree of communication and collaboration amongst all stakeholders. However, this communication and collaboration is more crucial during the early phases of problem finding, analysis and conceptual design, which contribute to the determination of 70–80% of production cost of the project (Schütze et al., 2003). The following subsections discuss and elaborate the characteristics of the early stages of building design process and the quality of communications within these phases.

14.2.1   The Conceptual Architectural Design Process

Since the early 1960s, many endeavours have been undertaken to explain or clarify the architectural design process and the genesis of design solutions (Lundequist, 1992). So far, diverse research agendas have been evident in the design thinking area (Akin and Lin, 1995). Some of these agendas dealt with the internal and external representations of designed objects (Akin, 1978); others with the issues of design generation (Cuomo, 1988), for example, the knowledge basis of the design thinking (Waldron and Waldron, 1988), the formulation of design problems (Akin et al., 1992) and the thought processes that apply to the learning (Schön, 1983). Yet, most others deal with refining the general descriptions of the design process offered by the initial group of studies (Akin and Lin, 1995). The first generation of the design methodologies dealing with design process as something chronological and linear are no longer taken into account (Moum, 2006). New generations of design studies generally acknowledge design as a learning domain in which experiences and basic techniques play a large role. However, Suwa *et al.* (1998) argued that the obtained skill in design process is rarely explicit so that even the experienced designers have difficulties in understanding their own design proficiency. As a remarkable effort in explaining the design process, Lawson (1997) critically emphasised that there was no obvious difference between problem and solution, analysis, syntheses or assessment in this procedure. Moum (2006) defined the architectural design as a concurrent learning process about the nature of the problem and the variety of the achievable solutions. Moum's taxonomy for all the activities of the whole design process comprises of four categories: generation of design solutions, communication, evaluation of design solutions and decision making. However, Lawson (1997) argued that the design problem is not easy to define or expose, as it is multi-aspect and iterative. According to Lawson (1997), design process starts with a basic idea in the designer's mind and proceeds as a complex and iterative process between problem and solution. In this manner, while the design team members are fulfilling the different requirements of the project, the primary idea 'materialises' into something to be considered as the conceptual base of the building project.

From a cognitive perspective, Schön (1983) asserted that every designer operates in a mental virtual world as an imitated simulation of the real world in practice; that is, during the conceptual design process, designers generate and develop design solutions by conducting diverse intellectual and physical tasks. However, the human mind is not strong enough to handle such a complicated task without any external aid. In this regard, Fish and Scrivener (1990) argued that the graphic forms provided by the external representations could facilitate the development of the useful ideas and concepts. Lawson (1997) further developed this discussion and asserted that the abstract external design representation could provide the designers with great manipulative and instantaneously analytical autonomy without wasting their time and resources. Lawson indicated that this wasting could be inevitable if the ideas had to be tested directly at the building site. Finally, Moum (2006) acknowledged this idea mentioning that the external design representation tools could significantly help the designers organise the design process and design ideas in a rational and logical way. Moum (2006) therefore asserted that studying the design communication culture and the new external design representation tools could still have great significance.

14.2.2   Communication Culture within the Architectural Design Process

Moum (2006) asserted that decisions made during the early conceptual building design stages were significantly affected by the transactions amongst different stakeholders. Every transaction comprises of the senders, the receivers, and the communication media. In essence, during the architectural design process, a transaction could comprise of an architect as the sender, who sends the design solution to the engineers as receivers. In this example, the design solution is encoded in a drawing as a form of symbolic language (transaction medium) to be understood by all design team members. In practice, even a client also might need to receive a copy of the design solution and decode the message to understand the solution. However, the success in decoding and encoding usually depends on both senders' and receivers' background knowledge (Kalay, 2004). Donath *et al.* (2004) therefore asserted that the success of interaction between designers and clients is dependent upon designers' competence in using some design media that is understandable by the clients.

Although conventional design tools and the symbolic design languages utilised in design disciplines are often compatible with its members' knowledge, the nuances of these are usually misunderstood by the clients and the designers from the other disciplines. Griffith *et al.* (2003) linked this misunderstanding to the tacit nature of the design knowledge. They asserted that a new generation of external representation media are needed to translate design knowledge into an explicit form, in order to be attained by all team members.

14.2.3   External Design Representation Tools

The early conceptual phases of the design process are characterised by fuzziness, coarse structures and elements, and the trial-and-error processes. Craft and Cairns (2006) asserted that searching for form and shape are the designers' principal goal during the early conceptual design stages. Here, the needs of correcting errors are the highest, so that the use of low-expenditure and sketch-like media is crucial. Cross (1999) asserted that during the conceptual design process, the designers' cognitions hinge around the relationship between internal mental processes and their expressions or the external representations through design media.

Cross (2007) argued that the famous architect, Santiago Calatrava, used sketching not only as a communication tool, but also as a means to orchestrate the mental thinking process and to develop the design ability. According to Cross (2007), acknowledging the dialogue or the 'conversation' that goes on between internal and external representations is a part of the recognition that design is reflective. Cross (2007) finally asserted that the designers have to have a medium to enable ill-defined ideas to be articulated, represented, reconsidered, revised and deployed. This supports an earlier claim by Frankenberger (1997) that since the design process is strongly influenced by feedback and dialogue, the communicative function of external representation tools is of great importance within the design profession. Finally, through a cognitive study, Schütze *et al.* (2003) ascertained that the external representations could serve designers as an aid for analysis, solution generation, evaluation, communication and external storage. They asserted that self-made sketch-like external representations could compensate for the shortcomings of the limited human memory regarding processing the complex design problems often encountered in global projects.

14.2.4   Conventional Computer Aided Design Tools

CAD tools are currently being more frequently used in drafting and modelling rather than early conceptual design stages. However, Kalay (2004) argued that a new generation of CAD tools should change the traditional entity of the conceptual design tasks as well. Moum (2006) asserted that such tool could have potential benefits for the conceptual designers, by helping them to generate a huge amount of highly realistic and professional representations of the design solution within a restricted time, hence transferring tacit design knowledge into an explicit form (Griffith *et al.*, 2003) in order to be attained by all other partners. However, according to Lawson (1997) – who ironically used the term 'Computer Aided Drafting' that such tools are yet to attain capabilities to support designing and intuitive thinking of humankind. In this regard, both Lawson (1997) and Griffith *et al.* (2003) argued that a new generation of more flexible digital media is needed, in order to provide the spontaneous reflections required during the conceptual design phases.

14.2.5   ICT and Virtual Reality

Cross (2007) asserted that design support systems should not just be designed to emulate or replace designers, acknowledging that design is also a pleasant and easy task for humankind. Cross (2007) argued that machines should only perform the tasks that are arduous and difficult for human beings. Kalay (2004) proposed the development of some expert Artificial Intelligence (AI) systems (as design agents) that can handle design operations on behalf of the designers. According to Kalay (2004), a design agent is a system, which can make a designer aware of inconstancy in construction checklists. However, in this approach the computers are merely design support tools, not designers. In this respect, VR is an example of emerging design support tools. VR has been defined as a 3D computer-generated alternative environment to be immersed in, for navigating around and interacting with (Briggs, 1996), or as a component of communication taking place in a 'synthetic' space, which embeds humans as its integral part (Regenbrecht and Donath, 1996). The justifications of VR systems usually includes a computer capable of real-time animation, controlled by a set of wired gloves and a position tracker, and use of a head-mounted stereoscopic display for visual output. For instance, Regenbrecht and Donath (1996) defined the tangible components of VR as a congruent set of hardware and software, with actors within a 3D or multi-dimensional input/output space, where actors can interact with other autonomous objects, in real time. Some other studies (Greenbaum, 1992; Yoh, 2001) defined VR as a simulated world, which comprises of some computer-generated images conceived via head-mounted eye goggles and wired clothing – thereby enabling the end users to interact in a realistic 3D situation.

Early studies that incorporated VR into the design profession used it as an advanced visualisation medium. Since as early as 1990, VR has been widely used in the AEC industry, as it forms a natural medium for building design by providing 3D models, which can be manipulated in real time and used collaboratively to explore different stages of the construction process (Whyte *et al.*, 2000). It has also been used as a design application to provide collaborative visualisation for improving construction processes (Bouchlaghcm, 1996). However, expectations of VR have changed during the current decade. According to Sampaio *et al.* (2008), it is increasingly important to incorporate VR 3D visualisation and decision support systems with interactive interfaces, in order to perform real-time interactive visual exploration tasks. This thinking supports the position that a collaborative virtual environment is a 3D immersive space, in which 3D models are linked to databases, which carry characteristics. This premise has also been followed through other lines of thought, especially in construction planning and management by relating 3D models to time parameters in order to design 4D models (Fischer and Kunz, 2004), which are controlled through an interactive and multi-access database. In similar studies, 4D VR models have been used to improve many aspects and phases of construction projects by:

1. developing and implementing applications for providing better communication amongst partners (Leinonen *et al.*, 2003);
2. supporting conception design (Petzold *et al.*, 2007);
3. introducing the construction plan to stakeholders (Khanzade *et al.*, 2007); and
4. following the progress of construction (Fischer, 2000).

14.3   Cognitive Approach to Design

Kan (2008) defined design as a series of decisions that expose the relationship of geometries, materials and performance. Kan clustered some of design activities as thinking and knowing (Cross, 2007), free-hand sketching and interactions (Lawson, 1997), social construction of design solutions (Minneman, 1991) and designing-by-making (Jones, 1970). Goldschmidt and Porter (2004) defined designing as a cognitive activity, which entails the production of sequential representations of a mental and physical artefact. Goldschmidt and Porter (2004) defined designing as a cognitive activity, which entails the production of sequential representations of a mental and physical artefact, and Tversky (2005) noted that when constructing the external or internal representations, designers are engaged in spatial cognition process in which the representations serve as cognitive aids to memory and information processing. Schön (1992) asserted that with the execution of action and reflection, each level of representation makes designers evolve in their interpretations and ideas for design solutions. Such cognitive approach to designing considers design media as something beyond a mere presentation tool. In this approach, reflections, which are caused by design media, are expected to either stimulate or hamper designers' creativity during design reasoning.

Visser (2004) discussed two main paradigms in the cognitive approach to designing: the '*symbolic information-processing*' (SIP) approach represented by Simon (1979), and the '*situativity*' (SIT) approach represented by Schön (1983). Visser described SIP as an approach, which deals with the designers, and the design problems emphasising the designers' rational problem solving process, and SIT as the one that depends on the designers' situational environment and context. Visser (2004) asserted that SIT approach defines designing as a reflective communication with the materials, which belong to design situation. Thus, Visser (2004) adopted a cognitive approach to design by focusing on individuals' activities, which are implemented in professional design projects. In the mentioned study, Visser is particularly concerned with dynamic aspects of designing focusing on the implemented activities by designers and the cognitive processes that they use. In this report, Visser criticised both SIT and SIP approaches, then integrated them into one comprehensive cognitive approach. Here Visser defined design process as an 'opportunistically organised' activity. Visser (2004) defined design as a process to define an artefact and to describe the characteristics, which can satisfy that artefact. Visser's asserted that design is the evolution of representations and there is no permanent hierarchy amongst representations of differing levels of idea abstraction. According to Visser, non-concrete problems cause interruptions, which do not hinder design quality, but can make some opportunities for reflection and improvement of practice. Visser argued that this reflection occurs by a mutual discovery process between the external representation and the designer's cognitive reasoning model.

In cognitive design studies, the other term for discussion is the role of visuo-spatial representations. According to Bruner (1973), external visuo-spatial representations monitor static and dynamic visuo-spatial characteristics of the design ideas. Burner (1973) asserted that deductive reasoning is the basis of the abstract cognition, which is something beyond mere recall from visual literature. In this regard, there are two steps for obtaining a clear understanding of '*design ideas*' and to enhance the '*design creativity*': to transform tacit design ideas into visuo-spatial representations and to make deduction from the external representations (Tversky, 2005). Whereas, from a design creativity perspective, the term 'creative' is used as a value for design artefacts (Kim and Maher, 2008). Yet, according to Visser (2004), in cognitive psychology discussions, creativity is often linked to a design activity, which also comprises particular procedures that have the potential to produce creative artefacts. Cross and Dorst (1999) defined a creative design procedure as a kind of non-routine designing activities that usually are differentiated from the others by the emergence of some considerable events or unanticipated novel artefacts. Kan (2008) asserted that the design process should be evaluated with the level of its creativeness, and that generating ideas and creativity should make common sense. On the other hand, Finke *et al.* (1992) acknowledged that creativity is a result of various mental processes, which totally lead design process towards the phase of creative insight and discovery. They posited that creativity should include initiative stages through which mental representations of '*pre-inventive structures*' are formed.

The *'Unexpected discoveries'* model by Suwa *et al.* (2000) is a more developed tool for measuring design creativity. They introduced unexpected discoveries as some perceptual activities of articulating tacit design semantics into visuo-spatial forms through unanticipated findings by the later inspections. Goldschmidt and Porter (2004) argued that unexpected discoveries (which happen when a designer perceives the depicted items) are some results of external representations that are more important than the representations *per se*. Suwa *et al.* (2000) considered unexpected discoveries as the stimuli that force design process to develop and evolve the 'solution-space'. They argued that there is an iterative interaction between development of the 'solution-space' and sparking new ideas about the 'problem-space'. 'Situative-invention' is another key factor for improving design process. According to Suwa *et al.* (2000), situative inventions refer to the actions in which a designer goes beyond the initial definitions of the problem space. In their explanation, by the situative-inventions, designers form new goals for the solution space to grab the significant parts of the design problem and go beyond a synthesis of solutions that suits the given requirements. On the other hand, Cross and Dorst (1999) posited the modelling of the design creativity as a co-evolution for both problem and solution spaces. Co-evolutionary design is an approach to 'problem solving' defined by Maher *et al.* (1996). In this approach, the design requirements and design artefacts are formed disjointedly while mutually affecting each other. Kim and Maher (2008) asserted that in this approach the changes of the problem could have an effect on the designer's insight into the solutions.

14.4   Virtual Reality Interfaces within Conceptual Architectural Design

The first generation of the design literature highlighted manual sketching using pencils and papers as the main abstract representation method for conceptual design phase. By the year 2000, its effectiveness was frequently appreciated by the scholars (Fish and Scrivener, 1990; Goldschmidt, 1994; Lawson, 1997; Kavakli *et al.*, 1998; McGown *et al.*, 1998; Purcell and Gero, 1998; Cross and Dorst, 1999; Rodgers *et al.*, 2000; Scrivener *et al.*, 2000). This advocacy was at the highest level when Schön and Wiggins (1992) highlighted the importance of freehand sketches as the indispensable media for designers to make reflective dialogue with their own ideas. However, the predisposition to manual sketching methodologies started to diminish with improvements in CAD tools, and their increasing use in the new complex global projects. The reviewed literature noted the increasing tendency for using CAD tools during the early conceptual architectural design process commencing after the year 2000. Scholars were impressed with its excellent capabilities, especially in advanced photorealistic visualisation of the projects (Madrazo, 1999; Marx, 2000). However, doubts about the effectiveness of CAD tools in handling early conceptual design stages started almost concurrently with the appreciations. For instance, Lawson (1997) ironically called such tools 'Computer Aided Drafting' rather than design. Other researches (Suwa *et al.*, 1998; Bilda and

Demirkan, 2003) doubted the usability of such tools, stating that although the CAD media have had a huge impact on the effectiveness of design groups, there are still some characteristics of designing that are exclusively related to freehand sketches. A study by Kwon *et al.* (2005) attributed this inadequacy to the limitation in the intuitive ideation capabilities of the current CAD media. Therefore, they posited that conventional CAD tools might not be desired during the conceptual design phases. Yet, due to the changes caused by the globalisation trends, conventional design representation media are no longer sufficient for handling engineering parts of the design process; the design of Boeing 777 (Dietrich et al., 2007) supports this idea. The difference in the needs for supporting both ill-defined and well-defined aspects of the emerging global design projects highlights the potential stark differences between the engineering documentation versus the intuitive conceptual design ideation by the architects. Currently, this difference makes designers use different design tools when dealing with the different phases of the design process. This transitional utilisation of different media could lead to potential loss of tacit design knowledge (Fruchter, 1998) within transitions of the interrupted process.

To address the aforementioned issues, a new generation of external design representation tools to compensate the shortcomings of both the conventional manual sketching tools and current CAD media, are needed. To fill this theoretical gap, VR as an emerging technology was considered appropriate. With a cognitive approach to the designing, the advanced features of tangible user interfaces (TUIs) and haptic devices, which could help designers to '*feel*' the virtual objects and include the touch sense in the design reasoning, are introduced. TUIs are an expression used to represent a technology, including digital information and physical objects to mimic an actual environment. Kim and Maher (2008) defined the characteristics of TUIs as similar to what is defined earlier for the graspable user interfaces by Fitzmaurice *et al.* (1995). Both studies defined five fundamental properties for such systems as space-multiplex for input and output systems, simultaneous control and manipulation of all components in the interface, high technology tools, spatially responsive digital tools and 3D re-configurability of the tools. According to Kim and Maher (2008), as opposed to a simple time-multiplexed technology that is used in ordinary input devices (e.g. a mouse), the main advantages of TUIs is the space multiplexing input technology that is able to control various functions at different times. One instance of such machinery is haptic technology. In computer science discussions, the term 'haptic' relates to the sense of touch. In other words, this is a technology that unites the user to a digital system, by simulating the sense of touch and applying force-feedback, vibrations and motions to the user (Basque Research, 2007). This physical interaction with the real world is the quality that Stricker *et al.* (2001) described as the technology that reinforces human's cognition by interaction with the physical world. Considering the aforementioned capabilities, this author believes that haptic technology can provide an advanced TUI for designers.

In haptic technology, the sense of touch is not limited to a feeling and it facilitates a real-time interactivity with virtual objects. According to Brewster

(2001), haptic technology is a space in the VR area, since it allows users to use their touch sense to feel virtual objects. Brewster argues that although touch is an extremely powerful sense, it has so far been abandoned in the digital world. Therefore, this study focused on the role of force-feedback facilitated by SensAble Technology TUIs in forming designers' spatial cognition and the effects of the proposed tangible VR 3D sketching interface on designers' collective cognitive and collaborative design activities.

Figure 14.1 illustrates the preceding theories for supporting VR TUIs as a solution for supporting the emerging global design processes. This presents the development of the conceptual architectural design process in order to enhance the integration of the whole building process besides developing and enhancing the current state of the design interfaces (Ibrahim and Pour Rahimian, 2010). This was expected to improve designers' cognition and collaboration during the conceptual architectural design process. This aim is intended to be achieved by developing a new design methodology based on Schön's (1983) 'reflective practitioner' theory and Fitts' (1964) 'motor learning' theory, then verifying its effectiveness based on the collected empirical data. The reflective practitioner theory argued that designers are in a mutual relationship with the external representations and getting reflections from them, whereas, the motor learning theory stated that the tangible interfaces could improve designers' cognitive actions. In this theory, the focus was on the integration of the designers' other senses (e.g. the sense of touch) with their visual sense.

The development of the conceptual architectural design process investigated the current state of the utilised conceptual design interfaces and the existing communication culture amongst the conceptual designers. According to Shuttleworth (2008), a qualitative case study research methodology is an appropriate research methodology for this kind of research that tests theoretical models by using them in 'real-world' situations. The quality of the designers' cognitive and collaborative actions in using VR 3D sketching as an instance of tangible VR design interfaces is also examined.

Design protocol analysis as a quantitative research methodology is predominantly used in the development of conceptual design process as the most prevailing research methodology for studying design processes (Cross et al., 1996). A qualitative case study research approach was undertaken at Universiti Putra Malaysia (Ibrahim and Pour Rahimian, 2010) based on ethnography for data collection and artefact analysis for data analysis was employed. The units of analyses that informed the development of the conceptual framework were design artefacts of the selected second-year architectural design studio, comprising 37 students and 4 studio mentors (Figure 14.2). The gatekeeper during the data collection was the Studio Master of the subject. The case study identified the characteristics of the current design media and the collaborative design culture within the conceptual architectural design process. Consequently, the recommendations of the case study helped to develop theoretical foundations of the study.

The purpose of the quantitative element of the study was to provide empirical evidence for the subjective view that proposed a VR-based 3D sketching interface could improve the designers' spatial cognition and

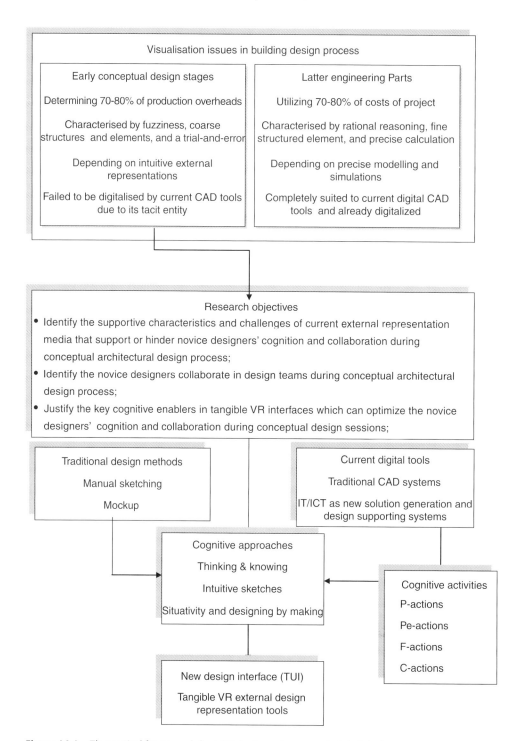

Figure 14.1   Theoretical framework for VR interfaces to support global design processes.

Figure 14.2   Traditional sketching (left) and 3D sketching (right) design settings.

collaboration. In conducting the experiment, a traditional sketching toolkit was developed as a baseline to be compared with the proposed 3D sketching design interface (Figure 14.2). The purpose was to reveal the cognitive and collaborative impacts of the proposed design system. Three pairs of fifth-year architecture students with relevant experience of traditional design and CAD systems were selected as participants for this experiment. During the experiment, protocol analysis methodology (Schön, 1983; Ericsson and Simon, 1993, Dorst and Dijkhuis, 1995, Lloyd *et al.*, 1995, Foreman and Gillett, 1997) was employed for data acquisition. The adapted methodology evaluated the designers' spatial cognition at four different cognitive levels, 'physical-actions', 'perceptual-actions', 'functional-actions' and 'conceptual-actions'. In addition, the designers' spatial cognition in two different collaborative levels, 'cognitive synchronisations' and 'gestures', were evaluated.

14.4.1   Conceptual Architecture Design Experiment: Results

The case study listed three dominant types of sketching (i.e. fully manual, mixed and fully digital) used by the students and their studio mentors. The study also employed four dependent variables and three independent variables in order to identify the supportive characteristics and challenges of the current external representation media. The dependent variables were solution quality, certainty of the correctness of the solution, total solution time and the experienced difficulty in design problem solving while the independent variables were fully manual (FM), mixed method (MM) and fully digital (FD) external design representation modes.

Based on the results from the selected sample, the variance analysis (ANOVA) revealed that the design solutions by the subjects using mixed traditional sketching and CAD modelling tools (MM) produced significantly higher solution quality compared to the other two groups. On the other hand, the fully manual sketching subjects had significantly higher solution quality compared to those subjects who solved the design problems completely in CAD environment. This study therefore posited that using the conventional CAD tools hindered the designers' creativity in the early conceptual design stages.

The analysis of the second dependent variable revealed a significant decrease in the certainties of the correctness of the solutions of FM subjects compared to the other two groups. The results also showed no significant difference amongst the three groups regarding the total time spent for creating their respective solutions. The aforementioned analysis therefore triggers doubts on the competence of the conventional manual sketching media in handling the complicated design stages. It implied that while the MM and FD groups used 3D prototyping techniques to ensure that various design parts fitted together, the FM group was unsuccessful in convincing designers in this regard.

The study applied subjective protocol to enable the difficulty experienced in the design problem solving to be evaluated. Based on the subjective protocol evaluation using the narrative stories transcribed from the recorded videotapes, results indicated that the subjects who had utilised mixed design media were able to pace their design processes with considerable lesser difficultly compared to the subjects from the other two groups. The observations also noted that the same subjects were able to manipulate free-hand sketches to solve design problems faster and easier. They were also able to use digital capabilities for solving the communicational problems, either within the design situations or with other designers. The results of the artefact analysis can be seen in Table 14.1.

From Table 14.1, the study found that amongst the three evaluated external design representation methods, the best tool comprises of both manual and digital tools. The observations and analytical results illustrated that neither manual sketching tools nor conventional CAD software applications are the better media for the conceptual design communications. This study posited that design semantic gets lost when the manual design tools fail in visualising complicated design ideas and design creativity diminishes when the arduous CAD software tools interrupt the designers' intuitive reasoning. The results supported the proposal for development of a tangible VR interface for filling the existing gap between the creative experimentation and the precise manufacturing-oriented modelling.

The experimental protocol analysis research identified key enablers in the VR 3D sketching interface that can optimise novice designers' cognition and collaboration during conceptual design sessions (Pour Rahimian and Ibrahim, 2011). This study employed a cognitive approach to design process to articulate all aspects of the utilised medium during the conceptual architectural design process. The traditional sketching method was selected as the baseline system to be compared with the proposed 3D sketching design methodology and to reveal the cognitive and collaborative impacts of the proposed design system. The experiment comprised of five main steps (van Someren et al., 1994):

1. to conduct the experiment;
2. to transcribe the protocols;
3. to parse process into the segments;

Table 14.1 Quality of the solution, total solution time, and certainty regarding the correctness of the solution (adapted from Ibrahim and Pour Rahimian, 2010).

Dependent variable (I) Type of sketching	N	Solution quality Mean Std. Dev	(J) Type of sketching	$D = J - I$	Sig.	Certainty of the correctness of the solution Mean Std. Dev	(J) Type of sketching	$D = J - I$	Sig	Total solution time Mean Std. Dev	(J) Type of sketching	$D = J - I$	Sig
Fully manual mode (FM)	14	3.38	MM	-.43564*	.03	2.38	MM	- 1.42***	.00	466.6	MM	170.07	.13
		0.47	FD	.47*	.03	0.40	FD	- 1.54***	.00	229.8	FD	88.49	.60
Mixed mode (MM)	13	3.81	FM	.43*	.03	3.80	FM	1.42***	.00	296.5	FM	- 170.07	.13
		0.34	FD	.91***	.00	0.37	FD	- 0.12	.74	168.7	FD	- 81.57	.66
Fully digital mode (FD)	10	2.90	FM	-.47*	.03	3.92	FM	1.54***	.00	378.1	FM	- 8.49	.60
		0.44	MM	-.91***	.00	0.38	MM	0.12	.74	262.7	MM	81.57	.60
Total	37	3.40	Total	f = 13.3***	.00	3.29	Total	f = 64.5***	.00	382.9	Total	f = 2.012	.15
		0.54				0.81				226.3			

* p < 0.05 (significant difference), ** p < 0.01 (very significant difference), *** p < 0.001 (absolutely significant difference).

4. to encode the segments based on the developed coding scheme; and
5. to analyse and interpret the encoded protocols.

In encoding the data collected and developing the hypotheses, the study categorised the designers' cognitive actions into five major action categories as physical, perceptual, functional, conceptual and collaborative. In interpreting the findings, the study relied on the observations of the designers' behaviours during the experiment and on the statistical analysis of the encoded design protocols. The encoded protocol data included the pairs' verbal accounts concurrently per experiment, hence providing adequate data for an empirical exploratory study. Therefore, the conclusion of this study relied more on the encoded protocol data rather the behavioural observations.

14.4.2   Conceptual Architecture Design Experiment: Discussion

The ethnography study (Pour Rahimian et al., 2008) and the artefact and protocol analyses on the ethnography findings (Ibrahim and Pour Rahimian, 2010) show that there is no better winner for the choice of the current media for external design representations. Both studies affirmed the inflexibility of the traditional geometric modelling tools in intuitive ideations. Moreover, both equally noted the shortcomings of the conventional manual sketching tools for further articulating design ideas, whereby it had difficulties in turning tacit knowledge into explicit knowledge (Griffith et al., 2003) for collaboration purposes. In other words, the results showed that neither manual sketching tools nor current CAD interfaces were the perfect media for emerging conceptual design communications. The study proposed that an alternative tool was needed to support the intuitive ideation, the precise manufacturing oriented modelling and the effortless design walkthroughs.

The results of the artefact and protocol analyses showed major barriers with conventional manual external design representation tools when designing complex design artefacts. This was due to their shortcomings in advanced visualisation as the design became further complicated. The current geometrical CAD modelling applications could not directly replace the aforementioned manual external design representation tools, given that the existing CAD software applications are unable to support certain intuitive design requirements of the conceptual design process. In short, the inflexibility of the ordinary arithmetical modelling software on one hand, and the restricted visualisation capabilities of the manual sketching tools on the other hand, increase the tendency for proposing a new generation of design representation media. In using the sketching metaphor, Levet et al. (2006) proposed the use of some design interfaces in which designers can swiftly produce a 3D prototype to exemplify the 3D object they have in mind. Kwon et al. (2005) considered this factor in order to improve computing performance for integrating the progress of the conceptual phase with the remaining design stages. There is an urgent need therefore to develop tangible VR interfaces that have some particular characteristics, as proposed by Levet et al. (2006) and Kwon et al. (2005).

The aim of the conducted experiment therefore was to provide objective and empirical evidence for the subjective view that the proposed sample tangible

 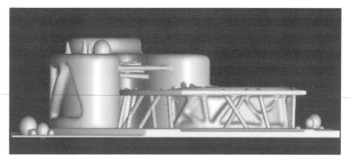

Figure 14.3   Sample 3D sketching design outcome (Pour Rahimian and Ibrahim, 2009).

VR interface, the 3D sketching interface, could improve the designers' spatial cognition and collaboration during the conceptual architectural design phase. The focus was on the designers' cognitive and collaborative actions that were developed. This provided an evidence to support that in 3D sketching sessions the increased integration between the physical actions and the mental perceptions could lead to occurrence of the epistemic actions to improve the designers' spatial cognition. In this regard, Kirsh and Maglio (1994) posited that the epistemic actions could offload the designers' mental cognition partly into the physical world, thus letting them have freer minds to create more design ideas. The tangible VR interface that is proposed has the potential to improve the designers' perception of visuo-spatial features, particularly in terms of unexpectedly discoveries of the spatial features and relationships. This has shown how association between mental cognition and the perception of the physical attributes can offload the mental load and stimulate creativity. According to Suwa *et al.* (2000), the occurrence of unexpected discoveries could lead to the occurrence of more situative inventions and consequently to more creativity.

In terms of functional-conceptual actions of the design process and utilisation of tangible VR design interface improved the participating three pairs of designers' problem finding behaviours as well as improving their co-evolutionary conceptions of perceptions. An example of this improvement is demonstrated by a sample 3D sketching design complied within three hours, as shown in Figure 14.3. Lastly, in terms of the collaborative activities, the tangible VR design interface is capable of motivating the participating three groups of designers to share more ideas together. Moreover, the study explained how the proposed tangible VR design interface was capable to change the type of conversations from ordinary clarifications to new proposals and arguments, which is evidence of more development of both problem and solutions spaces (Kim and Maher, 2008).

Emerging VR technologies (if developed and deployed appropriately) are capable of facilitating some senses beyond the visual aspects of the design artefact, by offering a new generation of 'promising' CAD tools, which embrace designers' cognition and collaboration during the conceptual architectural design process. This provides a theoretical basis for the development of cutting-edge collaborative virtual environments (Maher *et al.*, 2006) in architectural education and associated professional remits.

14.5   Technical Implications for Developing Tangible Virtual Reality Design Interfaces

Existing VR interfaces have ostensibly been formed, based on a single idea: creating 3D models and incorporating them with some pieces of information, so that both 3D models and information are editable through an interactive real-time interface (Pour Rahimian and Goulding, 2010). However, they differ from each other based on their architecture and the utilised methods for data creation and retrieval. Data creation and retrieval methods in VR interfaces can be investigated from two different perspectives, namely creating 3D bodies of constructional elements *per se* and defining characteristics of the elements.

Although creating 3D objects directly in VR environments is not impossible, this is usually created in CAD applications, since doing so in VR is often cumbersome and time-consuming. Consequently, current VR interfaces can be categorised considering how they convert CAD models into VR elements. In terms of transforming design elements from CAD into VR, there are three *de facto* approaches used by different practitioners. Whyte *et al.* (2000) noted three approaches for this translation as:

1. straightforward translation approach and importing the whole environment from CAD to VR;
2. library-based approach and putting the elements of construction in the library of VR environment, then calling them up when and where necessary; and
3. database-oriented approach with a central database for controlling the module characteristics.

Here the database uses both CAD and VR environments as graphical interfaces. Therefore, the third approach can be characterised as a combination of computer graphics and web programming.

Due to the ease of creating the environments by direct translation, most of the current VR systems in AEC projects use the straightforward translation method. Nevertheless, the new generation of interfaces have to migrate to the third paradigm in order to allow designers to see their artefacts throughout the design representation process. Moreover, this is the only way that can help participants share ideas from distance locations. VR interfaces in the AEC industry also vary based on the method of manipulating the objects within the environment and the adapted programming method in the VR interface. There are three major kinds of programming applications currently used by VR programmers:

1. 3D Application Programming Interfaces (APIs);
2. Virtual Reality Modelling Language (VRML) and 3D web technologies; and
3. recent commercialised object oriented VR programming packages.

In this respect, 3D Application Programming Interfaces (APIs) (e.g. Open GL and Direct 3D) are principal environments for VR programming in C++ and Visual Basic. Falling in the category of computer graphics, they are capable of either creating all models directly inside the space or/and importing them from CAD applications. They are perfect environments for advanced programmers for creating Win32 console applications, which are used in developing computer games; however, integration of such interfaces with web programming is difficult and often leads to failures in cases of complicated works.

Virtual Reality Modelling Language (VRML) and 3D web technologies in their first version were made as a division of Open Inventor, thereafter, having become the international standard for 3D web modelling. These applications provide a variety of facilities for manipulating immersive library-based web interfaces; however, they lack the capability of integration with inter-related databases, as they are not essentially database-oriented applications.

Recent commercialised object oriented VR programming packages contain built-in modelling environments for creating VR spaces directly or importing them from CAD applications. Such VR programming applications also contain logical libraries for defining behavioural links amongst the objects and simulating physical phenomena. Although the architecture of such applications is made based on APIs of C++, in some aspects they can offer a higher-level abstraction for programmers. Nowadays, there are three frontier commercial VR programming applications, namely Quest3D™, EON Reality™ and Virtools™. The outcomes of these applications are directly deployable into C++ and Visual Basic's web programming platforms (EON Reality, Inc, 2008). This makes them extremely flexible in terms of integrating VR programming (which is a part of computer graphics) with web programming and data mining. They also come with full Software Development Kits (SDKs) in order to help advanced programmers add some building blocks and prototypes to create rationales or behaviours that were not originally provided by the application. Besides, the SDKs let programmers integrate their interfaces with particular VR I/O devices, for example, Head-Mounted-Displays (HMDs) and data gloves. In this respect, it is now possible to employ tangible VR design interfaces as a new paradigm for simulating truly immersive AEC design collaboration – this technology could easily embrace a database-driven approach using structured modelling phases and API based programming for the development stages. By linking 3D objects to datasets through a web environment, such systems would therefore be able to optimise performance during detailed design stages, which contain collaboration among multiple designers.

14.6   Conclusion

A sequential case study and protocol analysis studies (Pour Rahimian, 2009) was reported in this chapter to present theoretical foundations for improving designers' cognition and collaboration during conceptual design process.

This has become necessary to improve and optimise the operational behaviour of design project teams. The purpose of the protocol analysis was to empirically evaluate and verify the role of 3D sketching using VR interface in facilitating integration between conceptual and engineering parts of building design process. This chapter identified:

1. issues and challenges of the multidisciplinary AEC teams in the emerging global project collaborations;
2. inherent characteristics of the conceptual design process and its external representation tools; and
3. the theoretical and technical requirements of the proposed tangible VR design interfaces as the future external design representation tools.

It is posited that by adopting a cognitive approach to design, the supportive characteristics of tangible VR interfaces to enhance design cognition and collaboration within the AEC practice and profession need further investigation. The introduced advanced digital interfaces could expedite knowledge transfer and stimulate 'creativity' and 'learning through experience' (Suwa et al., 1998) within the AEC education and the associated professional remits. It could also contribute towards transforming conceptual architectural design phase from analogue to digital format, hence linking it to the remaining digital engineering parts of the building design process. This process integration and digitisation could become the leading edge of the AEC enterprises through revolutionising their essential systems, for example, design representations within the interdisciplinary communications, tacit knowledge reuse via organisational memory, documenting and addressing the clients' needs, etc. Such improvements in dynamicity and flexibility of the communicational systems will therefore enable the AEC professionals to share and amplify design semantics throughout the project development life cycle and secure competitiveness of global projects via working collaboratively in geographically dispersed locations.

   This chapter presented how innovation and process improvement in the design process could be achieved by adapting existing VR technologies to enhance the conceptual architectural designers' cognition and collaboration. However, further research is recommended to reveal more technical and theoretical aspects of employing tangible VR design interfaces for multidisciplinary teamwork and utilising ICT in delivering of projects within time and budget. In order to progress the development of the conceptual and theoretical tangible VR design interface, there is a need for extensive research into various aspects of the protocol. Consequently, the of the recommended fields for future studies are development of tangible VR design interfaces in non-collocated collaborative design projects, investigation of the effects of the fully immersive interfaces on the designers' cognition and collaboration, and exploration of higher capabilities of VR interface tools in architectural design by developing customised interfaces based on open source programming applications, etc.

References

AIA (1993). Standard Form of Agreement Between Owner and Architect for Designated Services, AIA (No. B163): The American Institute of Architects (AIA).

Akin, Ö. (1978) *How do Architects Design? Artificial Intelligence and Pattern Recognition in Computer-Aided Design*. North Holland, NY. 65–104.

Akin, Ö. & Lin, C. (1995) Design protocol data and novel design decisions. *Design Studies*, **16**, 211–236.

Akin, Ö., Dave, B. & Pithavadian, S. (1992). Heuristic generation of layouts (HeGeL): based on a paradigm for problem structuring. *Environment and Planning B: Planning and Design*, **19**, 33–59.

Basque Research (2007). Using Computerized Sense Of Touch Over Long Distances: Haptics For Industrial Applications: ScienceDaily.

Bilda, Z. & Demirkan, H. (2003) An insight on designers' sketching activities in traditional versus digital media. *Design Studies*, **24**(1), 27–50.

Bouchlaghem, N., Thorpe, A. & Liyange, I.G. (1996) Virtual reality applications in the UK's construction industry. *CIB W78 Construction on the Information Highway, Bled*, 10–12 June.

Brewster, S. (2001). The Impact of Haptic 'Touching' Technology on Cultural Applications. Glasgow: Glasgow Interactive Systems Group, Department of Computing Science, University of Glasgow.

Briggs, J.C. (1996, 09-01-1996). The Promise of Virtual Reality. *The Futurist Magazine*, 30–31.

Bruner, J. S. (1973). Beyond the information given: studies in the psychology of knowing. Oxford, UK: W.W. Norton.

Cera, C.D., Reagali, W.C., Braude, I., Shapirstein, Y. & Foster, C. (2002) A collaborative 3D environment for authoring design semantics. *Graphics in Advanced Computer-Aided Design*, 22(3), 43–55.

Craft, B. & Cairns, P. (2006). *Work Interaction Design: Designing gor Human Work*. Paper presented at the IFIP TC 13.6 WG conference: Designing for human work. Madeira.

Cross, N. & Dorst, K. (1999) Natural intelligence in design. *Design Studies*, **20**(1), 25–39.

Cross, N. (2007) *Designerly Ways of Knowing*. Birkhäuser, Basel.

Cross, N., Christiaans, H. & Dorst, K. (1996) *Analysing Design Activity*. John Wiley & Sons, Inc, New York.

Cuomo, N.L. (1988) *A Study of Human Performance in Computer Aided Architectural Design*. PhD thesis, University of New York at Buffalo, New York.

Dietrich, A., Stephans, A. & Wald, I. (2007) *Exploring a Boeing 777: Ray Tracing Large-Scale CAD Data*. Paper presented at the Real-Time Interaction with Complex Models.

Donath, D., Loemker, T.M. & Richter, K. (2004) Plausibility in the planning process– reason and confidence in the computer-aided design and planning of buildings. *Automation in Construction*, **13**(2), 159–166.

Dorst, K. & Dijkhuis, J. (1995) Comparing paradigms for describing design activity. *Design Studies*, **16**(2), 261–274.

EON Reality, Inc., (2008) *Introduction to Working in EON Studio*. Available from http://www.eonreality.com

Ericsson, K.A. & Simon, H.A. (1993) *Protocol Analysis: Verbal Reports as Data*. Cambridge MIT Press, Cambridge, MA.

Finke, R. A., Ward, T. B., & Smith, S. M. (1992). Creative Cognition: Theory, Research, and Applications. Cambridge, MA: MIT Press.

Fischer, M. (2000) 4D CAD–3D models incorporated with time schedule, CIFE Centre for Integrated Facility Engineering in Finland. *VTT-TEKES, CIFE Technical Report*. Stanford University, Helsinki.

Fischer, M. & Kunz, J. (2004) The scope and role of information technology in construction. *CIFE Centre for Integrated Facility Engineering in Finland, Technical Report #156*. Stanford University, Helsinki.

Fish, J. & Scrivener, S. (1990) Amplifying the mind's eye: Sketching and visual cognition. *Leonardo*, **4**, 117–226.

Fitts, P.M. (1964) Perceptual–motor skill learning. In: ed. Melton, A.W., *Categories of Human Learning*. Academic Press, New York.

Fitzmaurice, G. W., Ishii, H., & Buxton, W. (1995). Bricks: laying the foundations for graspable user interfaces. Paper presented at the Proceedings of the CHI'95 Conference on Human Factors in Computing Systems New York.

Foreman, N. & Gillett, R. (1997) *Handbook of Spatial Research Paradigms and Methodologies*. Psychology Press, Hove, UK.

Frankenberger, E. (1997) *Arbeitsteilige Produktentwicklung:empirische Untersuchung und Empfehlungen zur Gruppenarbeit in der Konstruktion*. VDI, Düsseldorf.

Friedman, T.L. (2005) *The World is Flat: A Brief History of the 21st Century*. Farrar, Straus and Giroux, New York.

Fruchter, R. (1998) Internet-based web mediated collaborative design and learning environment. In: *Artificial Intelligence in Structural Engineering Lecture Notes in Artificial Intelligence*. Springer-Verlag, Berlin. 133–145.

Goldschmidt, G., & Porter, W. L. (2004). Design representation. New York: Springer.

Goldschmidt, G. (1994) On visual design thinking: the vis kids of architecture. *Design Studies*, **15(2)**, 158–174.

Goulding, J., Sexton, M., Zhang, X., Kagioglou, M., Aouad, Ghassan, F. & Barrett, P. (2007) Technology adoption: breaking down barriers using a virtual reality design support tool for hybrid concrete. *Construction Management and Economics*, **25(12)**, 1239 –1250.

Greenbaum, P. (1992). The lawnmower man. Film and video, 9(3), 58–62.

Griffith, T.L., Sawyer, J.E. & Neale, M.A. (2003) Virtualness and knowledge in teams: Managing the love triangle of organizations, individuals and information technology. *MIS Quarterly*, **27(2)**, 265–287.

Ibrahim, R. & Pour Rahimian, F. (2010) Comparison of CAD and manual sketching tools for teaching architectural design. *Automation in Construction*, **19(8)**, 978–987.

Jones, C.J. (1970) *Design Methods: Seeds of Human Futures*. Wiley, London.

Kalay, Y.E. (2004) *Architecture's New Media–Principles, Theories and Methods of Computer-Aided Design*. The MIT Press, MA.

Kan, W.T. (2008) *Quantitative Methods for Studying Design Protocols*. The University of Sydney, Australia.

Kavakli, M., Scrivener, S.A.R. & Ball, L.J. (1998). Structure in idea sketching behaviour. *Design Studies*, **19(4)**, 485–517.

Khanzade, A., Fisher, M. & Reed, D. (2007) Challenges and benefits of implementing virtual design and construction technologies for coordination of mechanical, electrical, and plumbing systems on large healthcare project. *CIB 24th W78 Conference*, Maribor, Slovenia, 205–212.

Kim, M.J. & Maher, M.L. (2008) The impact of tangible user interfaces on spatial cognition during collaborative design. *Design Studies*, **29(3)**, 222–253.

Kirsh, D. & Maglio, P. (1994) On distinguishing epistemic from pragmatic action. *Cognitive Science*, **18(4)**, 513–549.

Kwon, J., Choi, H., Lee, J. & Chai, Y. (2005) *Free-Hand Stroke Based NURBS Surface for Sketching and Deforming 3D Contents*. Paper presented at the PCM 2005, Part I, LNCS 3767.

Lawson, B. (1997) *How Designers Think – The Design Process Demystified*. Architectural Press, Oxford.

Leinonen, J., Kähkönen, K., Retik, A. Raja., Flood, R.A., William, I. & O'Brien, J. (2003) new construction management practice based on the virtual reality technology. *In: 4D CAD and Visualization in Construction: Developments and Applications*. A.A. Balkema Publishers, Amsterdam. 75–100.

Levet, F., Granier, X. & Schlick, C. (2006) 3D Sketching with Profile Curves. *LNCS*, **4073**, 114–125.

Lloyd, P., Lawson, B. & Scott, P. (1995). Can concurrent verbalization reveal design cognition? *Design Studies*, **16(2)**, 237–259.

Lundequist, J. (1992) *Prosjekteringsmetodikens teoretiska bakgrund*. KTH Reprocentral, Stockholm.

Madrazo, L. (1999) Types and Instances: A paradigm for teaching design with computers. *Design Studies*, **20(2)**, 177–193.

Maher, M. L., Poon, J., & Boulanger, S. (1996). Formalising design exploration as co-evolution: a combined gene approach. In J. S.Gero & F.Sudweeks (Eds.), Advances in formal design methods for CAD (pp. 1–28). Sydney: Key Centre of Design Computing, University of Sydney.

Marx, J. (2000) A proposal for alternative methods for teaching digital design. *Automation in Construction*, **9(1)**, 19–35.

McGown, A., Green, G. & Rodgers, P.A. (1998) Visible ideas: information patterns of conceptual sketch activity. *Design Studies*, **19(4)**, 431–453.

Minneman, S.L. (1991) *The Social Construction of a Technical Reality: Empirical Studies of the Social Activity of Engineering Design Practice*. Stanford University, Greater Manchester, UK.

Moum, A. (2006). A framework for exploring the ICT impact on the architectural design process. *ITCon, 11 Special Issue: The Effects of CAD on Building Form and Design Quality*. 409–425.

Petric, J., Maver, T., Conti, G. & Ucelli, G. (2002) *Virtual Reality in the Service of User Participation in Architecture*. Paper presented at the CIB W78 Conference, Aarhus School of Architecture, 12–14 June.

Petzold, F., Bimber, O. & Tonn, O. (2007) CAVE without CAVE: on-site visualization and design support in and within existing building, eCAADe 07. *25th Conference of Education and Research in Computer Aided Architectural Design in Europe*, Frankfurt. 161–168.

Pour Rahimian, F. (2009) *Cognitive Transformation Mediated by Digital 3D Sketching during Conceptual Architectural Design Process*. Universiti Putra Malaysia (UPM), Serdang.

Pour Rahimian, F., Ibrahim, R. & Jaafar, F.Z. (2008) Feasibility study on developing 3D sketching in virtual reality environment. ALAM CIPTA. *International Journal on Sustainable Tropical Design Research and Practice*, **3(1)**, 60–78.

Pour Rahimian, F. & Ibrahim, R. (2009) *Protocol Analysis Experiment Report for Developing VR Supportive Environment FRSB/EDI/TR2* (No. 2). Environmental Design Integration Group, Faculty of Design and Architecture, Universiti Putra Malaysia, Serdang.

Pour Rahimian, F. & Ibrahim, R. (2011) Impacts of VR 3D sketching on novice designers' spatial cognition in collaborative conceptual architectural design. *Design Studies*, 32(3), 255–291.

Pour Rahimian, F. & Goulding, J.S (2010) Game-Like Virtual Reality Interfaces in Construction Management Simulation: A New Paradigm of Opportunities. *9th International Detail Design in Architecture Conference 2010*. The Central University of Lancashire, Preston, UK.

Purcell, A.T. & Gero, J.S. (1998) Drawings and the design process: A review of protocol studies in design and other disciplines and related research in cognitive psychology. *Design Studies*, **19**(4), 389–430.

Regenbrecht, H. & Donath, D. (1996) *Architectural Education and Virtual Reality Aided Design (Vrad)*. uni-weimar.

Rodgers, P.A., Green, G. & McGown, A. (2000) Using concept sketches to track design progress. *Design Studies*, **21**(5), 451–464.

Sampaio, A.Z. & Henriques, P.G. (2008) Visual simulation of civil engineering activities: didactic virtual models, WSCG 2008. *16th International Conference in Central Europe on Computer Graphics, Visualization and Computer Vision*. Plzen, Czech Republic. 143–149.

Schön, D. A., & Wiggins, G. (1992). Kinds of seeing and their functions in designing. Design Studies, 13(2), 135–156.

Schön, D. (1983) *The Reflective Practitioner: How Professionals Think in Action*. Temple Smith, London.

Schütze, M., Sachse, P. & Römer, A. (2003) Support value of sketching in the design process. *Research in Engineering Design*, **14**, 89–97.

Scrivener, S.A.R., Ball, L.J. & Tseng, W. (2000). Uncertainty and sketching behaviour. *Design Studies*, **21**(5), 465–481.

Seng, W.K., Palaniappan, S. & Yahaya, N.A. (2005) *A Framework for Collaborative Graphical Based Design Environments*. Paper presented at the 16th International Workshop on Database and Expert Systems Applications.

Shuttleworth, M. (2008) *Case Study Research Design*. Podcast, available from http://www.experiment-resources.com/case-study-research-design.html (accessed 5 January 2009).

Simon, H. A. (1979). Models of thought (Vol. 1). New Haven, CT: Yale University Press.

Stricker, D., Klinker, G., & Reiners, D. (2001). Augmented reality for exterior construction applications. In W. Barfield & T. Caudell (Eds.), Augmented reality and wearable computers (pp. 53): Lawrence Erlbaum Press.

Suwa, M., Gero, J.S. & Purcell, A.T. (2000) Unexpected discoveries and S-inventions of design requirements: important vehicles for a design process. *Design Studies*, **21**(6), 539–567.

Suwa, M., Purcell, T. & Gero, J. (1998) Macroscopic analysis of design processes based on a scheme for coding designers' cognitive actions. *Design Studies*, **19**(4), 455–483.

Tversky, B. (2005) Functional significance of visuospatial representations. In: ed. Shah, P. & Miyake, A., *Handbook of Higher-Level Visuospatial Thinking*. Cambridge University Press, Cambridge. 1–34.

van Someren, M.W., Barnard, Y.F. & Sandberg, J.A.C. (1994) *The Think Aloud Method: A Practical Guide to Modelling Cognitive Processes*. Academic Press, London.

Visser, W. (2004) *Dynamic Aspects of Design Cognition: Elements for a Cognitive Model of Design*. France: Theme 3A-Databases, Knowledge Bases and Cognitive Systems, Project EIFFEL.

Waldron, M.B. & Waldron, K.J. (1988) A time sequence study of a complex mechanical system design. *Design Studies*, **9**, 95–106.

Whyte, J., Bouchlaghem, N., Thorpe, A. & McCaffer, R. (2000) From CAD to virtual reality: modelling approaches, data exchange and interactive 3D building design tools. *Automation in Construction*, **10**(1), 43–55.

Wojtowicz, J. (1994) *Virtual Design Studio*. Hong Kong University Press, Hong Kong.

Yoh, M. (2001) The reality of virtual reality. In: *Seventh International Conference on Virtual Systems and Multimedia (VSMM'01), Organized by Center for Design Visualization*. University of California Berkley, Berkley CA.

15

Virtual Planning and Knowledge-based Decision Support

Joseph H.M. Tah

15.1   Introduction

The construction industry is experiencing unprecedented change and dynamic conditions resulting from societal demands for low impact buildings and infrastructural assets with ever-increasing standards of performance, constantly diminishing environmental impacts, and steadily reducing costs of construction, operation and decommissioning challenges. This has fuelled the development of an increasing number of new methods, materials, technologies, processes and innovative practices aimed to improve buildings and communities, with respect to a multitude of sustainability performance considerations and indicators. In this respect, as the number of methods and technological options increases, so does the complexity and associated cost of choosing alternative combinations for any given situation. Therefore, informed decisions require the management of vast amounts of information and knowledge about the combinations of available options and the assessment of their performance. Given this, it is almost impossible to apply manual methods and physical prototypes to address these issues. Furthermore, it has been recognised that the use of computer-based virtual prototyping could offer some solutions to these challenges, and significant Research and Development (R&D) efforts have been made to develop various generations of virtual prototyping systems. These have evolved from early 3D CAD modelling software, through computer integrated environments ranging from 4D modelling, and more recently, to nD Modelling (Lee *et al.*, 2006), including the use of advanced virtual and augmented reality techniques with varying degrees of success.

Although significant progress has been made, further advances are needed as the current generation of virtual prototyping systems are still not 'smart'

Construction Innovation and Process Improvement, First Edition.
Edited by Akintola Akintoye, Jack S. Goulding and Girma Zawdie.
© 2012 Blackwell Publishing Ltd. Published 2012 by Blackwell Publishing Ltd.

enough to cope with the increasingly complex and dynamic nature of construction projects and problems facing the construction industry. This chapter posits that the next generation of virtual prototyping systems will need to draw heavily on multiple configurable knowledge sources in order to cope with the complexity that characterises construction projects. A brief examination of the complexity inherent in construction projects is therefore presented for discussion. An overview of research work related to virtual prototyping that has been undertaken to address these challenges and complexities is discussed, along with a selection of emerging technologies being developed to address this situation. On this theme, emerging developments in Building Information Modelling (BIM) (Eastman *et al.*, 2008) and Industry Foundation Classes (IAI, 2010) that facilitate inter-operability between disparate systems that support virtual prototyping are also examined. The case for the development and use of configurable knowledge bases to provide autonomous real-time decision support for virtual prototyping is introduced, along with a prototype system developed to support this. It is concluded that the current interest in BIM presents an opportunity to develop the next generation of virtual prototyping systems that include configurable knowledge bases. This would facilitate autonomous decision making, thereby rendering such systems more easy to use, and so encouraging wider adoption and use, particularly to support innovation processes in practice.

15.2   The Complex Nature of Construction Projects

It is widely acknowledged that by addressing the contemporary problems facing society, it is not just sufficient to be concerned with the physical built assets, but also the interplay between assets and the social, economic and environmental consequences that arise out of human activity. However, this presents significant challenges due to inherent complexities within and between the elements that constitute the physical built assets and the organisational systems used in their procurement, delivery, use and disposal. Construction projects are characterised by complexity, from their inception, through design, construction and operational life, to their disposal and recycling. For example in buildings, there are multiple interacting subsystems such as building materials and components, the structure, building fabric, building services, utilities, communication networks and the external environment. The procurement of these assets often involves complex interactions between multiple stakeholders and organisations operating in complex value and supply networks, often with conflicting perspectives on their creation, configuration and use. These organisational networks are characterised by social, economic, legal, regulatory and cultural interactions that strongly influence the management of the planning, design and production processes, which involves varying degrees of cooperation, competition and conflict. Thus, construction is a complex multi-disciplinary field that requires integrating expertise and input from various constituencies, with many people processing and exchanging complex heterogeneous information over complex human and communication networks with varying challenges and ever-changing constraints. The host of contemporary problems, together with the inherent complexities in the construction process when taken together, presents considerable challenges to decision makers. Existing conventional problem-solving methods are not always able to deal with this complexity adequately, given that the industry needs to constantly evolve and adapt in response to rapidly changing societal demands. It is for these reasons that there is a lot of interest in virtual prototyping, BIM, inter-operability using open semantic standards and knowledge-based decision support are discussed in the following sections.

15.3   Construction Planning and Virtual Prototyping

Construction planning plays a fundamental role in the process of construction management, and a good construction plan can often be used as a basis for developing the budget and schedule of work. Furthermore, it can also facilitate the formulation of correct strategies for guiding construction activities, and coordinating different construction processes. Therefore, developing a construction plan, (especially a robust one), is a critical task in construction management, given the complex nature of projects. A construction plan is normally represented using a bar chart or Gantt chart, which illustrates the relationship between construction activities and corresponding times, as well as logical constraints amongst the activities. Traditional off-the-shelf construction planning tools, such as Microsoft Project and Primavera Project Planner (P3), bring the convenience of specifying the plan tasks and generating static bar charts and network diagrams. However, they do not represent and communicate the spatial and temporal aspects of construction schedules effectively. Consequently, they do not allow project managers to create robust construction plans. The need for alternative approaches to creating construction plans have long been recognised, and virtual prototyping has emerged as one such approach. Virtual prototyping evolved in response to the requirement for designers and others involved in product development to have a better 'feel' for the end product at the earliest possible opportunity, without the expense of building a physical prototype or making the product itself. This is increasingly seen as necessary within the construction industry, and various technologies are emerging to support this. One such technology that is receiving wider acceptance in the construction industry is 4D CAD (3D CAD with the added dimension of time). This supports a kind of simulation approach to creating a construction plan. In 4D CAD, a construction plan is created and tasks are associated with the 3D building model. This essentially links the work breakdown structure (WBS) to the product breakdown structure (PBS). It allows users to generate a 3D dynamic construction sequence along the time dimension. Through this, a PBS-plus-WBS approach, potential conflicts in the construction plan can be visually identified and updated to formulate a feasible plan. The term '4D CAD technology' in construction was coined at the Centre for Integrated Facility Engineering (CIFE) at Stanford University (Collier and Fischer, 1996). This has been an area of active research due to its potential to facilitate the analysis of the multitude of factors and relationships amongst the logical, temporal, spatial and other dimensional information in construction.

In recent years, the possibilities offered by 4D CAD have spurred researchers into the exploration of the potential of virtual reality (VR) techniques in construction planning. Several prototypes of VR-based construction planning tools have been proposed by researchers. For example, Jaafari *et al.* (2001) developed a prototype application for the evaluation, visualisation and optimisation of construction schedules in a virtual reality environment and Waly and Thabet (2002) developed a Virtual Construction Environment (VCE), which allowed project teams to undertake rehearsals of major construction processes to examine various execution strategies before the start of the actual construction work. Similarly, Yerrapathruni *et al.* (2003) developed an immersive virtual environment (IVE) for generating and reviewing construction plans in a virtual environment and Kunz and Fischer (2005) developed a Virtual Design and Construction method for the simulation of the process of design and construction of buildings. Dawood *et al.* (2005) developed the VIRCON system to allow planners to rehearse and better understand project schedules through 4D visualisation and simulation. Other studies (McKinney-Liston, 1998; Kim *et al.*, 2001; Dawood *et al.*, 2002; Heesom and Mahdjoubi, 2004; Chau *et al.*, 2005; de Vries and Harink, 2007) have demonstrated related advancements adopting this notion. In addition, Li *et al.* (2008) presented cases of how virtual prototyping has been successfully adopted on a number of construction projects where it has integrated both in the design and construction processes, and a number of state-of-the-art virtual prototyping opportunities in construction are now emerging (Brandon and Kocatürk, 2008).

Commercial software for virtual prototyping has matured significantly in recent years. Current 4D modelling products and vendors include:

- ConstructSim™ from Common Point;
- fourDscape™ from Balfour Technologies;
- Innovaya Visual 4D Simulation from Innovaya;
- Navigator™ from Bentley Systems;
- PM-Vision™ from Construction System Associates
- the SmartPlant™ series from Intergraph; and
- Synchro from Synchro, Ltd.

Recognising the benefits of 4D-CAD, more and more construction projects used this technology during the construction process. However, although 4D modelling presents significant advantages over traditional planning tools, 4D models are often time-consuming to generate. Ostensibly, existing 4D systems rely on a plan to be pre-constructed and tasks individually associated with the PBS. This can be time-consuming, which limits the number of simulations and what-if scenarios that can be explored. Existing 4D models also tend to lack configurable data and knowledge bases of construction methods, procedures and resources, which can provide the context for rapidly generating alternative plans for different project scenarios. Although the experience and knowledge of construction managers cannot be wholly replicated in computer systems *per se*, the potential of computers to provide some decision support in automating or semi-automating the process of generating plans is not fully exploited in existing systems. However, the difficulty and cost of creating and using such models is currently limiting the widespread adoption of 4D systems in construction; but the evolution in Building Information Systems and semantic standards for interoperability could allow configurable knowledge bases to be developed to provide decision support for innovation.

15.4   Building Information Modelling

Building Information Modelling is one of the most promising developments in the construction industry. It allows an accurate virtual model of a building to be constructed digitally. When completed, the computer-generated model contains precise geometry and relevant data needed to support the construction, fabrication and procurement activities needed to realise the building. Autodesk (2010) claims to have coined the phrase BIM in 2002, and defined it as 'the creation and use of coordinated, consistent, computable information about a building project in design – information used for design decision making, production of high-quality construction documents, predicting performance, cost-estimating and construction planning, and, eventually, for managing and operating the facility.' Underpinning this is a BIM that has been defined as 'a digital representation of physical and functional characteristics of a facility'. As such, it serves as a shared knowledge resource for information about a facility forming a reliable basis for decisions during its life cycle from inception onwards. The basic premise of BIM is collaboration by different stakeholders at different phases of the life cycle of a facility to insert, extract, update or modify information in the BIM process to support and reflect the roles of those stakeholders. BIM is therefore a shared digital representation founded on open standards for interoperability (NIBS, 2010), which can facilitate information sharing amongst multiple project stakeholders (Figure 15.1). Further examples of BIM data sharing with multidisciplinary stakeholders can be seen in Figure 15.2.

In recent years, there has been a growing interest in the construction industry in the use of BIM tools. This has been brought about by the vendors of 3D CAD software that have re-positioned their software to exploit the BIM concept. Examples of these are:

- Revit and Architectural Desktop from Autodesk (Autodesk, 2010);
- ArchiCAD from Graphisoft (Graphisoft, 2010);
- MicroStation from Bentley (Bentley, 2010); and
- Tekla Structures by Tekla (Tekla, 2010).

However, these tools have their own internal models. These models are different from the information models implemented in applications used downstream in the construction process, for example, applications for

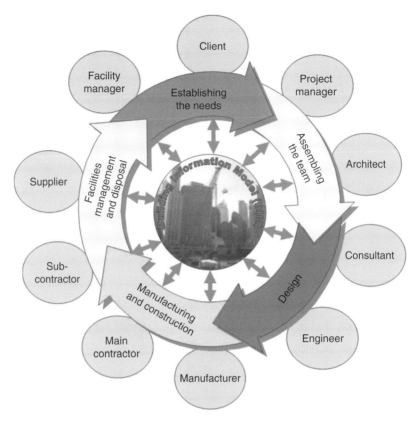

Figure 15.1 Stakeholder information sharing through BIM.

Figure 15.2 Examples of data shared through BIM by multidisciplinary stakeholders.

quantity take-off, cost estimation, construction planning, structural or environmental analyses, etc. This is an unfortunate consequence of the multidisciplinary nature of the construction industry, where historically the development of software has mirrored the specific requirements of individual disciplines, instead of taking a holistic view to facilitate integration and data sharing between project participants. This has led to the reproduction of data for different purposes, which is non-value adding and often error-prone. A study by NIST (Gallaher *et al.*, 2004) identified approximately $15.8 billion in annual interoperability costs in the US capital facilities industry, representing 1–2% of industry revenue. Gallaher *et al.* (2004) also acknowledged that this is likely to be only a portion of the total cost of poor interoperability, as the study did not cover all related costs. It has therefore long been recognised that the solution to this problem is the use on an open and neutral data format to ensure data compatibility across the different applications.

15.5   Interoperability and Industry Foundation Classes

The International Alliance for Interoperability (IAI) has been developing a specification for an international BIM standard, called the Industry Foundation Classes (IFC) since 1994 (IAI, 2010) to address the interoperability problem. The IAI has produced several versions of the specifications for IFC. For example, the IFC2x Platform was officially accepted as the ISO/PAS 16739:2005 in October 2005 (ISO, 2010). IFC is a comprehensive data representation of the building model; it is also a set of rules and protocols of how to define the data describing the building. In this respect, IFCs have been defined by the construction industry, which provides a foundation for the shared project model or BIM. This works through the specification of classes of components in an agreed manner that enables the development of a common language for construction. IFC-based objects allow construction professionals to share a project model, yet also allow each profession to define its own view of the objects contained in that model. IFC also enables interoperability amongst construction software applications, and software developers can use IFC to create applications that use universal construction objects based on the IFC specification. Applications that support IFC can allow members of a project team to share project data in an electronic format. This ensures that the data is consistent and coordinated. Furthermore, this shared data can continue to evolve after the design phase and throughout the construction and occupation of the building. The information generated by the project design team is therefore available to the building construction team and building facilities managers in an intelligent, electronic format through their IFC-compliant software. Numerous IFC-compliant tools have emerged in recent years, and some projects have already started to implement IFC-based applications in the design and construction process. In addition, some countries have adopted IFC specification as the basis of their national ICT strategy for the construction industry (BSA, 2010; Corenet, 2010; Vera, 2010). This has strengthened the status of IFC specifications as a *de-facto* standard and now it appears to be the only significant standard for inter-operable BIM. The CAD vendors previously mentioned now provide IFC-compliant import/ exchange interfaces to their software, to facilitate information exchange between the various BIM systems. Therefore, developments in BIM models and IFCs present a basis on which to explore the development of configurable knowledge bases to provide autonomous decision support for virtual prototyping and testing of innovative solutions for construction projects.

15.6   Knowledge-based Decision Support for Virtual Prototyping

Whilst advocating the development of configurable knowledge bases to support virtual prototyping, it has to be acknowledged that previous attempts at utilising knowledge-based systems techniques for decision support in project management are yet to make an impact in the construction industry. This can be attributed to the slow pace of technology adoption in the industry, and to some extent the lack of semantic data standards or ontologies on which to base this development. However, it is now timely to revisit the use of such techniques as part of BIM systems implementation that are now being widely adopted. The ensuing section presents a prototype system that was developed by the author to demonstrate the integration of 3D-CAD and project management for real-time 4D modelling (Tah *et al.*, 1997). The work centred on creating an information and knowledge continuum across the 3D CAD and Project Management interface. Initially, the goal was to establish the basis for information and knowledge continuum thinking by exploring the application of a combination of emerging techniques such as object-oriented modelling, knowledge-based systems, integrated project-model databases and integrated environments. A brief description of the work undertaken is presented as follows.

15.6.1   Semantically Enhanced Object Models

A comprehensive integrated project object model was developed to represent the data, behaviour and interrelationships between project information. It consisted of the following sub-system object models:

- a project object model;
- a structural system object model;
- a project task system object model;
- a resource system object model; and
- a productivity object model.

The project object model represented the top-level object classes, which pro-vides associations to the other sub-system models. The structural system object model represented information about a building's structural system, extending abstract structural element concepts into specialised concepts such as columns, beams, slabs with their associated material specifications. The structural elements are associated with the project task system object model via the task class, as they are constructed by tasks. This provides the link to project management. A construction project plan is often centred on the task class. The task class is sufficient for traditional planning systems, as is typical of current packaged project management software. However, the task class is predominantly too abstract for knowledge-level inferencing. Thus, the task class was specialised into classes that are more concrete. A task can be a 'work packet' or a 'unit of work' (UoW). The UoW class acts as a base class to specialised units of work, which for example in the case of work involving *in-situ* concreting could be 'fix reinforcement', 'erect formwork', 'place concrete', 'strike formwork', etc. A work packet consists of many units of work. The unit of work is significant for the purposes of knowledge representation and intelligent decision support. Through these methods, several knowledge bases are triggered to provide information and advice on the selection of methods of concreting, the selection of appropriate resources, and the retrieval and adjustment of productivity rates for task duration estimation. The resource system object model represents detailed information about construction resources, including labour, plant, materials, sub-contractors and resource group build-ups. The productivity object model represents different construction methods and their corresponding production rates for the individual units of work. The semantically enhanced object models provided the basis for developing a common language for representing information and an ontology for developing the knowledge models for knowledge-based development.

15.6.2   Knowledge Models

The following rules sets were produced from a knowledge elicitation exercise:

a) setting of the structural component material specifications;
b) elaboration of activities or tasks for each structural component type into units of work and work packets;
c) determining the amount of work in tasks;
d) selecting appropriate concreting methods;
e) selecting resources for concrete placement;
f) selecting resources for fixing bar reinforcement;
g) selecting resources for fixing mesh reinforcement; and
h) selecting resources for erecting formwork.

A plain English-like syntax was used to represent production rules conceptually. The names of classes and attributes in the object models were used in the syntax, where object instances participated in rules. The object models provided a basis and a common language for representing knowledge in the form of rules consistent with the information models. This helped form an information and knowledge continuum to facilitate the prototype development process.

15.6.3   The Prototype System

The object and knowledge models were used as the basis for developing a prototype system (Tah *et al.*, 1997). The MicroStation CAD and Microsoft Project packages were selected to provide CAD and project management services, as they both support Microsoft's Object Linking and Embedding (OLE) Automation capability. This allows systems to be developed inter-operably in real time. The object models were used to build an integrated database server that handled all the data required to be shared by all applications when inter-operating in real time. Two further software components (connectors/adapters), described in the following sections, were then built to provide the link between MicroStation and Microsoft Project.

A software component was built to map building object instances in MicroStation CAD and their interrelationships to the database server and vice versa. This component allows CAD and the database server to inter-operate in real time. It was developed to semantically enhance building objects in CAD to meet the object model's specification. In building this component, it was decided that the full drawing production facilities in MicroStation should be used for producing a drawing. This meant that this software component should be able to scan a MicroStation database to identify individual objects and create instances of objects and their inter-relationships in the integrated database server. A simple computation of the sizes of structural elements in terms of length, width, height and volume was also performed as part of the process of creating instances of objects in the database server. This software component was developed to allow bi-directional communication between MicroStation and the database server. In this respect, a software component was built to access and use the building design objects in the database server to generate construction project programmes using the KBS techniques based on the Units of Work concept presented earlier within MS Project. The KBSs were developed to retrieve instances of structural objects from the database server, and to automatically generate the units of work or work items and the work breakdown structure in Microsoft project. The full facilities of the Microsoft project were used to manually fine-tune the program, and the resulting program information was used to automatically update the database server. The intention was to use the fully populated database server to produce a simulation of the construction sequence in MicroStation using the start and finish times of activities.

The prototype was developed as an open software environment that allowed all applications to inter-operate through the integrated database server in real time. Figure 15.3 shows a working session with MicroStation running in the top left, and Microsoft project running in the bottom, and the software component that allows bi-directional communication between MicroStation and the database server on the top right-hand side. It depicts an object explorer to display the instances of the building objects in the database server. The database server and the software component linking Microsoft Project to the database runs in the background, and is therefore not shown.

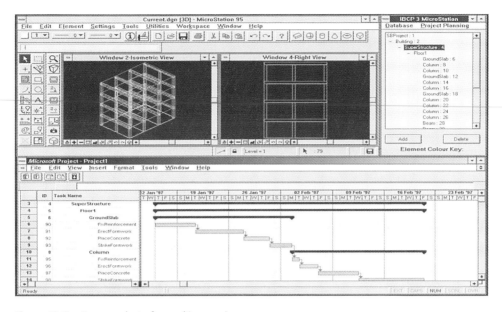

Figure 15.3   A screenshot of a working session.

This prototype demonstrated how complex behaviour could be handled through interactions amongst independent software components. The object classes presented here provide a powerful means for representing data and their interrelationships, to integrate disparate information and knowledge sources from the fragmented construction disciplines. The sharing of information through distributed processing amongst disparate disciplines provides the much needed coordination and communication of information, thereby minimising waste and errors and so resulting in significant reduction in costs. The functionality of the prototype was limited to the creation of construction programmes for the structural system of a building. In this respect, considerable effort concentrated on establishing the approach, as a significant amount of model development, knowledge acquisition and programming effort was required to produce a fully functioning system. This work contributed to the advancement of our understanding of how the next generation of virtual prototyping systems might be developed.

15.7   The Promotion of Innovation through Virtual Planning in Practice

Whilst significant work is still necessary to develop smarter knowledge-based virtual prototyping systems, virtual planning is already supporting innovation in modern construction management processes in practice. Recent improvements in technologies have significantly enhanced the ability to represent and visualise computer-generated information graphically, enabling professionals and practitioners to recognise patterns quickly and accurately. Visualisation enables experts and non-experts to understand the outputs resulting from often highly complex computations and information held in large-scale project and enterprise databases. It provides new ways of communicating options and choices across the multiple and diverse stakeholders involved in a construction project. Computer-graphics and visualisation is usually far better than text-based information for communicating with decision makers and involving them in innovation processes. This allows innovators to look for, and experiment with, new ideas in ways that were not previously possible.

Virtual planning is already influencing contemporary design and construction management practice, by allowing project teams to generate and test new ideas and introduce innovations in the design and construction process. In the UK, for example, construction firms such as Laing O'Rourke, Mace and VINCI, are already using virtual planning systems to explore alternative strategies to help pre-plan large-scale construction processes ahead of the commencement of on-site activities. These firms make extensive use of visualisation technologies to explore design options, cutting costs and reducing the time taken to deliver projects. Laing O'Rourke, for example, has established a Collaboration Centre with two digital prototyping theatres dedicated to virtual planning. This facilitates 4D modelling, simulation and visualisation of complex development and construction projects, from master-planning and regeneration schemes through to detailed structural, mechanical and electrical engineering, including construction sequencing and cost planning. This has enabled better decisions to be made when exploring alternative options for different designs and construction methods. The use of the facility was instrumental in the CLM team from three separate companies (Laing O'Rourke, CH2MHill and Mace) mounting a successful bid to become program managers for the construction work for the 2012 Olympics (Gann and Dodgson, 2008). The ability to produce fly-through simulations has proven invaluable for demonstrating virtual prototypes projects to clients.

Virtual planning can enhance the coordination of innovation processes and changes the traditional design and construction practice. This allows managers to take a wider and more strategic role in their capacity to visualise and understand the implications of alternative decision choices, which continues to improve with the convergence of emerging digital tools that support envisioning and innovation processes. Virtual planning has the potential to improve communication across professional, functional and organisational boundaries, thereby allowing professionals to search for better solutions to complex problems. This requires new management practices and organisational structures to enable collaboration across disciplines and between firms, and skills to broker, interpret, translate and recombine knowledge in multidisciplinary teams. It has been widely acknowledged that the benefits of any technological innovation accrue only when technological possibilities are aligned with clear management objectives, aligned to supportive organisational structures and skills. For example, the emergence of BIM has led large firms to establish new BIM management roles within their organisational structures, where for example, the BIM Manager oversees the successful exploitation of BIM (including virtual planning) and the coordination of input from the multi-disciplinary teams that contribute information to the BIM model. Therefore, the BIM Manager's role is pivotal, as it acts as a conduit to leverage strategy, especially how these parties should work together in the BIM environment.

15.8   Conclusion

The use of the object-oriented methodology provided the basis of a common language or an ontology for the representation of information for the development of the project-model database, the knowledge base and the connectors/adapters ensuring consistency across all applications and facilitating the development and integration process. The development of such systems is necessarily evolutionary and the availability of a widely accepted standards-based common language should inspire confidence in investing the enormous time and resources necessary to achieve real-world practical applications. Thus, the evolution and maturity of the IFCs as such a standard is viewed with a lot of interest. The potential of knowledge-based systems techniques to handle cross-application knowledge representation, data interpretation and transformation to support cross-application business processes and decision support to facilitate virtual prototyping is yet to be fully exploited. What is required is a framework within which to integrate multiple applications through a project model database with knowledge-based support to handle cross-application business processes. Preliminary work has provided a strong indication that multi-agent systems techniques could provide such a framework underpinned by Web Services technologies, but further work is still required (Tah, 2008). This, together with the current burgeoning interests in emerging developments in BIM models and IFCs, provides a strong basis for developing the next generation of virtual prototyping systems with configurable knowledge bases, in order to provide near real-time autonomous decision support. This would allow the modelling and simulation of complex construction projects in ways that are currently not possible. Furthermore, it would also provide multi-disciplinary project teams with easy-to-use collaborative virtual planning tools to support them in exploring alternative innovations in projects, and allow decision makers to select the best options for implementation.

References

Autodesk (2010) *Architectural Desktop and Revit by Autodesk*. Design software. Available from http://www.autodesk.com/ (accessed August 2010).

Bentley (2010) *MicroStation by Bentley*. Design software. Available from http://www.bentley.com/ (accessed August 2010).

Brandon, P. & Kocatürk, T. (2008) *Virtual Futures for Design, Construction and Procurement*. Blackwell Science, Oxford.

BSA (2010) *buildingSMARTalliance – National Institute of Building Sciences*. Available from http://www.buildingsmartalliance.org/index.php/about (accessed August 2010).

Chau, K.W., Anson, M. & Zhang, J.P. (2005) 4D dynamic construction management and visualization software: 1. Development. *Automation in Construction*, **14(4)**, 512–524.

Collier, E. & Fischer, M. (1996) Visual-based scheduling: 4D modelling on the San Mateo County Health Centre. *Proceedings of the 3rd ASCE Congress on Computing in Civil Engineering*. Anaheim, CA.

Corenet (2010) *Construction and Real Estate Network*. Available from http://www.corenet.gov.sg (accessed August 2010).

Dawood, N., Sriprasert, E., Mallasi, Z. & Hobbs, B. (2002) Development of an integrated information resource base for 4D/VR construction processes simulation. *Automation in Construction*, **12**, 123–131.

Dawood, N., Scott, D., Sriprasert, E. & Mallasi, Z. (2005) The *Virtual Construction Site (VIRCON) Tools: An Industrial Evaluation*, Itcon, vol. 10, Special Issue from 3D to nD modelling. 43–54. Available from http://www.itcon.org/ (accessed August 2010).

de Vries, B. & Harink, M.J. (2007) Generation of a construction planning from a 3D CAD model. *Automation in Construction*, **16(2007)**, 13–18.

Eastman, C., Teicholz, P., Sacks, R. & Liston, K. (2008) *BIM handbook – A Guide to Buildong Information Modelling for Owners, Managers, Designers, Engineers and Contractors*. John Wiley and Sons, Ltd, Chichester, UK.

Gallaher, M.P., O'Connor, A.C., Dettbarn, J.L & Gilday, L.T. (2004) *Cost Analysis of Inadequate Inter-operability in the US Capital Facilities Industry*. NIST GCR 04-867. Available from http://www.fire.nist.gov/bfrlpubs/build04/art022.html (accessed August 2010).

Gann, D. & Dodgson, M. (2008) Innovate with vision. *Ingenia*, 45–48.

Graphisoft (2010) *ArchiCAD by Graphisoft*. Design software. Available from http://www.graphisoft.com/ (accessed August 2010).

Heesom, D. & Mahdjoubi, L. (2004) Trends of 4D CAD applications for construction planning. *Construction Management & Economics*, **22**, 171–182.

IAI (2010) *International Alliance for Inter-operability*. Available from http://www.buildingsmart.com (accessed August 2010).

ISO (2010) *International Organization for Standardization ISO, Publicly Available Standards*. Available from http://www.iso.org/iso/en/CatalogueDetailPage.CatalogueDetail?CSNUMBER=38056 (accessed August 2010).

Jaafari, A., Manivong, K.K. & Chaaya, M. (2001) VIRCON: Interactive System for Teaching Construction Management. *Journal of Construction Engineering and Management*, **127**, 66–75.

Kim, W., Lee, H., Lim, H.C., Kim, O. & Choi, Y.K., (2001) visualized construction process on virtual reality. *Fifth International Conference on Information Visualisation (IV'01)*, **July**, 684.

Kunz, J. & Fischer, M. (2005) *Virtual Design and Construction: Themes, Case Studies and Implementation Suggestions*, CIFE Working Paper #097. Available from http://www.stanford.edu/group/CIFE/online.publications/WP097.pdf (accessed August 2010).

Lee, A., Wu, S., Marshall-Ponting, A., Aouad, G., Tah, J.H.M. & Cooper, R. (2006) nD modelling – A driver or enabler for construction improvement? *RICS Research Paper Series*, **5(6)**, 45.

Li, H., Huang, T., Kong, C.W., Guo, H.J., Baldwin, A., Chan, N. & Wong, J. (2008) Integration design and construction through virtual prototyping. *Automation in Construction*, **17(8)**, 915–922.

McKinney-Liston, K., Fischer, M. & Kunz, J. (1998) 4D annotator: a visual decision support tool for construction planners. In: ed. Wang, K.C.P., *Computing in Civil

Engineering, Proceedings of International Computing Congress, Boston, 18–21 October, ASCE, 330–341.

NIBS (2010) *National BIM Standards*. Available from http://www. buildingsmartalliance.org/index.php/nbims/about (accessed August 2010).

Tah, J.H.M., Howes, R. & Wong, H.W. (1997) Towards a concurrent engineering environment for integration of design and construction (CEE-IDAC). In: ed. Anumba, C.J. & Evbuomwan, N.F.O., *First International Conference on Concurrent Engineering in Construction*, The Institution of Structural Engineers, London. 206–215.

Tah, J.H.M. (2008) Future agent-driven virtual prototyping environments in construction. In: ed. Brandon, P. & Kocatürk, T., *Virtual Futures for Design, Construction & Procurement*. Blackwell Science, Oxford.

Tekla (2010) *Tekla Structure*.Design, software. Available from http://www.tekla.com (accessed August 2010).

Vera (2010) *Vera – Information Networking in the Construction Process 1997–2002*. Available from http://vera.vtt.fi (accessed August 2010).

Waly, A.F. & Thabet, W.Y. (2002) A Virtual Construction Environment for preconstruction planning. *Automation in Construction*, **12**, 139–154.

Yerrapathruni, S., Messner, J.I., Baratta, A. & Horman, M.J., (2003) Using 4D CAD and immersive virtual environments to improve construction planning. *Proceedings of CONVR 2003, Conference on Construction Applications of Virtual Reality*, Blacksburg, VA, 179–192.

16

E-readiness in Construction

Eric Lou, Mustafa Alshawi and Jack S. Goulding

16.1   Challenges Facing the Construction Industry

The construction industry is often hailed as the gauge of the global economy. From a value perspective, the industry contributes an estimated 9.8% towards the Gross Domestic Product (GDP) of the EU economy (Business Watch, 2005), and around 8% of GDP in the USA (Researchandmarkets, 2011). In this respect, the European Commission's Information and Communication Technology (ICT) Uptake Working Group report highlighted the importance of ICT-based innovation in bringing productivity improvements and competitive advantage to the industry (EC, 2006). This task force also reported a constant decline in labour productivity, which was mainly attributed to the lack of ICT-related investment since the mid-1990s. This evidence also highlights that higher productivity growth rates were observed in the USA and other world trade partners of Europe through the greater use/integration of ICT by all segments of the economy. However, industries have not been in a position to capitalise on the investment in terms of productivity growth (OECD, 2003). In this context, sustainability, competitiveness and growth of this vital sector of the economy can only be sustained through the pursuit of knowledge and innovation, the latter of which has historically been driven by rapid developments of ICT systems, specifically the ability to capture, store, analyse/manage and exchange data. Therefore, in an increasingly knowledge-based industry such as construction, it is vital to have early access to knowledge-based tools, together with an ICT infrastructure that can handle media-rich services; and in order to remain competitive, construction firms will have to fully embrace this technology (BERR, 2008).

Concerning ICT and the construction industry, whilst several success stories can be highlighted over the past decade, these have mainly focused

---

Construction Innovation and Process Improvement, First Edition.
Edited by Akintola Akintoye, Jack S. Goulding and Girma Zawdie.
© 2012 Blackwell Publishing Ltd. Published 2012 by Blackwell Publishing Ltd.

on technical operations such design, planning, estimating, etc. In addition, there has been continued growth in the uptake of collaborative environments. These applications are designed to manage and control project documents amongst partners, whilst also providing up-to-date information on their progress. However, although these applications can provide 'value' to projects, their actual role in achieving competitive advantage is not overtly documented in seminal literature. In addition, whilst a 'technology push' approach may bring about 'first comer' advantages to organisations, implementing ICT applications to create competitive advantage can only be leveraged by improving businesses processes in line with management objectives, and using ICT as the core enabler (Alshawi, 2007). Therefore, in today's economic climate, competitive advantage can be achieved by focusing on issues such as providing high-quality services and products with minimal cost, having the flexibility to predict and respond to market needs, or through the efficient management of resources. These can be realised by embracing ICT to enable streamlined business processes, which not only make organisations operate efficiently but also allow them to build their knowledge base in order to gain competitive advantage. Notwithstanding this, investment in IT-based business systems such as Building Information Modelling (BIM) collaborative environments (extranets), enterprise resource planning (ERP) and intelligent systems have not realised their full potential (Alshawi and Goulding, 2008). Furthermore, it is estimated that the worldwide cost of IT failure could account to US$6.2 trillion per year, or an estimated US$500 billion per month (Krigsman, 2009; Sessions, 2009) and similar failure stories are reported elsewhere (Alshawi, 2007; Burns, 2008; Business Wire, 2008; Krigsman, 2010). This situation also pervades the construction industry, for example, Salah (2003) identified that 75% of ICT investment in business-oriented systems did not meet there intended business objectives. Furthermore, some of these projects were abandoned, significantly redirected or 'kept alive' despite business integration failures. On reflection, the main attributes of these failures were rarely purely technical by origin, but more often than not related to organisational 'soft issues', which underpin the capability of organisations to successfully absorb ICT into their work practices. In this respect, Basu and Jarnagin (2008) noted that business executives did not fully recognise the full functionality and value of technology to the business, nor did ICT personnel possess an understanding of the business and its strategic objectives. This was reinforced through a survey of Chief Executives and Directors of construction organisations in the UK, where the results demonstrated a high level of awareness regarding the strategic benefits of ICT to achieve innovation and competitive advantage, but a lack of direction on how best to achieve these benefits in their organisations (Construct IT, 2008). This resonates with the findings by Goulding et al. (2007) regarding the importance of understanding technology adoption and diffusion issues, and with findings by Mata et al. (1995) concerning investment uncertainty.

16.2   Business Dynamics and Technology

Today's global business competition is forcing organisations to start using ICT, not just for performance improvement and cost reduction, but also to open up new markets and/or gain a niche advantage over their competitors (Coyne *et al.*, 2000; Davis and Walker, 2009; Toften and Hammervoll, 2009). In this respect, executives who hold a better understanding of ICT are more able to align ICT strategies to their business strategies in order to exploit and leverage innovative business processes. This alignment requires a careful and balanced approach between the level and complexity of the enabling technology, and the required level of process change (expected) within the organisation. However, achieving this balance can often be difficult, as it requires highly skilled professionals who fully appreciate the strategic needs of the business and the benefits and functionalities that advances in ICT could offer in order to leverage the business strategy (Alshawi, 2007). Therefore, the interrelationship between the dynamic nature of business and the supporting IT infrastructure can best be described in five layers, where:

- Layer 1 covers the business environment;
- Layer 2 identifies the ICT and business processes;
- Layer 3 identifies the package solutions;
- Layer 4 presents the enabling software; and
- Layer 5 identifies the hardware and communication technology infrastructure.

Whilst this five-layer approach can be used to influence the effective selection, development and implementation of ICT within organisations, it is also important to highlight the congruence and interrelationship of the four core elements (people and skills, business processes, IT infrastructure and work environment) on the successful implementation of ICT within organisations (Goulding and Alshawi, 2004). From a construction sector perspective, business dynamics embraces a wide range of disparate types and sizes of organisations, with a corresponding range of project scale and complexity. In this respect, the skills and resources needed range from highly skilled designers to non-skilled site workers. This skill set is used to deliver a project, which often includes teams from a range of different organisations and the complexity of each party's arrangements often causes communication problems (Cheng *et al.*, 2001, 2003). It is therefore important that organisations appreciate the importance of building ICT capability with this in mind, especially cognisant of internal and external communication and transfer protocols.

16.3   Building ICT Capability

Organisational capability requires the careful development and deployment of specific organisational competence to achieve business imperatives. In the context of ICT, organisational capability embraces many facets, not least highly flexible skills sets, an acute awareness of change, flexible management structures, well-articulated process improvement schemes, clear business goals and an advanced ICT infrastructure, which is aligned to deliver corporate goals (Brewer and Runeson, 2009; Ahuja *et al.*, 2010). For example, the competence of an organisation needs to develop in order to acquire the capability to strategically benefit from IT falls under four main elements, people, process, work environment and ICT infrastructure. These elements are highly interrelated, that is, developing competence in one element must be accompanied by improvement in the others. In this respect, these interrelationships need people with the necessary skills (and power) to implement process improvements. However, improvements often require management consent and approval, which requires organisations to instil an environment, which can facilitate the proposed change through activities such as motivation, empowerment and management of change. Therefore, the high level of integration between these elements can be enabled by a flexible and advanced ICT infrastructure. The first two elements (people and process) are the key to change and improvements (Lou and Alshawi, 2009), while the other two elements (IT infrastructure and work environment) can be seen as enablers, without which the first two elements could not materialise. Therefore, environments should be created where people are motivated, empowered and made aware of the expected change. They need to be ready to innovative, absorb new ideas and develop and implement them effectively. However, this requires business goals and improvement targets to be clearly articulated and communicated to employees, underpinned by strong support from senior management. In this respect, the time required for an organisation to build up ICT capability is therefore dependent on the level of maturity of the organisation in each of these elements. The following sections explain the relationship between these four elements in terms of achieving ICT capability in order to leverage business needs and secure competitive advantage.

16.4   Business Process and ICT

Business process and the ICT infrastructure within companies are ostensibly governed and augmented by the organisational culture and work practices. The more streamlined and effective these processes are, the more efficient the organisation is likely to be. Given this, careful alignment of ICT can help reshape business processes and facilitate the flow of information between processes (Coltman *et al.*, 2005; Siha and Saad, 2008; Yen, 2009). However, while ICT may be used to automate existing business processes, it is also important to note that automating inefficiently designed business processes will often end in failure (Dickinson, 1997). This is widely categorised as helping to 'do the wrong things faster'. Either way, the required process change is predominantly dependent upon how ICT is deployed, that is, implementing through a third-party product or developing and implementing a bespoke system. In the case of opting to engage a third-party product, business functions will often be 'challenged' by the functionalities of the

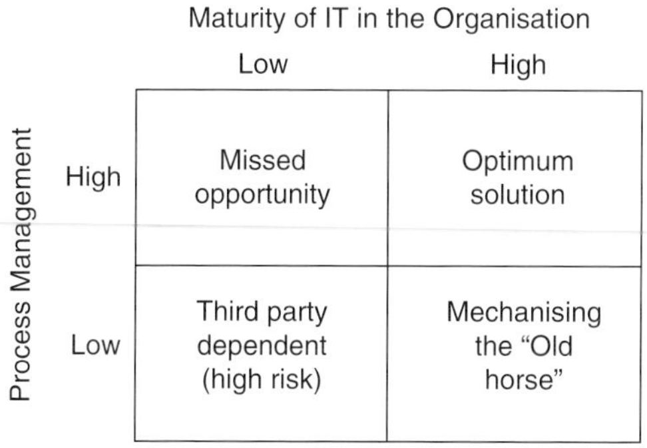

Maturity of IT in the Organisation

		Low	High
Process Management	High	Missed opportunity	Optimum solution
	Low	Third party dependent (high risk)	Mechanising the "Old horse"

Figure 16.1   Relationship between process and IT maturity (Alshawi, 2007).

proposed system. Therefore, the implementation process will require the organisation to change its business processes from its current practice to the one required by the ICT system. In the case of developing a bespoke ICT system, it is likely that the new system will mimic existing practices followed by the core business processes, with a slight change to the supporting business functions to accommodate the new technology. The following section examines the organisational capability needed to maximise the benefits of ICT investment (Figure 16.1). The nature of this process/maturity relationship can be separated into four quadrants. These quadrants link the level of maturity of organisations to manage process improvement with their maturity to utilise and manage ICT (Alshawi, 2007).

From an improvement perspective, innovation and improvement drivers are often initiated and led from within organisations by people (Wells, 2000; Dahlgaard-Park and Dahlgaard, 2010), but this can only be realised when employees are fully aware of the holistic processes involved in achieving sustainable business improvement leading to competitive advantage. Thus, this requires a shared congruence of organisational maturity, where people's readiness is aligned with organisation readiness, that is, they have the capability to change, and the mechanisms available to drive forward the strategic benefits; in essence, creating new core business functions that are enabled by ICT. However, this state of readiness requires investment in the ICT infrastructure, along with investment in people, and is often achieved over a considerable period of time, where systematic iterations of investment (technology/skills) evolve to reach the required a state of readiness. In this respect, there is an inverse correlation between investment and skills/benefits (Figure 16.2).

16.5   People and ICT

Organisational change often cites the paradigms of change in several layers, be it at individual, team, group or at corporate level. Ultimately, organisations need to be able to successfully adapt themselves to new situations and this cannot normally be accomplished without the influence and support of its constituent members (Gardner and Ash, 2003; Vakola *et al.*, 2004;

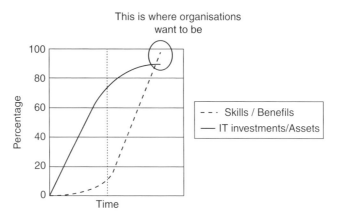

Figure 16.2 Relationship between IT investments, organisational benefits and time (Alshawi, 2007).

Brook, 2009). Thus, wherever the need for change is recognised within an organisation, and wherever the precise nature of such change is formulated, eventually it will be up to the 'people managers' to create the right environment for change to occur. This means that they will have a key role to play in the management of change in organisations. In this situation, organisations therefore need to find out how to release the 'creative energy', intelligence and initiative of people at every level, as resistance to change can be somewhat counterproductive. Consequently, in order for any business improvement to be successful, new behavioural patterns consistent within that of the business improvement initiative must be developed, or business performance will not improve. In this respect, effective change in organisational culture and structure is considered an essential ingredient of ICT augmentation, such that the role of people and organisational culture in delivering successful IT systems is paramount (Kennedy, 1994; Arendt, 2008). For example, Towers (1996) explained that managing change and people together is a major contributing factor to the success of process and IT, and Cooper and Markus (1995) stated that the inadequate treatment of the human aspect is the major cause of reengineering failure. Similarly, Kennedy (1994) highlighted that elements of human change management was 'the most difficult challenge', as employees often felt threatened by new processes and IT systems, which was supported by Arendt (2008), noting that the problem was predominantly down to a lack of proper knowledge, education and skilled owner-managers and employees within the enterprise.

Leadership of the ICT department can therefore be said to be paramount for the success of ICT planning, development, implementation and operation within an organisation. However, the precise role and remit of operations can vary considerably, due primarily to changes in both the technology and business challenges that currently face organisations. In this respect, Computer Sciences Corporation (CSC, 1996) identified six core IT leadership roles needed to execute an ICT initiative, these being: the Chief Architect, Change Leader, Product Developer, Technology Provocateur,

Coach, and the Chief Operating Strategist. Notwithstanding the importance of these, it is also important to acknowledge that the director of the ICT department may also be influenced by a range of issues, not least previous experience, technology predilection/aversion, level of incumbent support and rank/position within the organisation.

16.6   Business Process and Implementation

The work environment is often considered the main enabler of the process and people elements, and is often described as the *esprit de corps* of the organisation. This can be influenced by several factors, the most notable of which are committed leadership, empowerment of employees, communication, process vision development, project management and process-based team formation (Al-Mashari and Zairi, 2000; Damij, 2007). For example, leadership can play a vital role in directing change efforts towards success. The importance of leadership stems from its role in providing a clear vision of the future, communicating this vision and being able to involve people in all aspects of any transition. This includes the ability to motivate people and a holistic perspective and understanding of change. From an employee empowerment perspective, this can be an effective factor for influencing the success of ICT implementation, particularly where it promotes self-management and collaborative teamwork principles (Mumford, 1995; Keatin *et al.*, 2001). Furthermore, employees can become empowered when they are involved in deciding how work (change) should be approached, which technologies to use, and are given the chance to participate in the change and implementation process (Arendt *et al.*, 1995; Bovey and Hede, 2001). Another important facet is that of communication. Communication of change is a vital component of the implementation process, yet this is considered one of the most difficult aspects to achieve (CSC, 1994). The work environment also embodies process vision, as this embraces not only customers and competitors, but also the strategic direction of the company, and the integral links between strategy and process (Davenport, 1993). However, it is important not to lose sight of the importance of project management in the ICT implementation process, particularly concerning piloting (Hammer and Stanton, 1995). Piloting can help to identify potential risks and failures, the nuances of which embrace such issues as politics, through to investment decision making (Dekkers, 2008). The final ingredient within the work environment is that of teamwork. Teamwork has long been acknowledged as an important aspect in the implementation of change, the remit of which encapsulates managing work by making group decisions and coordinating activities, managing relationships by promoting trust, openness, and resolving conflicts, and finally, managing relationships with customers, suppliers and market partners. Thus, teamwork can leverage many advantages, such as facilitating interactions between functions, and speeding up the redesign process (Koufteros *et al.*, 2007; Loukis *et al.*, 2009) through to creating a learning environment in which team members are encouraged to share knowledge and expertise.

16.7   E-readiness

From an e-readiness perspective, it is imperative to implement the 'right' ICT solutions for the right processes, to the right degree, with the right timing. In this respect, organisations and employees need to be prepared and ready for the changes to come – the mantra of which is echoed in a myriad of Change Management literature. Therefore, striking the right balance is the main goal, and all organisations should be striving to achieve this concordance. Industry chiefs and leaders will therefore need to continuously assess and re-evaluate their business models and drivers. On this theme, successful e-business and e-commerce initiatives only take place if, and only if, emergent initiatives are built on robust foundations of readiness (Ruikar *et al.*, 2006; Waseda University, 2007; UN, 2008). Thus, the notion of e-readiness means different things to different people, in different contexts, and for different purposes. As a result, a large gap exists between ideas and concepts on one hand, and practical applications and implications, on the other (bridges.org, 2005; UN, 2008).

There is no single definitive definition for e-readiness, as different people describe and practice it differently. Therefore, the precise definition and meaning of e-readiness is still evolving. To provide a holistic overview, a few thoughts are outlined here to help this discussion. For example, The World Information Technology and Services Alliance (WITSA) stated that an e-ready country required consumer trust in e-commerce security and privacy along with better security technology, more trained workers and lower training costs, less restrictive public policy, new business practices adapted to the information age and lower costs for e-commerce technology (WISTA, 2004). However, the United Nations (UN) perceive e-readiness as the public sector e-Government initiatives of member states according to a weighted average composite index of e-readiness, based on website assessment; telecommunication infrastructure and human resource endowment (UN, 2008). The community assessment of e-readiness by the Center for International Development, Harvard University (CID, 2007) describes an e-ready society as one that has the necessary physical infrastructure (high bandwidth, reliability and affordable prices), has integrated current IT throughout businesses (e-commerce, local IT sector), along with communities (local content, many organisations online, IT used in everyday life, IT taught in schools), has Government (e-Government), has strong telecommunications competition, has independent regulation with a commitment to universal access and has no limits on trade or foreign investment. Furthermore, the Technology CEO Council (2005) views an e-ready community as being equipped with high-speed access in a competitive market, with constant access and application of IT in schools, Government offices, businesses, healthcare facilities and homes, user privacy and online security, and Government policies that are favourable to promoting connectedness and use of the network.

Following these themes, e-readiness can also be defined as the ability of an organisation or economy to deploy Internet-based computers and information technologies to transform traditional businesses into a new economy – an economy that is characterised by the ability to perform business transactions in real time – in any form, anywhere, anytime, and at any price (Bui *et al.*, 2002). Table 16.1 presents an overview of the worldwide

Table 16.1   E-readiness definitions.

Report	Definition of e-readiness
United Nations (UN) (2008)	This UN report assesses e-government readiness of Member States, according to a quantitative composite readiness of e-readiness based on website assessment; telecommunication infrastructure and human resource endowment.
Economist Intelligence Unit (EIU) (2009)	E-readiness is the "state of play" of a country's ICT infrastructure and the ability of its consumers, businesses and governments to use ICT to their benefit. The ranking allows governments to gauge the success of technology initiatives against other countries. It also provides companies that wish to invest in online operations with an overview of the world's most promising investment locations.
Center for International Development (CID), Harvard University (2007)	Readiness is the degree to which a community is prepared to participate in the Networked World. It is gauged by assessing a community's relative advancement in the areas that are most critical for ICT adoption and the most important applications of ICT. When considered together in the context of a strategic planning dialogue, an assessment based on these elements provides a robust portrayal of a community's readiness. The value to a community of assessing its readiness lies in evaluating its unique opportunities and challenges.
The World Information Technology and Services Alliance (WITSA) (2004)	The survey states that an 'e-ready' country requires consumer trust in e-commerce security and privacy; better security technology; more trained workers and lower training costs; less restrictive public policy; new business practices adapted to the information age; and lower costs for e-commerce technology.
McConnell International (2000)	E-readiness measures the capacity of nations to participate in the digital economy. E-readiness is the source of national economic growth in the networked century and the prerequisite for successful e-business.
Asian Pacific Economic Cooperation (APEC) (2000)	Readiness is the degree to which an economy or community is prepared to participate in the digital economy. Every economy, regardless of its level of development, presents a readiness profile on the global stage, composed of its national policies, level of technology integration, and regulatory practices. Readiness is assessed by determining the relative standing of the economy in the areas that are most critical for e-commerce participation. Six broad indicators of readiness for e-commerce are developed into a series of questions that provide direction as to desirable policies that could promote e-commerce and remove barriers to electronic trade.

definitions of e-readiness by leading international groupings, research groups and non-profit organisations. From this, it can be seen that the myriad of differences in e-readiness definitions raises the question: 'What is the most accurate definition for e-readiness?' The answer to this question is an ongoing debate, reflecting that there is no complete literature definition for e-readiness. In spite of all the differences in definitions and opinions, this chapter defines e-readiness as 'the degree to which an organisation may be ready, prepared, or willing to obtain benefits, which arises from the digital economy'. This definition reflects the importance of organisational soft issues such as business processes, management structures, change management initiatives, people and culture. Now it is important to note that e-readiness is increasingly being considered as part of organisational plans and business trajectories.

16.8   Organisational E-readiness in Construction

The construction sector is predominantly project-oriented, where teams of companies get together to design and construct a project, and the team is then disbanded after the project has been completed. Whilst it can be argued that the industry's main functions and processes are still relatively unchanged, there has been a challenge to improve performance and reduce costs using ICT as a lever of change (Marsh and Flanagan, 2000). However, although the potential to improve performance still exists, efforts have been hampered by:

1. the industry's structure;
2. the fragmented supply chain;
3. lack of investment in ICT;
4. limited ICT 'champions' who are able to understand innovation opportunities; and
5. limited support and empowerment of senior decision makers (Lou and Alshawi, 2009).

In this respect, two critical elements can significantly influence the level of ICT project integration, specifically:

1. *Process alignment*: the ability to align organisational processes with the proposed system's functionalities; and
2. *People*: the ability of employees to accept and adapt to the system.

Thus, the relationship between people, process and technology are common themes and enablers of e-readiness. These elements are highly interrelated, as developing competence in one element must be accompanied by improvement in the others. For example, process improvement is a competence that an organisation needs to develop in order to achieve the sought technology capability. By default, this element therefore needs people with the necessary skills and position power to implement process improvements – the mandate of which also embodies the creation of an environment that is conducive to, and can facilitate these proposed changes. This organisational context also embraces such levers as motivation, empowerment and the management of change. Thus, it is important to encourage and support the integration between people and process through a flexible and advanced technology infrastructure. Contextually therefore, the key elements of organisational e-readiness should consider the findings of e-readiness reports, rankings, assessments and measuring tools as 'building blocks for change'.

16.8.1   People

People are the pivotal drivers of a business, and the intellectual capital through which an organisation operates. As a collective force, the people element can be considered the most important asset for e-readiness success. However, people need leadership and a clear vision of the e-readiness future. Therefore, communicating this vision is exceptionally important (Hammer and Stanton, 1995), as they need to understand organisational processes and implement change where necessary. Leadership, empowerment and organisational culture are also important drivers of success in this respect (Lee-Kelley, 2002). One concept that has been particularly effective here is that of Organisational Learning, as this can be used to avoid repeating common mistakes (Oliver, 2009). Culture and society can also be seen as core levers of success, as this embodies organisational behaviour, which in turn contributes to organisational communication (Diefenbach, 2007). Consequently, it is important for organisations to align people (skills) to the right jobs, in order improve innovation and creativity (de Jong and Den Hartog, 2007). Therefore, the promotion of learning through people is fundamental to creating a sustainable ICT workforce that supports, underpins and integrates ICT into the organisation (MacPherson et al., 2005).

16.8.2   Process

Business process can be perceived as a core indicator of how an organisation functions. As a rule, the more effective business processes are, the more efficient the organisation tends to work. As Mulcahy (1990) observes, to be successful, construction organisations need to have clear objectives, and recognise the markets they wish to address, services they wish to provide, risk they may undertake, structures they will use, the environment they will operate within, the controls they will put in place and the returns they wish to achieve. These are core processes, therefore, capturing and disseminating knowledge provides a central organising theme for business change, which can be seen as a continuous process of adaptation and innovation – aligning external requirements with internal capabilities, and linking the two together (Nanoka and Takeuchi, 1995). However, it is important to acknowledge that processes are not typically isolated, as they more often than not are infused with people and technology elements. For example, processes involving project collaboration tend to embody the people involved in the project, and work processes associated with technology employed.

16.8.3   Technology

Within the construction sector, issues surrounding e-commerce, electronic transactions and collaborative environments are increasingly becoming more prevalent (4projects, 2011; BIW Technologies, 2011). Therefore, technology is a major factor of organisational e-readiness, especially as it embraces several core areas, from management and operation, through to standardisation, maintenance, forecasting and investment (APEC, 2000). Thus, ICT needs to be managed in order to ensure that new and emergent technologies are aligned with organisational strategic plans to leverage identifiable benefits. In this respect, this alignment can also help secure competitive advantage, using ICT as a core competency enabler (Construct IT, 2008). A key aspect of this is an organisation's ability to encourage users to engage in this process to procure the information infrastructure and develop management readiness (Boomer, 2006). These issues were reinforced by a strategic study into the thinking of UK AEC executives and IT/innovation directors, on their perception of investment for ICT-based innovation linked to competitive advantage (Construct IT, 2008). Research findings highlighted a series of 'missed opportunities', including people, process and the work environment – to develop core capabilities and be in a state of e-readiness (organisational) to effectively absorb technology into their work practices (to achieve innovation and competitive advantage). In this respect, the key enablers and subcategories can be seen in Table 16.2.

This work was supported by research findings from two scoping studies (Lou and Goulding, 2010) involving 40 domain experts, which identified that the People, Process and Technology issues also had core enablers – the ranking of which can be seen in Table 16.3. These findings identified that the industry did not perceive the solution to becoming e-ready as being predominantly technology-driven, but more through the engagement of leadership, to align change management issues to business processes and the strategic vision.

16.8.4   Means of Achieving E-readiness

Measures for assessing the performance and impact of ICT have predominantly focused upon post-investment/project appraisal decisions. This feedback helped executives learn from their ICT investment decisions. However, extant literature on ICT project failures indicates that many issues still need addressing. In this respect, it is advocated that organisational e-readiness models may be able to help here. The closest approach to this so

Table 16.2 People, process and technology (adapted from Lou and Goulding, 2010).

People	Process	Technology
Leadership & Empowerment	Business & Information Process	Connectivity & Reach
Culture & Society	Information Access & Connectivity	IT & Communication Infrastructure (Technology)
Human Capital & Skills	Security & Integrity	IT & Communication Infrastructure (Reliability)
Learning & Further Education	Policy & Vision	New Technologies
Promotion & Facilitation	Knowledge Sharing & Capture	New Investments
Change Management	Services & Support	Information Infrastructure & Management
Communication	Networked Economy	Interconnectability & Interoperability
Capacity Building	Web Measure & Services	Technology Transfer

Table 16.3 Ranking of the top five e-readiness core enablers (adapted from Lou and Goulding, 2010).

Rank	Enabler	Factor
1	Leadership & Empowerment	People
2	Change Management	People
3	Business & Information Process	Process
4	Policy / Strategy / Vision	Process
5	ICT Sharability / Interoperability	Technology

far is the VERDICT model (Ruikar, 2006), which presents e-readiness points/scores for organisations; but does not specifically address 'how' these organisation might improve their current situation, for example, through maturity level steps, etc. Other attempts include the General Practitioner IS Consultant Tool (Salah, 2003), which provides a generic e-readiness 'snapshot' of organisations. Notwithstanding this, the main objective of self-assessment is to produce an all-inclusive and quantifiable overview of organisational e-readiness that encapsulates the people-process-technology factors, as a well-planned and executed self-assessment (and follow-up action), can deliver significant benefits and help reinforce direction (Hillman, 1994). On this theme, self-assessment techniques are increasingly being used (Benavent *et al.*, 2005), reflecting the importance of these for self-positioning in the market.

In terms of providing a 'quantifiable' overview, maturity level rankings and benchmarking approaches can be used to provide a yardstick for executives to know 'where they are', and 'where they wish to be' in the future (Zairi, 1998), as best practice measures can often create breakthrough improvements (Andersen and Camp, 1995). However, in order to measure the current and expected organisational e-readiness capabilities, maturity-level techniques can be used (Galliers and Sutherland, 2003). For example, the Capability Maturity Model approach has been used successfully by many organisations in a variety of sectors to assess the relative maturity of practices (Rogers, 2009; Zwikael, 2009).

16.8.5   Case Study: E-readiness Construction Framework

Research findings from initial scoping studies identified that the main support mechanisms needed from a priority perspective was for e-readiness assessment tools to gauge organisational e-readiness maturity. Given this, an E-Readiness in Construction (ERiC) framework was developed for organisations to achieve e-readiness. ERiC is a self-assessment framework based on the concepts of maturity modelling. It provides a step-by-step guide for executives to evaluate their business holistically in order to secure e-readiness best practice. This can also be used to undertake benchmarking exercises in order to position themselves in the marketplace. A sample self-evaluation ranking questionnaire from ERiC covering Leadership and Empowerment can be seen in Figure 16.3.

This e-readiness framework presents a solution for construction industry executives to evaluate their state of e-readiness. From an operational perspective, one of the fundamental parameters needed from the outset is that of transparency. Executives therefore need to undertake an organisation-wide audit through this model. This includes strategic and operational issues, from executives and senior management, through to middle management and functional levels. The framework questionnaire consists of five factors, specifically Leadership and Empowerment, Organisational Change Management, Business and Information Process, Organisational Policy and ICT Interoperability. Each factor is evaluated through five separate questions with five maturity-modelled statements. The framework can also include additional weightings to create additional flexibility.

The results from this framework are displayed in a 'web' diagram, which identify the maturity of the five framework factors (Figure 16.4). Each level is characterised by conditional requirements for an organisation to achieve. The maturity levels show a sequential development, from an initial level with basic requirements (Level 1), through to a maximum maturity level (Level 5), categorised as the optimum performance level. The operationalisation of this approach follows the principles of Sarshar et al. (2004), where progression from one level to the next represents a step change in maturity. In this respect, organisations in Level 5 are classified as 'Comprehensively Ready'. at Level 4 'Advanced', Level 3 'Intermediate', Level 2 'Restrictive' and at Level 1 'Obsolete'. The questionnaire results are analysed and presented based on the average score of each sub-factor. Organisations are then provided with factors for improvement, along with corresponding action plans.

Factor

Maturity level

Leadership & empowerment

Involvement	Hands-on	Open-door	Filtered	Restrictive	Non-existent
High-involvement leaders view employees at all levels as true partners - such practices allow the organisation to tap into the creativity and energy of their employees to an extent that is not possible with traditional forms of management.	Top management is involved in developing, participating, delivering and collaborating in new IT ideas/projects from its beginning; and providing an equal platform for communication/debate between all levels in the organisation.	Top management shares initial IT ideas and projects, and employees are free to participate and comment, only then the final decisions are made.	View of employees for new IT ideas/projects is sought through internal IT teams. Team leader will filter the ideas and discussions before passing to top management.	Top management establishes new IT ideas/projects – employees' feedback are welcomed.	An autocratic regime is practised – top management provide the idea/strategy, and the operational people executes – there is a clear divide between the management and operational levels in the organisation.

Example (Collaborative Environments/Extranets implementation)

	Hands-on	Open-door	Filtered	Restrictive	Non-existent
	Top management and employees participate in open forum discussions of the possibility extranets implementation – tapping into the experiences and knowledge of employees before decision is taken by top management to implement extranets in the organisation.	Top management and employees participate in open forum discussions of the possibility extranets implementation – tapping into the experiences and knowledge of employees before decision is taken by top management to implement extranets in the organisation.	Top management and employees participate in open forum discussions of the possibility extranets implementation – tapping into the experiences and knowledge of employees before decision is taken by top management to implement extranets in the organisation.	Top management and employees participate in open forum discussions of the possibility extranets implementation – tapping into the experiences and knowledge of employees before decision is taken by top management to implement extranets in the organisation.	Top management and employees participate in open forum discussions of the possibility extranets implementation – tapping into the experiences and knowledge of employees before decision is taken by top management to implement extranets in the organisation.

Notes					Score

Figure 16.3 Questionnaire sample.

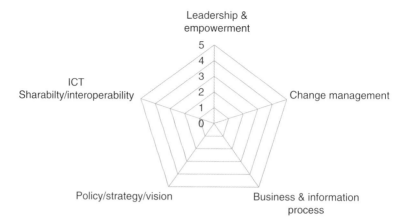

Figure 16.4   ERiC framework: five core factors.

16.9   Conclusion

Whilst globalisation has fundamentally changed businesses, the construction industry has not readily absorbed the integration and implementation of ICT into its business to improve innovation and productivity. It has invested heavily in ICT, and several success stories can be cited, including design, planning, costing, etc. However, there is little evidence to support success in areas such as generating competitive advantage, opening up new market leverage opportunities, applying high levels of innovation, etc. One of the principal reasons behind this is an open acknowledgement that the industry needs executives with a strategic understanding of ICT in order to align and exploit strategies to leverage innovation and secure competitive advantage. This requires organisations to build ICT competence, with agile and flexible skill sets that are aligned to business deliverables – the precursor of which requires flexible management structures, well-articulated process improvement schemes, clear business goals and leaders fully conversant in Change Management. However, it is also important to recognise at this juncture that the prevailing level of organisational culture can work as both a lever and a barrier to success. In this respect, the role and importance of people and organisational culture should not be underestimated.

Sustained successful e-business initiatives traditionally only occur if they are based upon robust foundations. This requires companies to have effective measures and systems in place to design, develop, deliver and evaluate these. For example, e-readiness initiatives embrace a myriad of factors, from business processes, through to management, people and culture. These elements are all interrelated, and more often than not, improving one area in isolation will not ordinarily procure success without improvements in other areas being made at the same time. On this theme, research findings identified three pivotal areas existed, specifically People, Process and Technology. These were recognised as the main underpinning factors, with leadership and empowerment, managing business and information processes, and change management seen as the main enablers. In this respect, an e-readiness framework was presented for discussion. This framework enables construction executives to critically evaluate their organisation and position against five levels of maturity: Leadership and Empowerment, Organisational Change Management, Business and Information Process, Organisational Policy and ICT Interoperability. Approaches of this nature are posited as being increasingly important in the construction industry, as organisational e-readiness can be seen as a conduit for leveraging business excellence. However, for this to happen, organisations will have to radically re-think their People, Process and Technology relationships (internally and externally), as these symbiotic relationships will govern the success or otherwise of any approach adopted.

References

4projects (2011), Available from *http://www.4projects.com/OurProduct/Integration.aspx* (Date accessed 20/12/2011)

Ahuja, V., Yang, J. & Shankar, R. (2010) IT-enhanced communication protocols for building project management. *Engineering, Construction and Architectural Management*, **17**(2), 159–179.

Al-Mashari, M. & Zairi, M. (2000) Revisiting BPR: A holistic review of practices and development. *Business Process Management Journal*, **6**(1), 10–30.

Alshawi, M. (2007) *Rethinking IT in Construction and Engineering: Organisational Readiness*. Taylor and Francis, Oxford.

Alshawi, M. & Goulding, J. (2008) Organisational e-readiness: Embracing IT for sustainable competitive advantage. *Construction Innovation: Information, Process Management*, **8**(1),

Andersen, B. & Camp, R.C. (1995) Current position and future development of benchmarking. *The TQM Magazine*, **7**(5), 21–25.

APEC (2000) *E-Commerce Readiness Assessment Guide*. Asia-Pacific Economic Co-operation (APEC) Readiness Initiative, Electronic Commerce Steering Group.

Arendt, C.H., Landis, R.M. & Meister, T.B. (1995) *The Human Side of Change*, Part 4. IIE Solutions, 22–27.

Arendt, L. (2008) Barriers to ICT adoption in SMEs: how to bridge the digital divide? *Journal of Systems and Information Technology*, **10**(2), 93–108.

Basu, A. & Jarnagin, C. (2008) How to tap it's hidden potential. *Wall Street Journal* Available from http://online.wsj.com/article/SB120467900166211989.html (accessed 6 January 2011).

Benavent, F.B., Ros, S.C. & Moreno-Luzon, M. (2005) A model of quality management self-assessment: an exploratory research. *International Journal of Quality & Reliability Management*, **22**(5), 432–451.

BERR (2008) *Supporting Innovation in Services, Department for Business*. Enterprise and Regulatory Reform, Crown copyright, URN 08/1126, London.

BIW Technologies (2011), Available from *http://www.biwtech.com/cp_root/h/Applications/Performance/201/?lang=* (Date accessed 20/12/2011)

Boomer, L.G. (2006) The 10 rules of technology management. *Accounting Today*, **20**(3), 20–23.

Bovey, W.H. & Hede, A. (2001) Resistance to organizational change: the role of cognitive and affective processes. *Leadership and Organization Development Journal*, **22**(8), 372–382.

Brewer, G. & Runeson, G. (2009) Innovation and attitude: Mapping the profile of ICT decision-makers in architectural, engineering and construction firms. *International Journal of Managing Projects in Business*, **2**(4), 599–610.

Bridges.org (2005) *E-readiness assessment: Who is Doing What and Where?* Cape Town. Available from http://www.bridges.org/files/active/0/ereadiness_ whowhatwhere_bridges.pdf (accessed 28 February 2008).

Brook, I. (2009) *Organisational Behaviour: Individuals, Groups and Organisation*, 4th edn. Financial Times/Prentice Hall, Essex, UK.

Bui, T.X., Sebastian, I.M., Jones, W. & Naklada, S. (2002) *E-Commerce Readiness in East Asian APEC Economies – A Precursor to Determine HRD Requirements and Capacity Building*. Asia Pacific Economic Co-operation (APEC), Telecommunication and Information Working Group.

Burns, M. (2008) Enterprise software survey 2008. *CA Magazine*, **September**, 14–15.

Business Watch (2005) *The European e-Business Market Watch, Sector Report No. 08-II, ICT and Electronic Business in the Construction Industry*, IT adoption and e-business activity in 2005, European Commission, Enterprise and Industry Directorate General.

Business Wire (2008) *Survey Cites Dissatisfaction with Incumbent ERP Vendors*. Available from www.sys-con.com/node/594952 (accessed 11 January 2009).

Chan, A.P.C., Chan, D.W.M. & Ho, K.S.K. (2003) Partnering in construction: Critical study of problems for implementation. *Journal of Management in Engineering. (ASCE)*, **19**(3), 126–135.

Cheng, E.W.L., Li, H., Love, P.E.D. & Irani, Z. (2001) Network communication in the construction industry. *Corporate Communications: An International Journal*, **6**(2), 61–70.

CID (2007) *Readiness for the Networked World. A Guide for Developing Countries, Information Technologies Group*. Center for International Development (CID), Harvard University.

Coltman, T.R., Devinney, T.M. & Midgley, D.F. (2005) Strategy content and process in the context of e-business performance. In: ed. Baum, B., *Strategy Process (Advances in Strategic Management)*. Emerald Group Publishing Ltd, UK. 349–386.

Construct IT (2008) *Strategic Positioning of IT in Construction: An Industry Leaders' Perspective*. Construct IT for Business, Salford, Greater Manchester, UK.

Cooper, R. & Markus, M. (1995) Human Reengineering. *Sloan Management Review*, 36 (Summer), 39–50.

Coyne, K.P., Buaron, R., Foster, R.N. & Bhide, A. (2000) Gaining advantage over competitors. *McKinsey Quarterly*, June. Available from http://mkqpreview1. qdweb.net/PDFDownload.aspx?ar=1057 (accessed 20 August 2010).

CSC (1994) *State of Reengineering Report, Computer Sciences Corporation*. CSC Index Research, London.

CSC (1996) *New IS leaders*. Computer Sciences Corporation, CSC Index Research, London.

Dahlgaard-Park, S.M. & Dahlgaard, J.J. (2010) Organizational learnability and innovability: A system for assessing, diagnosing and improving innovations. *International Journal of Quality and Service Sciences*, **2**(2), 153–174.

Damij, N. (2007) Business process modelling using diagrammatic and tabular techniques. *Business Process Management Journal*, **13**(1), 70–90.

Davenport, T. (1993) *Process Innovation: Reengineering Work through Information Technology*. Harvard Business School Press, Boston, MA.

Davis, P.R. & Walker, D.H.T. (2009) Building capability in construction projects: a relationship-based approach: Engineering. *Construction and Architectural Management*, **16**(5), 475–489.

de Jong, J.P.J & Hartog, D.N. (2007) How leaders influence employees' innovative behaviour. *European Journal of Innovation Management*, **10**(1), 41–64.

Dekkers, R. (2008) Adapting organizations: the instance of business process reengineering. *Systems Research and Behavioral Science*, **25**, 45–66.

Dickinson, B. (1997) Knowing that the project clothes have no emperor. *Knowledge and Process Management*, **4**(4), 261–267.

Diefenbach, T. (2007) The managerialistic ideology of organisational change management. *Journal of Organizational Change Management*, **20**(1), 126–144.

EIU (2009) The 2009 e-readiness rankings. The Usage Imperative. *Economist Intelligence Unit (EIU) Research Reports*. Available from http://graphics.eiu.com/pdf/E-readiness%20rankings.pdf (accessed 7 September 2009).

European Commission (2006) *ICT Uptake, Working Group 1, ICT Uptake Working Group draft Outline Report*, October 2006. Brussels.

Galliers, R.D. & Sutherland, A.R. (2003) The evolving information systems strategy. In ed. Galliers, R.D. & Leidner, D.E., *Strategic Information Management*, 3rd edn. Butterworth Heinemann, Oxford.

Gardner, S. & Ash, C.G. (2003) ICT-enabled organisations: a model for change management. *Logistics Information Management*, **16**(1), 18–24.

Goulding J.S., Sexton, M.G., Zhang, X., Kagioglou, M., Aouad, G. & Barrett, P. (2007) Technology Adoption: Breaking Down Barriers Using a Virtual Reality Design Support Tool for Hybrid Concrete. *Journal of Construction Management and Economics*, **25**(12), 1239–1250.

Goulding, J.S. & Alshawi, M. (2004) A Process-Driven IT Training Model for Construction: Core Development Issues. *Journal of Construction Innovation*, **4**(4), 243–254.

Hammer, M. & Stanton, S. (1995) *The Reengineering Revolution*. Harper Collins, New York.

Hillman, G.P. (1994) Making Self-assessment Successful. *The TQM Magazine*, **6**(3), 29–31.

Keating, C.B., Fernandez, A.A., Jacobs, D.A. & Kauffmann, P. (2001) A methodology for analysis of complex socio-technical processes. *Business Process Management Journal*, **7**(1), 33–50.

Kennedy, C. (1994) Reengineering: the human costs and benefits. *Long Range Planning*, **27**(5), 64–72.

Koufteros, X.A., Nahm, A.Y., Cheng, E.T.C. & Lai, K-H. (2007) An empirical assessment of a nomological network of organizational design constructs: From culture to structure to pull production to performance. *International Journal of Production Economics*, **106**(2), 468–492.

Krigsman, M. (2009) *IT Project Failure. Worldwide cost of IT failure: $6.2 Trillion*. Available from http://blogs.zdnet.com/projectfailures/?p=7627&tag=col1;post-7695 (assessed 6 January 2011).

Krigsman, M. (2010) *Understanding Marin County's $30 million ERP failure. ZDNet*. Available from http://www.zdnet.com/blog/projectfailures/understanding-marin-countys-30-million-erp-failure/10678?tag=nl.e539 (assessed 13 September 2010).

Lee-Kelley, L. (2002) Situational leadership. Managing the virtual project team. *Journal of Management Development*, **21**(6), 461–476.

Lou, E.C.W. & Alshawi, M. (2009) Critical success factors for e-tendering implementation in construction collaborative environments: people and process issues. *Journal of Information Technology in Construction (ITcon)*, **14**, 98–109.

Lou, E.C.W. & Goulding, J.S. (2010) The pervasiveness of e-readiness in the global built environment arena. *Journal of Systems and Information Technology*, **12**(3), 180–195.

Loukis, E., Pazalos, K. & Georgiou, S. (2009) An empirical investigation of the moderating effects of BPR and TQM on ICT business value. *Journal of Enterprise Information Management*, **22**(5), 564–586.

MacPherson, A., Homan, G. & Wilkinson, K. (2005) The implementation and use of e-learning in the corporate university. *Journal of Workplace Learning*, **17**(1/2), 33–48.

Marsh, L. & Flanagan, R. (2000) Measuring the costs and benefits of information technology in construction. *Engineering and Architectural Management*, 7(4), 423–435.

Mata, F.J., Fuerst, W.L. & Barney, J.B. (1995) Information technology and sustained competitive Advantage: A resource-based Analysis. *MIS Quarterly*, **19**(4), 487–505.

McConnell International (2000) *Risk E-Business: Seizing the Opportunity of Global E-Readiness*. McConnell International and WISTA publishing.

Mulcahy, J.F. (1990) Management of the Building Firm. *Proceedings CIB 90, Joint Symposium on Building Economics and Construction Management*, Sydney. 11–21.

Mumford, E. (1995) Creative chaos or constructive change: business process reengineering versus socio-technical design. In: ed. Burke, G. & Peppard, J., *Examining Business Process Reengineering: Current Perspective and Research Directions*. Kogan, New York, 192–216.

Nanoka, I. & Takeuchi, H. (1995) *The Knowledge Creating Company – How Japanese Companies Create the Dynamics of Innovations*. Oxford University Press, Oxford.

OECD (2003) *Comparing Labour Productivity Growth in the OECD Area: The Role of Measurement*. OECD Statistics Working Paper 2003/5. OECD Statistics Directorate, Paris.

Oliver, J. (2009) Continuous improvement: role of organisational learning mechanisms. *International Journal of Quality and Reliability Management*, **26**(6), 546–563.

Researchandmarkets (2010) US Construction Sector: An Analysis. Available from http://www.researchandmarkets.com/reports/335700/us_construction_sector_an_analysis (assessed 6 January 2011).

Rogers, G.P. (2009) The role of maturity models in it governance: a comparison of the major models and their potential benefits to the enterprise. In: ed. Cater-Steel, E., *Information Technology Governance and Service Management: Frameworks and Adaptations*. IGI Global, New York. 254–265.

Ruikar, K., Anumba, C.J. & Carrillo, P.M. (2006) VERDICT-An e-readiness assessment application for construction companies. *Automation in Construction*, **15**, 98–110.

Salah, Y. (2003) IT *Success and Evaluation: A General Practitioner Model*, PhD Thesis, Research Institute for the Built Environment (BuHu), University of Salford, Greater Manchester, UK.

Sarshar, M., Haigh, R. & Amaratunga, D. (2004) Improving project processes: best practice case study. *Construction Innovation: Information, Process, Management*, **4**(2), 69–82.

Sessions, R. (2009) *The IT Complexity Crisis: Danger and Opportunity*. ObjectWatch, Inc, Houston, TX.

Siha, S.M. & Saad, G.H. (2008) Business process improvement: empirical assessment and extensions. *Business Process Management Journal*, **14**(6), 778–802.

Technology CEO Council (2005) *Freeing Our Unused Spectrum*. CEO, Washington, DC.

Toften, K. & Hammervoll, H. (2009) Niche firms and marketing strategy: An exploratory study of internationally oriented niche firms. *European Journal of Marketing*, **43**(**11/12**), 1378–1391.

Towers, S. (1996) Reengineering: middle managers are the key asset. *Management Services*, 40(12), 17–18.

United Nations (2008) *UN E-Government Survey 2008. From E-Government to Connected Governance*. Department of Economic and Social Affairs, Division for Public Administration and Development Management, United Nations Online Network in Public Administration and Finance (UNPAN), United Nations Publications, New York.

Vakola, M., Tsaousis, I. & Nikolaou, I. (2004) The role of emotional intelligence and personality variables on attitudes toward organisational change. *Journal of Managerial Psychology*, **19**(2), 88–110.

Waseda University (2007) *e-Government Ranking 2007, Waseda University Institute of E-Government*. Graduate School of Global Information and Telecommunications Studies, Japan.

Wells, M.G. (2000) Business process reengineering implementations using Internet technology. *Business Process Management Journal*, **6**(2), 164–184.

WITSA (2004) *The WITSA Public Policy Report 2004*. World Information Technology and Services Alliance (WISTA), Geneva.

Yen, V.C. (2009) An integrated model for business process measurement. *Business Process Management Journal*, **15**(6), 865–875.

Zairi, M. (1998) *Effective Management of Benchmarking Projects. Practical Guidelines and Examples of Best Practice*. Butterworth-Heinemann, Oxford.

Zwikael, O. (2009) Critical planning processes in construction projects. *Construction Innovation: Information, Process, Management*, **9**(4), 372–387.

17

Building Information Modelling

Umit Isikdag, Jason Underwood and Murat Kuruoglu

17.1   Introduction

Building Information Modelling (BIM) has become a key research area for addressing problems relating to the sharing of information and collaboration throughout the life cycle of a building. The importance of BIM stems from having an open interchange of information across platforms and a transferable record of building information throughout a life cycle of a building. BIM is a general term used for defining the information management process throughout the life cycle of a building, which mainly focuses on enabling and facilitating the integrated way of project flow and delivery, through the collaborative use of semantically rich 3D digital building models in all stages of the project and building life cycle. BIMs are semantically rich shared 3D digital building models that form the backbone of the BIM process.

The traditional nature of the Architecture Engineering and Construction (AEC) industry involves bringing together multi-disciplines/practitioners in a one-of-a-kind project, which requires a high amount of collaboration and coordination. The AEC industry is therefore an information intensive industry, the work practices of which are 'document-centric' with construction project information being captured predominately in documents, such as CAD drawings, specifications, blueprints and so on. Therefore, information is ostensibly distributed amongst the various multi-disciplinary teams involved in the project as disparate documents. This document-oriented nature of the industry combined with the lack of interoperability between applications has resulted in significant barriers to communication and collaboration between the stakeholders; which in turn, has significantly affected the efficiency and performance of the industry (Latham, 1994; Egan, 1998, 2002). The software applications used in the

Construction Innovation and Process Improvement, First Edition.
Edited by Akintola Akintoye, Jack S. Goulding and Girma Zawdie.
© 2012 Blackwell Publishing Ltd. Published 2012 by Blackwell Publishing Ltd.

AEC design processes have until the last 15 years been focused on automating 2D drawing board techniques, structural analysis and time/cost management activities, while model-based information management and interoperability was mostly disregarded. Gallaher *et al.* (2004) indicated that US$15.8 billion was lost annually in the USA Capital Facilities Industry due to the lack of interoperability. In order to tackle the problems related to information integration and interoperability, recent Research and Development (R&D) has focused on:

1. facilitating collaboration of teams over the Internet. for supporting virtual AEC enterprises;
2. developing tools for generating semantically rich 3D digital building models; and
3. utilising distributed platforms and databases for managing building information.

Having evolved from the term Building Product Modelling (i.e. the realisation of the Product Modelling paradigm for the AEC industry), BIM has now become a key research area in addressing the problems related to the sharing of information and collaboration in AEC industry. For example, Underwood and Isikdag (2010) noted that in the late 1990s BIM was prescribed as a remedy for overcoming the illness of 'Data Interoperability' between applications in the AEC industry. However, today it apparent that this 'magic remedy' has evolved to cure many more issues than it was originally prescribed for, where BIM is now promising to be the new facilitator of integration, interoperability and collaboration towards reshaping the traditional AEC industry. BIM is now also widely accepted as a new and promising information management process and strategy for integrating the processes and stakeholders of an AEC project. BIMs are intelligent digital building models that can either reflect the 'to-be' or 'as-is' situation of a project/building, and resides at the centre of the BIM strategy.

This chapter focuses on BIM as a process and strategy of information management based on using BIMs to store and manage building information in all stages of an AEC project (from conception to demolition/disassembly). Currently, the implementation of the BIM paradigm is achieved by enabling information exchange of several applications supporting BIMs by the use of agreed models (i.e. schema standards), such as Industry Foundation Class (IFC) and CIS/2. In this context, BIM can be realised in many different areas and on different levels, including facilitating the design phase of the project life cycle, and depending on the environment in which they are used, they can have different functions such as linking macro (outdoor) and micro urban spaces (i.e. buildings) to facilitate information sharing between software applications. They can also store building information throughout the life cycle of a building, including procurement related tasks in the building life cycle, whilst supporting the simulation of construction processes (i.e. in nD). In addition, BIM can help facilitate the integration of information systems, provide real-time and on-demand building information over the Internet, and enable advanced analysis, whilst supporting the design and construction of environment friendly/energy efficient buildings, including facilitating emergency response operations.

Recent research demonstrates many examples on the different uses and applications of BIMs for fulfilling a variety of requirements in the building life cycle. A few key examples are explained in Rebolj *et al.* (2010) and Spearpoint (2010), who identified how 4D to nD simulation applications can be facilitated using BIMs and countries such as Singapore have used BIM in earnest. BIM can also facilitate the design of energy-efficient buildings towards addressing sustainability and reduction in CO_2 emissions issues (Hua, 2010; Solis and Mutis, 2010). In fact, the use and functions of BIMs as intelligent digital models is not limited to these few areas – BIM is firmly developing as the solid information management strategy for the AEC industry. This chapter focuses on introducing BIM as an innovative approach to information management for AEC processes. It develops the different viewpoints on BIM, from sharing and exchange, through to implementation planning. Latter sections of this chapter present case studies on different BIM implementations/practices, in Turkey and UK, for discussion and reflection.

17.2   Background

BIM has emerged as an evolution from *de-facto* standards of drawing exchange formats, for example, Drawing Exchange Format (DXF) through to Building Product Models (BPM), which in the main are based on ISO 10303 technologies. Most of the BPMs adopted product-modelling concepts by being enablers of communication, exchange, sharing, interpretation and processing of information. In parallel with the wider recognition of IFC and CIS/2 models in the AEC industry, BPMs have more recently been referred to as BIMs, as most of the BPMs (including IFC and CIS/2) are defined using the ISO 10303 standard, which uses the term information modelling in its model definition language EXPRESS. The term BIM is highly recognised with distinguished efforts of valuable R&D groups in academia and industry, and with wide support of buildingSMART (formerly the International Alliance for Interoperability). Today, IFC and CIS/2 are the key schema standardisation efforts in the area of BIM, and IFC is the effort of IAI/buildingSMART whose goal is to specify a common language for technology to improve the communication, productivity, delivery time, cost and quality throughout the design, construction and maintenance life cycle of buildings. For example, in 2005, IFC became an ISO Publicly Available Specification (as ISO 16739). In addition, CIS/2 is a widely adopted open standard for the digital exchange and sharing of the engineering information relating to a structural steel framework.

17.2.1   Overview on Building Information Modelling

BIM has been defined as a new way of creating, sharing, exchanging and managing the information throughout the entire building life cycle (NBIMS, 2007). The NBIMS initiative categorised BIM in three ways:

1. a product (an intelligent digital representation of a building);
2. a collaborative process which covers business drivers, automated process capabilities and open information standards use for information sustainability and fidelity; and
3. a facility of well-understood information exchange, workflows and procedures, which teams use as a repeatable, verifiable, transparent and sustainable information-based environment used throughout the building life cycle.

The US General Services Administration BIM Guide (2006) indicated that the information in a BIM catalogues the physical and functional characteristics of the design, construction and operational status of the building. This information spans a number of disciplines and application types. BIM therefore integrates this information into one database in a consistent, structured and accessible way. The importance of BIMs stems from having an open interchange of information across platforms and a transferable record of building information throughout a building life cycle. Isikdag *et al.* (2007) identified the definitive characteristics of BIMs as being object-oriented, data rich/comprehensive, in 3D, spatially-related and rich in semantics. Cerovsek (2010) underlined five standpoints for BIM analysis, which can be interpreted as viewpoints to BIM. From these standpoints, BIM can be viewed as:

1. a Model
2. a Modelling Tool
3. Communicative Intent
4. Individual Project Work, or
5. Collaborative Project Work.

From this perspective, Cerovsek (2010) describes a 'BIM' as a digital representation of an actual building for project communication over the whole building-project life cycle. This includes the physical, tangible appearance of a building from a time standpoint, which can be represented by three model categories, 'as-it was', 'as-it-is' or 'as-to-be'. Furthermore, a 'Building Information Model Schema' can be defined as a non-linguistic data structure that describes abstractions of generalised properties of a collection of states of information about buildings to be used in project communication. Underwood and Isikdag (2010) defined BIM as the information management process throughout the life cycle of a building, from conception to demolition, which mainly focuses on enabling and facilitating the integrated way of project flow and delivery, through the collaborative use of semantically rich 3D digital building models in all stages of the project and building life cycle. The authors interpreted BIMs as the set of semantically rich shared 3D digital building models that form the backbone of the BIM process.

17.2.2   BIMs, Interoperability, Integration

In Information Systems, interoperability is the ability for two systems to understand one another and to use the functionality of one another. The word 'interoperable' indicates that one system can perform an operation of the other system as if it is the other system. The case of interoperability equates to loose system integration. As mentioned by Panetto and Molina (2008), integration goes beyond interoperability as it involves functional dependence. While interoperable systems can also function independently, an integrated system loses significant functionality if one of the systems is interrupted. Two integrated systems are inevitably interoperable, but two interoperable systems are not necessarily integrated. The literature review in the area of software integration highlights two main levels of integration/interoperability as Data/Information Level and Application/Service Level (Hohpe and Woolf, 2003; Linthicum, 2003; Erl, 2004; Imhoff, 2005; Ruggiero, 2005). Data/Information Level Interoperability refers to the situation where software can perform functions using the data produced by other software that is interoperable with the first one. In fact, if two systems are considered as integrated in the data level, then one must use the data produced by the other system to function properly. From the perspective of software integration and interoperability, BIMs can be categorised as standard data/information models, which act as enablers of Data Level Interoperability. In other words, software applications exchange BIMs or interact with a shared BIM residing in a database to perform operations over the building information that is generated, processed or stored by another application.

17.2.3   Sharing and Exchange of BIMs

From an ISO 10303 viewpoint, in the information exchange scenario one software system maintains the master copy of the data internally and exports a 'snapshot' of the data for others to use. In information sharing, there is a known master copy of the data, and Create/Read/Update/Delete (CRUD) operations can be performed over the data by multi-user interactions. A BIM schema represents the logical data structure (or data model) containing all entities, attributes and relationships in the model. The model is then created by software and stored in physical files or databases. It is therefore possible to exchange physical files representing the entire set of model classes or a model subset between different applications. BIMs can be stored in a shared database where CRUD operations can be performed by multiple applications. This type of shared data stores are referred to as Shared Databases or Building Information Model Servers (e.g. Express Data Manager, 2010). BIMs can be accessed over the web for querying and visualisation purposes, where applications automating web-based querying and visualisation of BIMs are known as Building Information Model Servers (BIMServer, 2010). The information stored in BIMs can be accessed by using web-services (web interfaces to data and applications). Two definitive characteristics of web services are Loose Coupling and Network Transparency (Pulier and Taylor, 2006). Web services are *loosely coupled*, that is, when a piece of software has been exposed as a web service, it is simple to move it to another computer as the service functions are independent of the client application that is using the service. On the other hand, as web services' consumers and providers send messages to each other using open Internet protocols, web services offer total *network transparency* (i.e. the location of the web service does not have an impact on its function). Stateless Web Services (i.e. Representational State Transfer – REST) are built upon resources (i.e. anything that is available digitally over the web), their names (identified by uniform indicators, i.e. URIs), representations (i.e. metadata/data on the current state of the resource) and links between the representations. In Stateless Web Services, each client request is treated as an independent transaction, and the service does not store information regarding the client that makes the request. From the perspective of BIMs, services can be used to query the models.

17.2.4   Implementation Planning

Implementation planning often has to be performed prior to any BIM implementation, the nuances of which involve the development of an implementation/execution plan in order to make full use of the exchange and sharing of building information based on the developed implementation/execution plan. A valuable information resource of this area is the *BIM Execution Planning Guide* (2010) developed by Pennsylvania State University. This guide can be viewed as a tool that provides process maps on accomplishing several design, analysis and construction processes for using BIMs. A BIM Project Execution Plan is therefore developed in the early stages of a project, and updated as additional participants are added to the project. It is also monitored, updated and revised (as needed) throughout the implementation phase of the project. The plan defines the scope of BIM implementation on the project, identifies the process flow for BIM tasks, defines the information exchanges between parties and describes the required project and company infrastructure needed to support the implementation. The guide provides a four-step procedure to develop a detailed BIM Execution Plan. These steps are:

1. identifying the appropriate BIM goals and use on the project;
2. designing the BIM execution process;
3. defining the BIM deliverables; and
4. identifying the supporting infrastructure needed to successfully implement the plan.

Each organisation or project can follow its own route of BIM implementation, as each individual AEC project tends to have its own unique characteristics. Thus, the BIM implementation effort can be seen as being distinct. However, in order to understand how BIMs can be implemented, it is useful to investigate previous real-life experiences. Following this, the implementation of BIM real-life case studies from two different countries, specifically Turkey and the UK, are presented for discussion.

17.3   A Case Study on the Implementation of BIM

Various case studies on the implementation of 4D modelling for visual representation of construction processes (with or without BIM) have been reported in literature (Collier and Fischer, 1995; Haymaker and Fischer, 2001; Kam and Fischer, 2002; Yerrapathruni, 2003; Heesom and Mahdjoubi, 2004; Gao et al., 2005; Faddoul et al., 2006; Gao and Fischer, 2008; Hsieh, 2009; Mahalingam et al., 2010). This case study explains the use of BIM in the process of visualisation (i.e. 4D modelling) for a real-life large-scale AEC project completed in Turkey. This case study used BIM to generate a 4D model of a football stadium. The Galatasaray Türk Telekom Stadium project is based in Istanbul (Turkey), the project of which consisted of a stadium, an indoor sports hall and shopping centres, and areas assigned for concerts, meetings and congress. The project encompassed a structure that included technical characteristics that were different from other stadiums in Turkey, in terms of architecture and technology. The groundwork of the construction commenced in December 2007. The stadium spans a horizontal axis of 190 m and a longitudinal axis of 228 m, with a ground area of approximately 43,000 m². Cast concrete and pre-cast concrete construction technology was used in the construction of the main parts of the structure. The stadium can accommodate 52,647 people and it was formally opened in January 2011. The work schedule for the stadium construction was prepared by the contractor's planning group using Primavera Project Planner (P3) software. When the work schedule was being prepared, the project was considered in six main phases covering the construction of six blocks of the stadium. Each construction block was named A, B, C, D, E and F, and each main phase (block) was divided into three sub-phases within itself (to simplify the scheduling process). The main purpose of this segmentation was to define an efficient production planning strategy by dividing the work schedule into phases and sub-phases. The BIM for this stadium was generated using BIM applications (i.e. AutoCAD and Revit) and transferred into Synchro 4D Modelling software. Synchro was then used to link the work schedule and 3D geometric representation of the building provided by the BIM (Figure 17.1).

In the object linking process, Synchro considers the building elements available in the BIM as material resources, which can be linked with the tasks in the work schedule. In the linking process, any incorrect resource assignment can be deleted from the resources based on the warnings issued by the software or by user perception. In every stage of the linking process, new tasks may be assigned to resources. Following the object linking procedure, the software prepares the visualisation and simulation of the construction process (i.e. the 4D model of the construction). In the 4D visualisation of the construction processes, the state of the construction can be observed for any given date and from various angles. Figure 17.2 presents four different states of the construction starting from early dates through to completion.

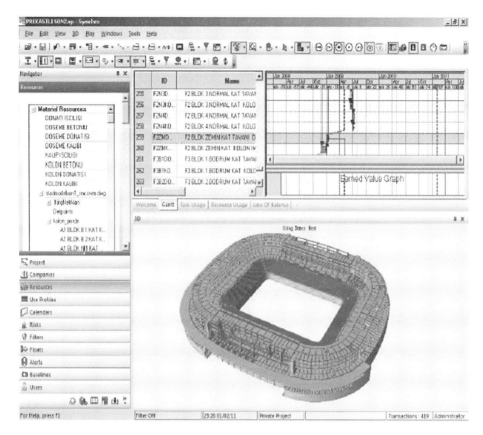

Figure 17.1 Linking work schedule with 3D geometric representation.

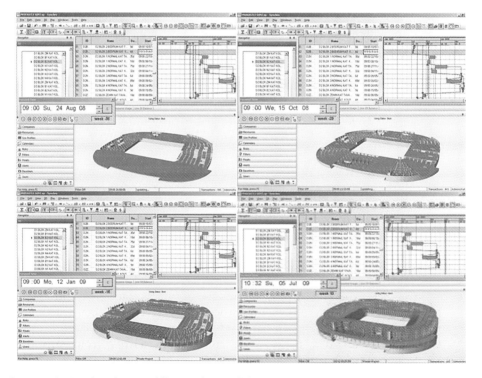

Figure 17.2 Synchro showing different phases of the project.

The strategic analysis conducted with the contractor organisation indicated that the IT knowledge level of users, their BIM related experience and technical support from software vendors could play a critical role in the successful implementation of BIMs for 4D visualisation. Problems encountered during the BIM-based 4D visualisation process have often been acknowledged as being related to the time spent on establishing the 4D model, along with reporting difficulties. This study revealed that BIM-based 4D visualisation was not well recognised in Turkey, due mainly to the lack of focus in Turkish AEC industry regarding BIM technologies and process visualisation of AEC projects. The next section explores how well BIM technologies were received and recognised in the UK AEC industry.

17.4   Building Information Modelling in the UK

BIMs can be seen as interoperability enablers. In this respect, the interoperability requirements of construction include the drawings, procurement details, environmental conditions, submittal processes and other specifications necessary for building quality. Proponents advocate that BIM can be utilised to bridge the information loss associated with handing a project from the design team, to the construction team and the building owner/operator, by allowing each group to add to and reference back to all information they acquire during their period of contribution to the BIM. This interoperable element is vital for a number of reasons. First, there is little evidence to indicate that any single tool can contain all the data relating to the entire building for its entire life. This was reinforced by Smith and Tardif (2009), who stated that:

- The entire building life cycle of business process and workflows is too complex to be modelled effectively in one system.
- Business processes vary too much across the industry.
- A single project model would involve too much change to existing information management infrastructure and business processes to support viable migration paths from existing workflows to new ones.
- The cost and technical challenges of such a system would be prohibitive.

Therefore, an open non-proprietary exchange method for data and processes is an essential element of any BIM system. Observations of all of the BIM definitions seem to lead to significant consistency, but all of these fall short of a clear, concise definition that could be placed before a client or chief executive. This has proved to be the biggest downfall in the UK market, as the industry has failed to recognise the importance of this. However, the BIM market is still gaining momentum.

17.4.1   The Evolutionary Adoption of BIM

The industry is now faced with a new economic situation that both threatens and offers real opportunities for exploitation. For example, it is important to observe how the industry has evolved to this point since moving from

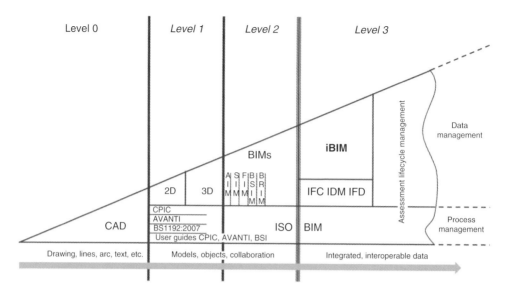

Figure 17.3 BIM Evolutionary Ramp – construction perspective (adapted from Bew *et al.*, 2008). Reproduced by permission of Mark Bew & Marilyn Richards.

drawing boards in the 1970s and 1980s, towards facilitating industry-wide adoption, where the BIM vision can be seen in a strategic way. This can be observed as a pseudo-Darwinian evolution (Bew *et al.*, 2008; Bew and Underwood, 2010). The 'evolutionary ramp' model (Figure 17.3) recognises that all forms of asset data managed in a collaborative way form part of what could loosely be called a BIM, and in the context of the model outlined, this would indicate anything beyond Level 1.

An important theme in the development of any implementation and adoption strategy is the separation of the management of data and process (Figure 17.3). The BIM levels presented within the 'evolutionary ramp' have been identified to enable a simple identification for the level of BIM a business or project is using. Tangible savings have been identified at each level, as progress is made along the evolutionary ramp. The level definitions can be seen as:

- *Level 0:* Unmanaged CAD, probably 2D with paper as the most likely data exchange mechanism.
- *Level 1:* Managed CAD in either 2D or 3D format using BS1192:2007, with a collaboration tool providing a common data environment and possibly some standard data structures and formats. Commercial data is managed by standalone finance and cost management packages with no integration.
- *Level 2:* Managed 3D environment held in separate discipline 'BIM' tools with attached data. Commercial data is managed using Enterprise Resource Planning (ERP), while integration is based on proprietary interfaces or bespoke middleware that could be regarded as 'pBIM'. The approach may also utilise 4D program data and 5D cost elements.

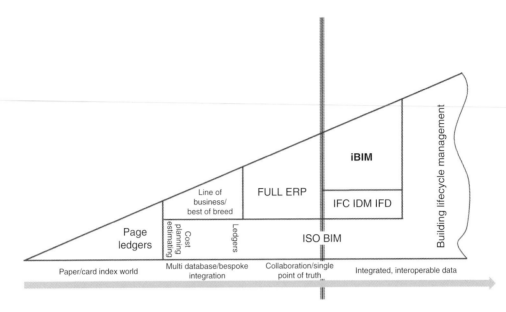

Figure 17.4 BIM evolutionary ramp – commercial systems (adapted from Bew *et al.*, 2008). Reproduced by permission of Mark Bew & Marilyn Richards.

- *Level 3:* Fully open process and data integration enabled through IFC/IFD. This is managed by a collaborative model server that could be regarded as iBIM or integrated BIM, potentially employing concurrent engineering processes.

Later stages where there are fully interoperable models will need new technologies to deliver the concept, possibly using Atomic or Federated BIM, to enable effective large data model sharing. This may need advanced Identity Lifecycle Management systems to control access and security. As can be seen from these definitions, 3D CAD data or model data alone is only a small part of the story, and tools that create data and enable processes to act on that data are the vital differentiator in the world of true BIM. Clearly, all of these tools have a similar evolution to create, save and transact processes over asset data. In this respect, leading players in ERP and environmental markets have been active in enabling their products with features to enable BIM-like operation. The model in Figure 17.4 shows the parallel route the 'commercial systems' (ERP) market is taking, developing more functionality that could support moving BIM into the core ERP toolset. This is also evident in the consolidation at the top end of the market, with large players such as Oracle acquiring businesses such as Symmetry and Primavera to give object viewing, planning/project control and life-cycle management capabilities.

While there are common characteristics in definitions with respect to the people, process and technology aspects, the wide variety of definitions that exist have emanated from the perspectives and vested interests of individual industry stakeholders. However, it is important to note that none of these definitions are from the practising UK industry itself. Consequently, this has significantly contributed to what appears to be a distinct lack of understanding and awareness of BIM across the UK industry, with BIM currently meaning different things to different stakeholders. The implications could potentially stall progress towards a significant uptake and critical mass of industry-wide adoption. Therefore, to drive industry-wide BIM adoption, there needs to be a concerted effort towards agreeing and establishing a universal, industry-wide definition, along with benefits statements to facilitate greater awareness and increased coherency across the sector. This also requires the different dimensions of People, Processes and Technology to be considered as part of the holistic solution.

17.4.2   Selected Cases on Evolutionary Ramp

BIM evolution can be seen as a journey of building data and process sets, which improve as the organisation progresses. The benefits that can be accrued at each evolutionary step can be seen through a number of case studies, towards incremental improvement, and it is anticipated that the step to stage three will yield even more significant returns. The examples described below have been selected as representative of each of the various stages of evolutionary development described by the 'ramp'. The case studies presented below have been selected to demonstrate the characteristics of how the various projects have approached key issues of technology, contracts, training, etc, to deliver significant improvements in performance over their traditionally run projects at different stages of BIM evolution.

17.4.2.1 Level 'o' BIM Adoption

In terms of BIM adoption and potential benefits, the focus is towards the top of the evolutionary ramp and level of BIM adoption, therefore no case studies at level 0 are presented. However, it worth reflecting on the figures quoted by the research into the US market by the National Institute of Standards and Technology (NIST) in 2004 in relation to the UK market. NIST conservatively estimated the cost of poor interoperability to be $15.8M, which represents around 2% of the industry's revenue. By extrapolating this to the UK market, which employs 2.1 million people in 250,000 companies, and is responsible for 8.2% of Gross Value Added, this demonstrates a proportionally larger number based on a more complex, bespoke market.

17.4.2.2 Level '1' BIM Adoption

The vast majority of businesses are currently focused on seriously attempting to move towards the Managed CAD Environment level of the model. Leading businesses such as Laing achieved good results in early production pilots during the early 1990s using these techniques, and the work they pioneered subsequently resulted in the development of BS1192:2007 and Avanti Processes. This section provides a case study to demonstrate a BIM implementation at BIM Level '1'.

17.4.2.3 Basingstoke Festival Place

The Festival Place Shopping Centre in Basingstoke was a £110 M redevelopment involving a complex jigsaw of rebuilding parts of the old shopping centre without disturbing any remaining shops in Festival Place. The project was a client-led design team project with the contractor (Laing) employed on a traditional JCT Design and Build contract. Laing decided to build a full 3D model from the design team's drawings/documentation. The project also made use of a bespoke in-house collaboration system for the model, including document and project information control; although no model-based data management was attempted. All errors and clashes were identified in the model using Navisworks software, from which RFIs (Request for Information) were raised in the project as a result. The RFI procedure is used to confirm the interpretation of a detail, specification or note on the construction drawings or to secure a documented directive or clarification from the architect or client that is needed to continue work. For example, an RFI raised by the general contractor that has been answered by the client or architect and distributed to all stakeholders is generally accepted as a change to the scope of work, unless further approval is required for costs associated with the change. It is therefore common and accepted practice for a subcontractor or supplier to use an RFI to state concerns related to the omission or misapplication of a product, in order to seek further clarification of the building owner's intended use or the building official acceptance of the specified product. It is also acceptable for the subcontractor to use an RFI to call attention to an inferior product that may not meet the building owner's needs, with discretion to recommend a better/more correct product (Blanker, 2011).

As the RFIs raised in this project were not fully dealt with, the budget increased by approximately 25%. Some of this growth was due to client changes, but some £10 M was attributed to design ambiguity, which represented 9% of the final construction cost, that could have been avoided. It was concluded that a more appropriate way would be for the design supply chain, rather than the contractor, to collaborate and develop the 3D models as part of the normal design development and data management activity. It was further opined that models, built as a secondary activity on projects where the design teams do not have the required skills, are also effective in reducing cost and risk. Therefore, it was acknowledged that investment in building the model was likely to be approximately 0.5–1.5% of the project cost, with a reduction of 50% of the normal package growth of 25%. equating to 10% of the final construction cost for material cost only.

17.4.2.4 BIM Level '2' BIM Adoption

Level '2' is the managed 3D environment, where commercial data is managed by ERP. This section introduces a case study on the BIM implementation at Level '2'.

17.4.2.5 Enfield Town Centre Project, London

One of the first projects to employ the standards supported by Avanti was a multi-use development in Enfield, North London, designed by Reid Architecture and funded by the client ING Real Estate. The £25 M scheme of largely retail/restaurant/health club users also included a civic facility incorporating council offices, a library and a public theatre. The main contractor, Costain, took the decision to adopt the Avanti Standards and operate in a shared model environment, that is, Costain's bespoke extranet iCosnet, while the collaborative model was built using active (.dwg) files. The team therefore exchanged and built live model files, the orientation, origin point and scaling of which were completely aligned, which avoided the necessity of swapping fragments of digital information. The process of restructuring the previously completed CAD work to achieve this alignment required considerable concentration and effort, but the benefits far outweighed the initial reticence. Subsequent coordination of the work of the subcontractors, all of whom were obliged to participate in using the system, was vastly simplified. Bourne Steel also used the model data, together with Revit software, as the basis for its own detailed modelling. The 3D expression of the shared model was facilitated, managed and updated by a third-party company employed by Costain. They assisted the team with the coordination activity, focusing on 3D clash detection along with also providing construction process simulations (4D), which were used by Costain to improve the efficiency of construction programming. The models were also used as the basis for photo-realistic renderings for marketing, which were built using 3D object tools including Autodesk Architectural Desktop, Tekla, Multisuite and BS Link, and brought together for coordination in Navisworks. Considerable cost savings were noted by adopting this BIM approach. In addition, appreciable benefits were experienced by the entire team associated to the clarity, simplicity and efficiency of use of the model for everyday processing of information, design creation and data exchange. For example, the team claimed that at least 20 weeks were saved in the time needed to prepare information for issue to others.

17.4.2.6 BIM Level '3'

Currently, there are no true end-to-end implementations of 'iBIM', as no commercial environments are available to implement the IFC data exchange, in addition to the Information Delivery Manual (IDM) process models still being developed. However, this does not mean that great things have not yet been achieved with the technology in the specific line of business, including the environmental market. Developments in this area are progressing rapidly and efforts are in place to bring this approach to market.

17.5   Innovation through BIM

The philosophy that lies behind BIM stems from four dimensions in relation to the management of building information, and these have been articulated and agreed by the industry over the last two decades. These dimensions can be summarised as enabling i) model-based management of ii) shared building information, which provides iii) meaningful data about a building/ facility in a iv) standardised way. Four newly emerging dimensions in the management of building information towards transforming from BIM to BIM 2.0 focuses on enabling an i) integrated environment of ii) distributed information, which is always iii) up-to-date and open for iv) derivation of new information. In this respect, innovation through BIM can be investigated in two dimensions, specifically innovation through the implementation of BIM and innovation through the use of BIM together with newly emerging technologies, such as cloud computing, sensor network, web services, etc).

17.5.1   Innovation through the Implementation of BIM

Innovation through BIM can be achieved by:

1. increasing the organisational and industrial adoption rate of BIM;
2. focusing on BIM maturity;
3. moving towards lean and green construction;
4. benefiting more from process simulation and monitoring by use of BIMs; and
5. facilitating processes in urban management and disaster response fields, by bridging the gap between building and geoinformation.

17.5.1.1 Adoption

The implementation of product modelling approaches and BIMs has been the subject of research for nearly 30 years in the AEC industry. In fact, the low interest and investment in ICT and the lack of strategic perspective on the use and implementation of this have prevented the successful adoption of collaborative systems and interoperable information models within the industry. In addition, the move from CAD-based thinking to the vision of BIM is more difficult, as it involves a shift in fundamental data management philosophy. In a similar manner to the move from old accounting packages to ERP systems, this transformation includes the formal management of processes on a consistent, repeatable basis (Bew and Underwood, 2010). It also has to be acknowledged that implementation can often be difficult to achieve, and the lack of mature process management tools and methodologies for projects have made this transition more confusing. In addition, the industry's approach to contracts, training and education will need greater attention if it is to deliver this operating model and approach. In this respect, as BIM adoption is most likely to occur in phases, considerable effort will need to take place in order to move from one phase into another. For example, in the evolutionary approach to implementing BIM, organisations must be realistic about their current capability and maturity to undertake this evolution. Industry-wide adoption should therefore consider the bigger picture when contemplating BIM adoption. The focus of studies on industry-wide adoption is towards the positioning of BIM adoption across disciplines in relation to their status and future expectations, based on such factors as tools, people and processes, and Gerrard et al. (2010) provides a bird's eye view of this. In addition, efforts to measure the extent to which BIM has been implemented nationwide provides an indication of the industrial uptake of BIM, which can be used to position the BIM maturity of organisations in the context of national BIM adoption levels. This positioning is important, as the adoption levels are affected by the environment and the incumbent barriers facing organisations (Underwood and Isikdag, 2010).

17.5.1.2 Maturity

A key area in successful BIM adoption is developing the necessary capabilities and core competence concerning organisational readiness, as this can play a significant role in the absorption and diffusion of ICT within enterprises. The level of absorption and diffusion can also affect the rate of success in ICT implementation and subsequent uptake. If BIM is considered as a set of new technology and methodologies supporting information management in the AEC industry, then maturity in terms of implementing and using BIM (technology and methodologies) is critical to the success of a BIM implementation. Frameworks for measuring BIM maturity can greatly facilitate organisations in positioning themselves against their competitors in terms of technological, methodological and process maturity. Such a maturity framework is explained in Succar (2010), where a five-level BIM-specific maturity index is developed to measure the BIM maturity of organisations.

17.5.1.3 Lean and Green

In recent years, the AEC industry has begun to adopt lean production principles that have been successfully applied in manufacturing – the approach of which is known as Lean Construction. The aim of Lean Construction is to enable continuous improvement of all construction processes in the building life cycle, starting from design through the demolition of the building (Solis and Muits, 2010). Process improvement is carried out to maximise 'value' in construction (i.e. building production) and minimise the cost of production. On the other hand, the industry is also embracing initiatives to build more 'environment-friendly' buildings, along with reducing its own carbon footprint to address global concerns on environmental issues. The use of shared building models, together with collaborative environments (i.e. BIM methods), can contribute to leaner and greener construction. For example, this would minimise the need for travelling to project meetings, and the intelligent BIMs for design optimisation (i.e. in CO_2 emission analysis, etc) could play an important role in the design and operation of environmentally friendly buildings. Finally, in terms of leaner construction, the Integrated Project Delivery (IPD) approach (a US perspective on the future of project life-cycle management and project delivery) covers a lot of process changes/improvements and information management over shared digital models, which in turn could help in the process changes required by IPD (IPD, 2007).

17.5.1.4 Process Simulation and Monitoring

Construction process simulation has been facilitated through visualisation for over 10 years. This approach involves the visual simulation of construction processes is known as 4D CAD/4D Modelling. The efforts in this area are making more use of 3D CAD models, and in recent years, BIMs have superseded 3D CAD models in the visual simulation of construction processes. In this respect, analysis such as clash detection, can now be completed using BIM software. BIMs are also used in monitoring the construction progress, where activity progress can be monitored directly using a combination of data collection methods based on the BIM, especially on the 4D model of the building (Rebolj *et al.*, 2010).

17.5.1.5 Building and Geo-information Integration

There is significant value in integrating BIMs and GeoInformation Systems into a single system (Peters, 2010). The differences between CAD applications and GeoInformation Systems have generated barriers of integration of different representations of buildings in AEC and urban models. For example, the need to integrate geometric models and harmonise semantics between the two domains could tackle interoperability issues (Van Oosterom *et al.*, 2006; Isikdag and Zlatanova, 2008). In addition, Song *et al.* (2010) reviewed the benefits of integrating BIM with urban-scale contextual data, and indicated that a range of stakeholders such as building contractors, estate agents, city management and the public sector, etc could benefit from the integration of BIMs and (3D) GIS. Wang and Hamilton (2010) also provided information on the design and development of an integration framework for BIM and geospatial information using the Building Feature Service (BFS) – a service defined to retrieve building information similar to OGC web services, used for retrieving geospatial information. Support for semantic queries in BIMs can also enable users to extract partial spatial models from a full building model (Borrmann and Rank, 2010). Connectivity relationships in BIMs also have significant importance in terms of correct mathematical representation of the building (Paul, 2010). If these relations are structured correctly, the transformation of geometric information between BIMs to GeoInformation models can easily be validated.

17.5.1.6 Emergency Response

Emergency response operations indoors require a high amount of geometric, semantic and state information related to the building elements. Until recently, Egress Models used in building evacuation have mainly been based on 2D floor plans. This situation is now about to change, as the required level and amount of information can now be transferred from BIMs. In addition, BIMs are now used in fire simulations. For example, Spearpoint (2010) identified how the IFC building information model could be used to transfer building geometry and property data to fire simulation models, and suggests improvements to the IFC model with respect to the needs of fire engineering.

17.5.2   Innovation through BIM and Emerging Technologies

Innovation through BIM can be achieved through several newly emerging technologies, namely: i) Web Services, ii) Cloud Computing and iii) Sensor Networks.

17.5.2.1 Web Services

A current trend in the software industry is towards enabling interoperability over web services. In fact, the AEC industry has yet to benefit from the service-oriented approaches as the focus is still data integration-oriented. As indicated in the recent Construction Informatics roadmaps (ROADCON, 2003; Strat-CON, 2007), BIM-based web services are likely to be the catalysts of information integration and interoperability in the near future. The use of BIM servers is now increasing with open-source implementations such as BIMServer (BIMServer, 2010). London *et al.* (2010) explained a framework developed to facilitate multi-disciplinary collaborative BIM adoption through the informed selection of a project specific BIM approach and model servers. The framework consisted of four interrelated key elements, including a strategic purpose and scoping matrix, work process mapping, technical requirements for BIM tools and model servers, and a framework implementation guide. The authors concluded that the future BIM approaches would require shared models in model servers to be linked with external systems in a heterogeneous environment.

17.5.2.2 Cloud Computing

The term 'cloud computing' involves the use of the Internet (i.e. the cloud) for managing highly scalable and customisable virtual hardware and software resources, which are provided as services. Cloud computing can be disaggregated into three segments, specifically 'Software as a Service (SaaS)', 'Platform as a Service (PaaS)' and 'Infrastructure as a Service (IaaS)'. Cloud computing also involves the virtualisation of storage environments to form a virtual data centre, available over the Internet. The AEC industry will benefit from cloud computing, mostly by making use of the SaaS approach and data centre virtualisation. Applications used in various stages of the building's life cycle can work within a distributed environment (i.e. offered as software services), where the information backbone of the construction project or building (i.e. the BIMs) would reside in a virtual data centre (and offered as a data service).

17.5.2.3 Sensor Networks

Recent developments in the field of BIM have shown that they can be successful in presenting semantic information about building elements along with their geometric representation. In essence, by querying a BIM, one can understand that a rectangular prism visualised in CAD is not just a simple rectangular box, but a column made up of concrete, residing in the second floor of a residential building; specifically, the information contained in a BIM is meaningful. Although information within BIMs can be meaningful, this becomes 'stateless' after the construction of the building is completed. In this situation, up-to-date building information should be provided by sensors, or by a network of sensors, which monitor the building. In the context of BIM, distributed sensors and Wireless Sensor Networks are increasingly likely to be used to monitor conditions such as temperature, gas levels, pollutants, humidity, state of doors and windows (e.g. being open/closed and so on), room occupancy, etc. The information provided by the sensors, when integrated with the building information infrastructures, therefore would become valuable for transforming building information into meaningful full-state information.

17.6   Conclusion

Since the advent of the personal computer, the last 40 years has witnessed an 'information/digital' revolution. This is comparable with that of the industrial revolution, which has shaped our world and becoming embedded in our culture at a personal, business and global level. This has facilitated considerable change and improvements to the AEC industry. For example, the historic approach to document management and lack of interoperability between software applications created significant barriers to communication and collaboration between the various stakeholders, which in turn adversely affected efficiency and performance. BIM is a key area that could offer important opportunities to revolutionise the sector by enabling seamless processes that support the complete life cycle of the facility. BIM is an emerging information management strategy that could also resolve problems relating to information sharing and collaboration, from enabling and facilitating integrated project flow and delivery, through the collaborative use of semantically rich 3D digital building models in all stages of the project and building life cycle. BIM is currently being employed across the globe on a variety of projects, from housing through to high prestige projects, and clients are increasingly aware of the potential of BIM to deliver real value beyond the conventional design and construction phases.

The BIM momentum is now changing the AEC industry in a way that is likely to cause a paradigm shift in the overall construction process towards intelligent construction and intelligent buildings. However, this paradigm shift in the overall construction process can only be achieved by a combination of:

1. more efficient information management;
2. further automation of processes;
3. better collaboration between stakeholders; and
4. enabling leaner and greener construction for producing eco-sensible buildings.

The requirement for consistency and accuracy can be realised using BIM strategy, which makes it *sine-qua-non* for enabling the paradigm shift towards intelligent construction and intelligent and eco-sensible buildings. BIM strategy is therefore likely to play a key role in enabling efficient team working, better collaboration and better coordination (especially in complicated projects), along with adding value by automating most of the information management processes. The next paradigm shift is likely to occur when offsite and rapid manufacturing become more common, and production processes become leaner and more value-oriented. In this respect, as offsite manufacturing and design processes become leaner and value-oriented (with the support of BIM), this will provide inertia to move the industry and associated industries from design-produce/build, to produce-design-assemble types of construction.

References

Bew, M. & Underwood, J. (2010) Delivering BIM to the UK Market. In: ed. Underwood, J. & and Isikdag, U., *Handbook of Research on Building Information Modelling and Construction Informatics: Concepts and Technologies*. IGI Global.

Bew, M., Underwood. J., Wix, J. & Storer, G. (2008) *Going BIM in a Commercial World*. 7th European Conference on Product and Process Modelling (ECPPM),. Sophia Antipolis, France, 10–12 September.

BIM Execution Planning Guide (2010) *BIM Project Execution Planning Guide.* Available from http://www.engr.psu.edu/bim/download/ (accessed 11 Jan 2011).

BIM Server (2010) *Open Source BIM Server*. Available from http://www.bimserver. org (accessed 5 January 2010).

Blanker (2011) *Definition of RFI*. Available from http://blanker.org/request-information

Borrmann, A. & Rank, E. (2010) Query support for BIMs using semantic and spatial conditions. In: Underwood, J. & Isikdag, U., *Handbook of Research on Building Information Modelling and Construction Informatics: Concepts and Technologies.* IGI Global.

Cerovsek, T. (2010) A review and outlook for a 'Building Information Model' (BIM): A multi-standpoint framework for technological development. *Automation in Construction*, article in press doi:10.1016/j.aei.2010.06.003

Collier, E. & Fischer, M. (1995) *Four-dimension Modelling in Design and Construction*. CIFE Technical Report No. 101, CIFE Stanford University, CA.

Egan, J. (1998) *Rethinking Construction*. Department of Environment, Transport and Regions (DETR), UK.

Egan, J. (2002) *Accelerating Change: a report by the Strategic Forum for Construction* (Chaired by Sir John Egan). Strategic Forum for Construction, London.

Erl, T. (2004) *Service-Oriented Architecture: A Field Guide to Integrating XML and Web Services*. Prentice Hall, London.

Express Data Manager (2010) *Object Database Server*. Available from http://www. jotne.com/products.41332.en.html

Faddoul, M., Messner, J., Hagan, S. & Morrell, J. (2006) *Revolutionary Scheduling: The Practical Application of 4D Technology from a CM Perspective*. CMAA 2006 Spring Leadership Forum, Philadelphia.

Gallaher, M.P., O'Connor, A.C., Dettbarn, J.L. & Gilday, L.T. (2004) *Cost Analysis of Inadequate Interoperability in the US Capital Facilities Industry. NIST Publication GCR 04-867*. Available from http://www.bfrl.nist.gov/oae/publica-tions/ gcrs/04867.pdf)

Gao, J. & Fischer, M. (2008) *Framework & Case Studies Comparing Implementations & Impacts of 3D/4D Modelling Across Projects*. CIFE Technical Report #TR172. Stanford University, Stanford.

Gao, J., Fischer, M., Tollefsen, T. & Haugen, T. (2005) *Experiences with 3D and 4D CAD on Building Construction Projects: Benefits for Project Success and Controllable Implementation Factors*. Available from http://itc.scix.net/data/ works/att/w78-2005-D3-1-Gao.pdf (accessed 5 September 2010).

Gerrard, A., Zuo, J. Zillante, G. & Skitmore, M. (2010) Building Information model-ling in the Australian architecture engineering and AEC industry. In: ed. Underwood, J. And Isikdag, U., *Handbook of Research on Building Information Modelling and Construction Informatics: Concepts and Technologies*. IGI Global.

Panetto, H. & Molina, A. (2008) Enterprise integration and interoperability in man-ufacturing systems: Trends and issues. *Computers in Industry*, 59(7), 641–646.

Haymaker, J. & Fischer, M. (2001) *Challenges and Benefits of 4D Modelling on the Walt Disney Concert Hall Project*. CIFE Working Paper No. 64. CIFE, Stanford, CA.

Heesom, D. & Mahdjoubi, L. (2004) Trends of 4D CAD applications for construc-tion planning. *Construction Management and Economics*, 22, 171–182.

Hophe, G. & Woolf, B. (2003) *Enterprise Integration Patterns: Designing, Building, and Deploying Messaging Solutions*. Addison Wesley, Reading, MA.

Hsieh, C. (2009) Four-Dimensional Computer-Aided Drafting – Current Status and Potential Business Applications. In: ed. Rao, M., *Proceedings of University of*

Southern Mississippi, Decision Science Institute Conference 2009, 344–351. Available from http://www.swdsi.org/swdsi2009/Papers/9K03.pdf

Hua, G.B. (2010) A BIM based application to support Cost Feasible 'Green Building' concept decision. In: ed. Underwood, J. & Isikdag, U., *Handbook of Research on Building Information Modelling and Construction Informatics: Concepts and Technologies*. IGI Global.

Imhoff, C. (2005) Understanding the Three E's of Integration EAI, EII and ETL. *DM Review Magazine* 1023893-1.

IPD Working Definition (2007) *A Working Definition: Integrated Project Delivery*. Available from http://www.ipd-ca.net/images/Integrated Project Delivery Definition. pdf (accessed 5 January 2010).

Isikdag, U. and Zlatanova, S. (2008) Towards defining a framework for automatic generation of buildings in CityGML using building information models. In: ed. Lee, J. & Zlatanova, S., *3D Geo-Information Sciences*. Springer LNG&C, Berlin.

Isikdag, U., Aouad, G., Underwood, J. & Wu, S. (2007) Building Information Models: A review on storage and exchange mechanisms. In: ed. Rebolj, D., *Proceedings of 24th W78 Conference :Bringing ITC Knowledge to Work*. Maribor, 135–144.

Kam, C. & Fischer, M. (2002) PM4D final report. *CIFE Technical Report No. 143*. CIFE, Stanford, CA.

Latham, M. (1994) *Constructing the Team. Joint Review of the Procurement and Contractual Arrangements in the UK Construction Industry, Final Report*. HMSO, London.

Linthicum, D.S. (2003) *Next Generation Application Integration: From Simple Information to Web Services*. Addison Wesley, Reading, MA.

London, K., Singh, V., Gu, N., Taylor, C. & Brankovic, L. (2010) Towards the development of a project decision support framework for adoption of an integrated building information model using a model server. In: ed. Underwood, J. & Isikdag, U., *Handbook of Research on Building Information Modelling and Construction Informatics: Concepts and Technologies*. IGI Global.

Mahalingam, A., Kashyap, R. & Mahajan, C. (2010) An evaluation of the applicability of 4D CAD on construction projects. *Automation in Construction*, **19**(2), 148–159.

NBIMS (2007) *National Building Information Modelling Standard Part-1: Overview, Principles and Methodologies*. US National Institute of Building Sciences Facilities Information Council, BIM Committee. Available from www.wbdg.org/pdfs/NBIMSv1_p1.pdf

Paul, N. (2010) Basic topological notions and their relation to BIM. In: ed. Underwood, J. & Isikdag, U., *Handbook of Research on Building Information Modelling and Construction Informatics: Concepts and Technologies*. IGI Global.

Peters, E. (2010) BIM and geospatial information systems. In: Underwood, J. & Isikdag, U., *Handbook of Research on Building Information Modelling and Construction Informatics: Concepts and Technologies*. IGI Global.

Pulier, E., Taylor,H. (2006) Understanding Enterprise SOA, Manning Publications, Greenwich, USA.

Rebolj, D., Babic, N.C. & PodBreznik, P. (2010) Automated building process monitoring. In: ed. Underwood, J. & Isikdag, U., *Handbook of Research on Building Information Modelling and Construction Informatics: Concepts and Technologies*. IGI Global.

ROADCON (2003*) ROADCON Final Report*. Available from http://cic.vtt.fi/projects/roadcon/docs /roadcon_finalreport.pdf (accessed: 22 September 2007).

Ruggiero, R. (2005). Integration Theory, Part 1. Infomanagement Direct, Online at http://www.information-management.com/infodirect/20050812/1034584-1.html?pg=1 (accessed November, 2011)

Smith, D. & Tardif, M. (2009) *Building Information Modelling: A Strategic Implementation Guide*. John Wiley & Sons, Inc, New York.

Solis, J.L.F. & Mutis, I. (2010) The idealization of an integrated BIM, lean, and green model (BLG). In: ed. Underwood, J. & Isikdag, U., *Handbook of Research on Building Information Modelling and Construction Informatics: Concepts and Technologies*. IGI Global.

Song, Y., Boghdan, J., Hamilton, A. & Wang, H. (2010) Integrating BIM with urban spatial applications: A VEPS perspective. In: ed. Underwood, J. & Isikdag, U., *Handbook of Research on Building Information Modelling and Construction Informatics: Concepts and Technologies*. IGI Global.

Spearpoint, M. (2010) Extracting fire engineering simulation data from the IFC. In: ed. Underwood, J. & Isikdag, U., *Handbook of Research on Building Information Modelling and Construction Informatics: Concepts and Technologies*. IGI Global.

Strat-CON (2007) *Strat-CON Final Report*. Available from http://cic.vtt.fi/projects/stratcon/stratcon_ final_report.pdf (accessed 16 March 2008).

Succar, B. (2010) Building information modelling maturity matrix. In: ed. Underwood, J. & Isikdag, U., *Handbook of Research on Building Information Modelling and Construction Informatics: Concepts and Technologies*. IGI Global.

Underwood, J. & Isikdag, U. (2010) *Handbook of Research on Building Information Modelling and Construction Informatics: Concepts and Technologies*. IGI Global.

US General Services Administration BIM Guide (2006) *GSA BIM Guide Series 01 3D-4D-BIM Overview*. Available from http://www.gsa.gov/bim

Van Oosterom, P., Stotter, van, J. & Janssen, E. (2006) Bridging the worlds of CAD and GIS. In: ed. Zlatanova, S. & Prosperi, U, *Large-scale 3D data integration – Challenges and Opportunities*. Taylor & Francis, Abingdon, UK. 9–36.

Wang, H. & Hamilton, A. (2010) BIM integration with geospatial information within the urban built environment. In: ed. Underwood, J. & Isikdag, U., *Handbook of Research on Building Information Modelling and Construction Informatics: Concepts and Technologies*. IGI Global.

Yerrapathruni, S. (2003) *Using 4 D CAD and Immersive Virtual Environments to Improve Construction Planning*. Pennsylvania State University.

18

Industry Preparedness: Advanced Learning Paradigms for Exploitation

Jack S. Goulding and Farzad Pour Rahimian

18.1 Introduction

The European Union (EU) Architecture, Engineering and Construction (AEC) sector is one of the largest industrial employers, encompassing more than 2 million enterprises and approximately 12 million employees. This represents 9.8% of the EU's Gross Domestic Product (GDP) as it employs over 7.1% of the workforce (Business Watch, 2005). However, well-documented problems relating ostensibly to failures in communication and information processing still exist (Latham, 1994; Egan, 1998). These failures have contributed to an increased proliferation of adversarial relationships between the different parties involved in a project (Forcade *et al.*, 2007), which has also affected the veracity of design information (Fruchter, 1998; Cera *et al.*, 2002) within the project life cycle. In essence, the nature and complexity of communication within AEC projects has changed significantly over the last ten years, especially with advances in technology and the increased prevalence of web-based project collaboration technologies and project extranets. Within the AEC sector, Information and Communications Technology (ICT) has revolutionised production and design (Cera *et al.*, 2002), which has led to dramatic changes in terms of labour and skills (Fruchter, 1998). However, it is also important to acknowledge that the capabilities of such applications (and implementation thereof) can be used to predict the cost and performance of optimal design proposals (Petric *et al.*, 2002) or mitigate/reduce risks prior to construction (Goulding, 2007). Consequently, the success of AEC projects is highly dependent upon the type, level and quality of the innovative communication exchange between various disciplines involved in the design and implementation phases.

Construction Innovation and Process Improvement, First Edition.
Edited by Akintola Akintoye, Jack S. Goulding and Girma Zawdie.
© 2012 Blackwell Publishing Ltd. Published 2012 by Blackwell Publishing Ltd.

Acknowledging these issues, project teams are also continuing to face real and signification problems and challenges regarding heterogeneous project extranet systems (Alshawi and Ingirige, 2003; Ibrahim and Pour Rahimian, 2010). The problem here is that the industry is experiencing confusion as to how to manage project information in order to meaningfully support the decision-making processes. This is the point where Fruchter (2004) suggested the need for digital integration (creation, retrieval and management) in order to prevent tacit knowledge loss and miscommunication amongst various parties from different disciplines. In this respect, recent innovation in Virtual Reality (VR) technologies and AEC decision-support tools have matured sufficiently to enable tele-presence engagement to occur in collaborative environments. In this respect, several opportunities are now available, including improved immersive interactivity with haptic support that can enhance users' engagement and interaction (Rahimian and Ibrahim, 2011). These advanced interfaces are also expected to leverage new training systems within the AEC sector (Fruchter, 1998), to address the shortcomings of 'typical' learning models that often provide trainees with only general instructions (Laird, 2003), which is especially important when considering the different approaches available (Clarke and Wall, 1998). This is even more apparent with VR, as learning environments are now able to provide enhanced interactivity that incorporate innovative experiential learning approaches (Alshawi et al., 2007, Goulding et al., 2012).

This chapter extends the findings of previous studies in this area, with specific emphasis on learning and learning styles. Moreover, it explains how learning support tools can be used to assist the decision-making process in construction, the potential of applying Game Theory in training environments to non-collocated design teams, approaches for mitigating the problem of 'compartmentation' of knowledge (Mole, 2003) and new ways for enhancing actor engagement. A VR interactive learning environment prototype, based on offsite production (OSP) and Open Building Manufacturing (OBM), is presented for discussion within the AEC sector.

18.2   Learning and Training Developments and Opportunities

The AEC sector has witnessed unprecedented levels of technological change over the last 20 years. This has enabled decision makers to fundamentally re-think and reflect on the ways that they have historically done business in the past, especially in light of the new challenges and opportunities ahead. Examples of this can been seen through the increased use, proliferation and maturity of ICT, in such areas as Enterprise Resource Planning, Building Information Modelling (BIM) and Computer Integrated Construction. These developments are now enabling organisations to capture, store and reuse knowledge from organisational settings in a structured, strategic and more coherent way. Moreover, e-readiness initiatives are now starting to become a focal point for discussion (Lou and Goulding, 2010), along with the enhanced effectiveness of ICT, especially across multidisciplinary collaborative teams (Lam et al., 2010). However, whilst it is acknowledged that ICT has revolutionised and transformed business operations within the sector, it is increasingly important to acknowledge that there is also a need to engage appropriately trained and skill personnel to leverage these developments, through new business models. This requires the appropriation of 'appropriate' training and education programmes that are designed specifically to meet these needs. In this respect, the way in which training and education is delivered, managed and assessed has also changed significantly over the last 20 years. For example, pedagogical solutions can now be matched to learning outcomes, and the delivery of online learning has matured to such an extent that interoperability, scalability, adaptability and mass-customisation are now becoming mainstream solutions. However, whilst advocates of virtual learning environments, and advanced learning management systems often extol the virtues of e-learning *per se*, they often fail to articulate the limitations of such systems, especially concerning the 'personalisation' of the learning process and incompatibility with current pedagogic needs (Alshawi *et al.*, 2006). Learning therefore needs to be carefully managed in order to maximise the effectiveness of Organisational Learning (Argyris and Schon, 1978; Senge, 1990; Huber, 1991), and through the development of organisational intellectual capital (Nonaka and Takeuchi, 1995; Edvinsson and Malone, 1997; Sullivan, 1999; Ward and Daniel, 2005).

18.2.1   Learning Pedagogy

Current e-learning literature encompasses several pedagogical theories, from Constructivism, through to Systems theory and Behaviourism. These are ostensibly based on the underlying principles of pedagogical and andragogical learning (Knowles, 1980), where andragogy relates to '*man*' and pedagogy relates to '*child*' (Davenport, 1993). This also includes experiential issues (Kolb, 1984) through to reflection and learning (Jarvis *et al.*, 2003). Each of these theories have their relative strengths and weaknesses, for example, Bloom's Taxonomy identified classification levels of intellectual behaviour associated in the learning process (Bloom, 1956), whereas Piaget (1952) identified four stages of logical and conceptual learning growth and Gagne (1965) determined the 'Conditions of Learning'. It is therefore important to appreciate the subtle nuances of these, especially the context through which these are to be applied, and the learner profiles involved.

From an e-learning perspective, Conole *et al.* (2004) noted that there was little evidence to link new learning models to pedagogical effectiveness, especially regarding mapping pedagogical approaches to specific characteristics of learning. However, emergent technological developments and innovation approaches are now creating new opportunities through increased understanding on learning styles and learner preferences (Weber and Brusilovsky, 2001; Dimitrova, 2003). Similarly, the effectiveness of virtual learning is a promising avenue that is still unfolding (Stonebraker and Hazeltine, 2004; Davis and Wong, 2007) and meta-cognitive techniques are now starting to facilitate individual preferences concerning inquiry learning and reflective thinking (Wen *et al.*, 2004). These developments are likely to present instructors with powerful knowledge-based learning environments that can be tailored to suit individual learner needs (to blend learning content with learner styles).

18.2.2   Learning Styles and Models

Technology-enhanced learning is often associated with improving the efficiency and cost-effectiveness of learning practice in order to meet learners' needs (Wang and Hannafin, 2005; Bouzeghoub *et al.*, 2006). This is area is still evolving, especially in e-Learning (Sampson *et al.*, 2002; Clark and Meyer, 2011), with several definitions and terminologies being offered (often interchangeably), which adds to this confusion (Syed-Khuzzan *et al.*, 2008). However, from a learning styles perspective, much of what is described can more easily be explained in terms of informative and behaviourist approaches to learning (Conole *et al.*, 2004), underpinned by pedagogical techniques aligned to domain-specific knowledge (Govindasamy, 2002). Pedagogical principles can be seen as backbone theories that govern content and delivery approaches. It is therefore important to acknowledge the role of technology as a key enabler of existing and new pedagogical concepts (Pahl, 2003).

Given these issues, there are a variety of different models that can characterise learning styles (Coffield *et al.*, 2004; Karagiannidis and Sampson, 2004). Whilst some of these have attracted conceptual and empirical problems (De Bello, 1990; Pashler *et al.*, 2008), the philosophical foundations of these are still considered to enhance learning, particularly when the instructional process accommodates the various learning styles of learners (Kim and Chris, 2001). However, it also has to be acknowledged that learners often come from a variety of different backgrounds (especially in the AEC sector), which by default has a wide variety stakeholders, with differing profiles, learning styles, preferences, etc. Thus, when considering the type and level of skill sets needed, it is particularly useful to try and accommodate learners' learning experiences, in order to make learning as personalised as possible (Vincent and Ross, 2001). This is especially important, as the adoption of a 'one-size-fits-all' approach to learning has generally been considered as being ineffective (Watson and Hardaker, 2005). Cognisant of this, 'There needs to be a deliberate and documented choice of model, which reflects a broad awareness of the field and which will allow for results and outcomes to be dealt with within a clear a clear conceptual framework' (Cassidy, 2004).

18.2.3   Learning through Game Theory

Game theory was first introduced as an alternative solution to conventional approaches to learning. Turocy *et al.* (2003) defined Game theory as a method of studying 'conflict and cooperation' of multiple participants, especially where their actions are co-dependent. Turocy *et al.* (2003) proposed this theory as a common protocol for formulating, structuring, analysing and understanding strategic scenarios. Based on this protocol, a game was described as a 'formal strategic situation' encompassing interactions of various agents. More intrinsically, Nisan *et al.* (2007) identified that Game theory could be used to model complex situations with multiple parties and outcomes using algorithms. However, Goulding *et al.* (2007) noted that educational training tools needed to 'engage' learners by putting them in the role of decision makers in order to 'push' them through challenges, hence enable different ways of learning and thinking through frequent interaction and feedback (to engender and create real-world context). Similarly, Wellings and Levine (2010) suggested redesigning existing text-based lessons into problem-based learning processes, noting that this was possible using immersive visualisation and simulation environments, which embedded games and interactive interfaces (to increase engagement), and Thai *et al.* (2009) asserted that educational digital games offered opportunities to empower trainees' engagement, to help transform teaching and learning into a new stage. However, within the milieu of game-based learning environments, it is equally important to acknowledge the formal transference of learning and the context of this learning (Gross, 2007). There is also a need to distinguish between playing and learning – especially a need '...to appreciate how content, skills and attitudes are not just transferable, and neither merely a question of interaction between player and computer game' (Egenfeldt-Nielsen, 2007).

18.3   Virtual Reality Systems

Virtual Reality (VR) has been defined as a 3D computer-generated alternative environment, which facilitates high degrees of immersion, navigation and interaction (Briggs, 1996; Yoh, 2001; Pour Rahimian *et al.*, 2008), or a component of communication that takes place in a 'synthetic' space (Regenbrecht and Donath, 1996, Sampaio *et al.*, 2010). The various components of VR systems usually include a computer capable of real-time animation, controlled by a set of wired gloves and a position tracker, along with a head-mounted stereoscopic display as visual output. For instance, Regenbrecht and Donath (1996) defined the tangible components of VR as a congruent set of hardware and software, with actors engaged in a 3D or multi-dimensional input/output space; where the actors interact with autonomous objects in real time.

18.3.1   Virtual Reality in Architecture, Engineering and Construction

Over the last 30 years, ICT systems have matured sufficiently to enable construction organisations to fundamentally restructure and enhance their core business functions. Sampaio and Henriques (2008) asserted that the main objective of using ICT in the construction field was to support the management of digital data; specifically to convert, store, protect, process, transmit and securely retrieve datasets. They acknowledged VR as an important stepping stone for data integration in construction design and management, as this was seen as being able to present holistic information about buildings (e.g. size, material, spatial relationships, mechanical and electrical utilities, etc) through one single conduit. Similarly, Zheng *et al.* (2006) proposed the use of VR to reduce time and costs in product development, and also to enhance the quality and flexibility of providing continuous support during the development life cycle. In this respect, early studies incorporating VR into the design stage used this predominantly as an advanced visualisation medium. However, since as early as 1990, VR has been widely used in the AEC industry for creating 3D design visualisation models that can be manipulated in real time and used to explore different stages of the construction process (Whyte *et al.*, 1998), especially for improving construction processes in collaborative environments (Bouchlaghem *et al.*, 2005). However, the expectations of VR have changed significantly over the last decade. For example, Sampaio and Henriques (2008) noted that it was increasingly important to incorporate VR 3D visualisation and decision support systems with interactive interfaces, in order to perform real-time interactive visual exploration tasks. This resonates with thinking that supports the increased use of collaborative virtual environments in which 3D models are linked to databases, which carry enhanced characteristics and metadata (knowledge on data). For example, construction planning and management can now link to 3D models along with time parameters (known as 4D), in order to design 4D models (Fischer and Kunz, 2004), which are controlled through interactive and multi-access databases. VR models can therefore be used to improve many aspects and phases of construction projects by:

1. providing better communication amongst partners (Leinonen *et al.*, 2003);
2. enhancing design creativity (Rahimian and Ibrahim, 2011);
3. improving coordination (Khanzade *et al.*, 2007);
4. aligning directly with construction processes (Fischer, 2000); and
5. integrating with BIMs to enhance data integration (Xie *et al.*, 2011).

18.3.2   Virtual Reality as a Training Medium

VR applications are increasingly being used as a part of the teaching and learning process, as this can contribute to the trainees' professional future by developing learning activities beyond what is available through conventional training systems (Zudilova-Seinstra *et al.*, 2009). With respect to educational issues in the AEC industry, Sampaio *et al.* (2010) asserted that interaction with 3D geometric models could lead to active learner thoughts (which seldom appear in conventional pedagogical conditions). Moreover, Juárez-Ramírez *et al.* (2009) noted that when learning is augmented to 3D modelling, VR could lead to better communication in the process of AEC training. Moreover, Gomes and Caldeira (2004) identified new opportunities for enabling trainees and trainers to exchange information about specific domains, in order to enhance interaction and learn collaboratively. However, VR training environments are still evolving and presently are not capable of delivering some learning outcomes. This is primarily down to the ways in which technology is aligned to systems, processes, learning outcomes and pedagogical delivery mechanisms. It is therefore argued that educational training tools need to 'engage' learners by putting them in the role of decision makers and 'pushing' them through challenges, hence enabling different ways of learning and thinking through frequent interaction and feedback, and connections to the real-world context (Goulding *et al.*, 2007). In this regard, Wellings and Levine (2010) suggested focussing on problem-based learning processes, where training is matched to real-world problems. They asserted that this was possible using immersive visualisation and simulation environments, which increased the engagement of trainees and trainers. Similarly, Thai *et al.*, (2009) noted that educational digital games offered opportunities for empowering trainees' engagement, to help transform teaching and learning into a new stage. In this respect, ACS (2009) summarised the benefits of interactive game-like immersive training environments as:

1. exploring knowledge by clicking on objects with linked information;
2. strengthening education by providing a repository of aids, tools, etc, associated with 3D objects;
3. offering collaborative workspaces, e.g. 3D informal discussion forums;
4. providing traditional instructor-based education via a distance delivery method; and
5. creating simulated learning by modelling a process or interaction that closely imitates the real world in terms of outcomes.

It is therefore argued that paring instructional content with game features is more likely to engage users in the learning process (to help achieve the desired instructional goals). This follows similar thinking where instructional programmes that incorporate certain features or characteristics from gaming technology are used to trigger a cycle of user judgment or reaction (often based on enjoyment or interest), to maximise persistence or time on task, in order to achieve the desired learning outcomes (Garris *et al.*, 2002).

18.4   Case Study

This section presents the rationale and rubrics for the development of a VR construction site simulator. This simulator was designed to help learners appreciate the subtle nuances and differences facing construction professionals that wished to adopt offsite manufacturing concepts into traditional work practices. The development aim of this simulator was to capture and present 'real life' OSP issues facing construction professionals in an innovative way that proactively stimulated actor engagement, to provide meaningful solutions. In this respect, a real-life construction project was used to govern the authenticity of the learning environment. The simulator was therefore purposefully designed to allow 'things to go wrong', thereby allowing 'learning through experimentation' or 'learning by doing'. The 'scenes' within the simulator take place on a virtual construction site, and the target audience is aligned to construction professionals (e.g. project managers, construction managers, architects, designers, suppliers, etc). Thus, the construction site (within the simulator) was used as a domain through which unforeseen issues and problems (caused through upstream decisions, faulty work, etc) could be enacted, so that real implications could be better appreciated in respect of time, cost, resources, etc. The simulator was developed not purely to solve OSP problems *per se*, but rather to allow things to 'go wrong' in order to demonstrate the implications of decisions taken by learners. In this context, the simulator was designed to:

- *Maximise learner autonomy*: allowing learners to make all decisions (independently);
- *Stimulate engagement*: where interactivity is promoted, supported by feedback on decisions taken, and their implications, e.g. cost, time, resources, health and safety, etc;
- *Engender personal reflection*: where users can defend their decisions, especially on means for mitigating future problems in areas, such as:

1. OSP strategies, e.g. Design for Manufacture Logistics and Assembly (DFMLA);
2. new business processes, procurement/contractual arrangements, project management, quality assurance, etc;
3. impact on Health and Safety procedures;
4. supply chain integration;
5. new manufacturing technologies, etc.

The simulator was created using scenarios with predefined learning objectives. In this respect, learning was driven by problems, encountered in the VR environment, and aligned to the following design contextual constraints:

- All scenarios/scenes had to occur within the virtual construction site simulator;
- Learners would adopt the sole role of a construction manager;
- Training would engage instructional signposting techniques (e.g. interactive messaging, multi-decision criteria, time-dependent options, etc); and
- Learning outcomes would be aligned to OSP working practices only.

18.4.1 Simulator Development Framework

The simulator development framework encompasses four main activities, specifically: Training Objectives; Development Scenario; VR Environment; and Validation (Figure 18.1). This framework required extensive input from domain experts selected from the construction industry in order to help secure relevance and govern authenticity of these stages.

The training objectives underpinning the simulator were gathered from a synthesis of seminal literature covering the potential risks and threats facing OSP in general, and Open Building Manufacturing in particular. This knowledge was seen as fundamental, as learners would need to understand how different stakeholders deal with the implications of problems; and consequently, help them to appreciate how these could be mitigated for future practice. In this context, the following seven core risks were identified:

1. late design changes;
2. loss of factory production, or production capacity;
3. unpredictable planning decisions and designs that are not suited to offsite manufacturing;
4. issues associated with tolerances;
5. suppliers' failure to deliver on time;
6. manufacturer bankruptcy; and
7. issues associated with alternative manufacturer selection.

The scenarios were developed to expose learners to new working conditions and issues that they were likely to face on real construction projects employing OSP concepts. It was therefore deemed important to challenge learners to think about the underlying causes of these problems, rather than just reacting to the problems themselves. This philosophy was used to 'provoke' learners to think proactively about future OSP projects. In this context, the main scenario was based on identifying possible problems and issues that could be faced – colloquially referred to as problem 1, problem 2, etc (Figure 18.2). For each of these problems, there were a

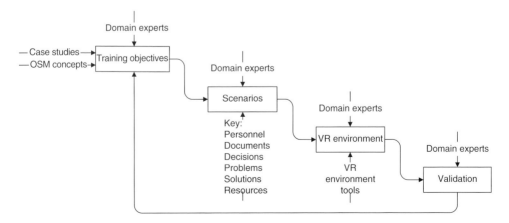

Figure 18.1 VR environment development framework.

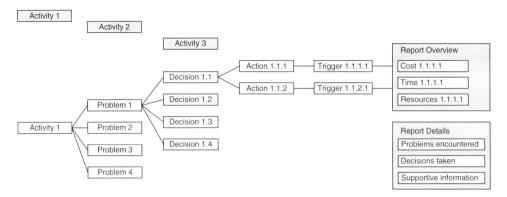

Figure 18.2   Scenario implementation concept.

number of possible decisions with associated actions. Depending on the action chosen, the programme schedule, along with corresponding costs, time and resources are affected.

These scenarios were used to simulate how OSP operates in real life; which allowed learners to think 'how' and 'why' things go wrong, and consequently why OSP may end up being more expensive than traditional ways of working and thinking. The rationale here was to stimulate learners to think about the holistic process and implications regarding the decisions they take (rather than just the immediacy of the options available). In this respect, a debriefing session was identified as a conduit for questioning learners further on the decisions they took in the learning environment. This approach was seen as a way to help distinguish between 'being immersed' within the environment and the process of critical reflection that often takes place outside the VR environment (De Freitas and Oliver, 2006).

To run a scenario, various information and data has to be input into the system in order to help populate the scenario. This data includes:

a) construction site type, location, constraints and layout;
b) project type and primary use of the building (e.g. commercial or residential), budget allocated, type of structure, special layout and planning, etc;
c) manufacturer type, in terms of scope, capacity, location, costs associated and maintenance;
d) equipment hire, in terms of size, capacity, assembly rate, required labour, hire rate, etc; and
e) work plan and associated possible interruptions/problems, including manufacturing options.

This information was 'mirrored' from the case study project and mapped into a relational database within the simulator. A schematic representation of this workflow approach can be seen in Figure 18.3.

The first stage of the prototype development established the requirements and priorities in an ontological structure. The generic structure and content for the knowledge objects were then assigned wrappers using metadata.

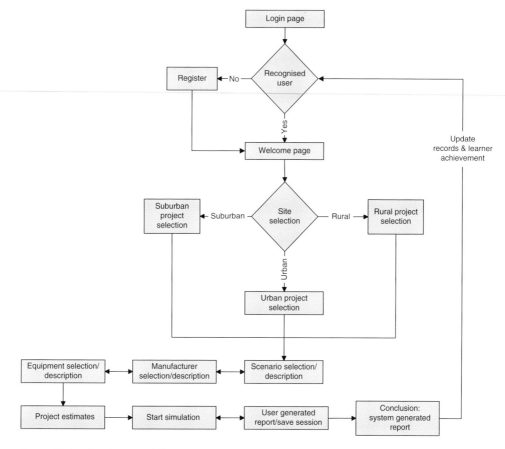

Figure 18.3   Simulator workflow.

The next stage established object classes and their hierarchy to satisfy multiple abstractions, and compliance with extranet metadata. The final stage developed the user interface to comply with Human-Computer Interaction protocols and accessibility guidelines, etc, in order to provide the system with a robust and reliable structure. The developed system included simulated scheduling of the project; association of the 3D models and building blocks with project life cycle; supply chain analysis monitoring for each building block or activity; management of delays in material delivery and a final breakdown for project costs and labour.

18.4.2   System Architecture

Existing VR interfaces have ostensibly been formed based on one single idea-creating 3D models and incorporating them with some pieces of information so that both 3D models and information are editable through an interactive real-time interface (Pour Rahimian and Goulding, 2010).

Consequently, data creation and retrieval methods in VR interfaces can be investigated from two different perspectives, namely, creating 3D bodies of constructional elements, and then defining the characteristics of these elements. However, whilst creating 3D objects directly in VR environments is not impossible, these are normally created in CAD applications; since doing so in VR is often cumbersome and time-consuming. In terms of transforming design elements from CAD into VR, there are three *de facto* approaches used by different practitioners. Whyte *et al.* (2000) noted three approaches for this translation as being:

1. *Straightforward translation approach*: importing the whole environment from CAD to VR;
2. *Library-based approach*: putting the elements of construction in the library of the VR environment, then 'calling' them as necessary; and
3. *Database-oriented approach*: using a central database for controlling the module characteristics.

In this study, the database-oriented approach was selected, as this enables learners to gain access to the system from multiple remote locations.

From an environment development perspective, there are three major kinds of programming applications currently used:

1. 3D Application Programming Interfaces (APIs);
2. Virtual Reality Modelling Language (VRML) and 3D web technologies; and
3. Commercialised object oriented VR programming packages.

From the above, 3D Application Programming Interfaces (e.g. Open GL and Direct 3D) are the principal environments for VR programming in C++ and Visual Basic. Falling into the category of computer graphics, they are capable of either creating the models directly inside the space, and/or importing them directly from CAD applications. Anecdotally, VRML and 3D web technologies in their first iteration were made as a division of Open Inventor; thereafter have become the international standard for 3D web modelling. These applications provide a variety of facilities for manipulating immersive library-based web interfaces, but lack the capability of integrating with interrelated databases, as these are ostensibly not database-oriented applications. On this theme, recent commercial object oriented VR programming packages contain built-in modelling environments for creating VR spaces directly, or importing them from CAD applications. Such VR programming applications also contain logical libraries for defining behavioural links amongst the objects and simulating physical phenomena. Although the architecture of such applications is made based on APIs of C++, in some aspects they can offer a higher-level abstraction for programmers. Nowadays, there are three frontier commercial VR programming applications, namely: Quest3D™, EON Reality™ and Virtools™. These applications are directly deployable into Visual C++ and Visual Basic web programming platforms (EON Reality Inc, 2008). This

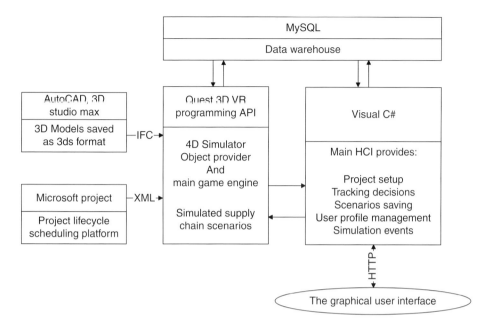

Figure 18.4   Simulator architecture.

makes them extremely flexible in terms of integrating VR programming with web programming and data mining. They also come with full Software Development Kits (SDKs) in order to help programmers add building blocks and prototypes to create rules and behaviours that were not originally provided by the host application.

This study implemented a database-driven approach, using structured modelling phases and API-based programming for the development stages. This was adopted primarily because 3D objects can be linked to datasets through the web environment along with schedules of activities (4D visualisation) – the product of which can be used to visualise changes in real time, which can be used to enhance learning outcomes. In this respect, all elements and components of the construction site were modelled in either AutoCAD™ or 3D Studio Max™, two of most the popular modelling software applications used today, and the scenarios were scheduled in MS Project™. All VR programming tasks were performed in Quest3D™ environment, whilst the Active Server Pages (ASP.Net™) web development tool using C#™ programming language was employed for developing the user interface. Finally, a MySQL™ database that was compatible with both programming environments was installed on a server in order to track, manage and transmit user data. The adapted Unified Software Development Process (Jacobson et al., 1999) was used to iteratively test, diagnose and troubleshoot the prototype regarding functionality, compliance, grouping, integration, maintenance, version control and validation. The simulator architecture arrangements can be seen in Figure 18.4.

18.4.3   3D Modelling in AutoCAD and 3D Studio Max

AutoCAD and 3D Studio Max were employed as the main geometrical modelling platforms. Different construction site elements including both permanent (e.g. structural/architectural element and building blocks) and temporary (e.g. scaffolds and site barriers) constructional objects were created in these environments. Models for construction machinery and equipment (e.g. tower cranes and trucks) were downloaded from CAD forums and websites. In order to optimise system performance, the project created the models at a primitive level (e.g. beams, columns, walls, building blocks) and left the final assembly for Quest 3D as the 4D simulation tool to regenerate and replicate single models, as geometrical arrays do not place as much demand on physical storage space or require as much graphical memory. For the same reason, the project used bitmap graphics to give the illusion of secondary details, rather than creating them in 3D models. The bitmaps also helped the simulation look more realistic by visualising texture of different materials (e.g. concrete, wood, etc). Ultimately, the models were converted into the 3DS file format, in order to be readable by Quest3D.

18.4.4   Project Scheduling Environment

MS Project™ was used for scheduling project activities based on the developed scenarios. For each scenario, the temporal relationships amongst different assembly tasks were planned from the commencement of the project to the day of completion. This approach also included the inclusion of random occurrences relating to 'unexpected' problems and interruptions during the project life cycle for learners to solve (to minimise repetition). These schedules formed the essential basis for data regarding sequence amongst different assembly activities, duration of different constructional tasks and their commencement time and finishing time. Finally, the schedules were converted to MySQL databases, which are accessible for both C# and Quest3D programming tools. Figure 18.5 shows a sample project array within Quest3D that is imported from MySQL, which identifies the details of assembly of various building blocks in terms of the 3D position of the building block, sequence of assembly, exact delivery time, and the machinery involved in transportation and assembly of it. Since this data is stored in

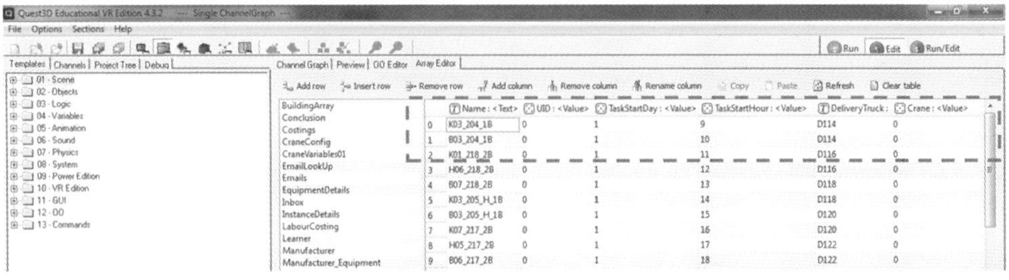

Figure 18.5   Sample project array within Quest3D, imported from MySQL database.

a relational database, the pieces of information regarding costs and labour for both building blocks and equipment were also automatically associated to the tasks. This made it possible for learners to modify or update scheduling data through the ASP.Net interfaces.

18.4.5   Data Warehouse Environment

MySQL™ was selected as the platform to host the databases of this simulator, as it is the only application that provides Quest3D with a SDK. This is also accessible the in MS Visual Studio environment through Devart DotConnect™ and ADO.Net™. A relational database comprising of 44 different tables was created in MySQL to manage information regarding manufacturers, equipment, labour and costs associated with different tasks, schedules for different scenarios, etc. This enhanced flexibility provides learners with full control of project data, both through web forms, and through the project simulation environment.

18.4.6   Main Human Computer Interaction Interface

The simulator provides ASP.Net web forms as the initial Graphical User Interface to transfer messages from users to the system, in order to gather data regarding specifications of the desired construction process. Using these web forms, learners are able to control the type and sequence of construction tasks and make appropriate decisions; for example, on the type and level of machinery involved in the project. The system also provides learners with estimates of project costs and time, then proceeds to the simulation of the project. The simulation is generated in the ASP.Net environment as an embedded object by 'calling' an external object exported by Quest3D. Based on the collective decisions made by the learner, the system calculates the total costs of the project and provides comparison between the results obtained, compared with optimal solutions. The system then generates a detailed report on the performance of learners. Figure 18.6 presents a sample screenshot of this report, together with the supportive C# code derived from the project database.

18.4.7   Programming Environment

The geometrical 3D models of the constructional elements were imported by Quest3D™ in order to provide the basic entities and building blocks of VR programming. Quest3D is an Object Oriented programming platform in which the programming logic is formed through interconnection of Logical Building Blocks. The structure of the program is comprised of four main components, specifically:

1. static 3D models of construction site, including tower cranes, trucks, land, surroundings and supporting elements (which do not change from one learner to another, e.g. scaffolding);

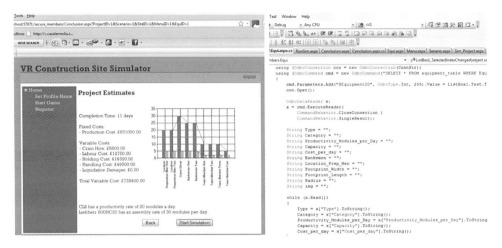

Figure 18.6 Design interface (Left) and C# code-behind (right).

2. building blocks as the dynamic 3D models of project;
3. project schedules for controlling all events of assembly and delivery process; and
4. monitoring tools for controlling project time and resources.

Static elements were directly generated using 3DS and bitmap files, which appear at particular locations. However, tower cranes and trucks were programmed to perform the desired animations at certain points of time. The modules of all static objects were directly connected to the project interface, except additional tower cranes and trucks that may (or may not) have been hired by the learners. In these two cases, the modules were connected to the interface through an IF Toggle Channel, in order to 'call' the entities based on the preferences of the learner. In terms of assembling dynamic objects, the system relies on project schedule imported from MySQL database (Figure 18.5), for the assembly sequence and project time. Based on the given sequence, the system checks for assembly permission for each module, and if the project time coincides with the time allocated to any module, the program runs the related animations in order to deliver and assemble that module. However, at certain intervals, some random interruptions are programmed to occur in the project sequence, so the implementation process can only continue after the learner has responded to the prompt, and sent the right email to the appropriate recipient. In this case, a Trigger Channel reinitiates the performance of system, subject to the delivery of these emails to the database. Delays in making the right decisions therefore result in an increased project cost and completion time. Finally, with respect to project scheduling and monitoring tools, the system relies on system arrays formed and updated based on MySQL databases and particular algorithms, which control project time and retrieve and visualise project cost in real time. In essence, the simulator interface provides learners with a 4D simulated environment and an IPod-type interface for showing project statistics in real time (Figure 18.7).

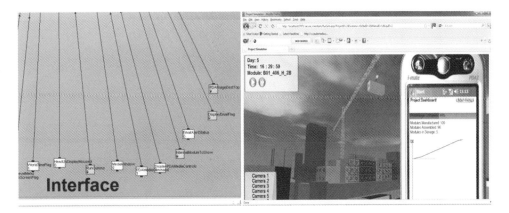

Figure 18.7   Project interface in channel and design view.

18.4.8   Case Study Discussion

This construction site simulator was designed as an interactive learning environment for learners to appreciate the subtle differences required with offsite production, as opposed to the more traditional approach to construction. It was developed as a web-based simulation tool using both non-immersive and immersive pages, in order to allow learners to experience new OSP challenges based on real-life AEC projects using simulated scenarios. The Graphical User Interface was designed to be as simple and straightforward as possible with respect to data input, and was developed to be accessible through any standard web browser (to provide users with login account details and other criteria, e.g. selection of available construction sites, projects, contractors, equipment, scenarios, etc). All choices made by the learners, as well as their registration data, are automatically recorded in a MySQL database, which is also accessible through the immersive application for project simulation. After completing the initial decision-making process through the interactive ASP.Net Web Forms, learners are able to commence the training session, starting with a 'walkthrough' to experience and appreciate the complexity of the project. At this stage, the application provides users with a summary of the project and contract, and runs the simulation of the project within an immersive and interactive environment developed in Quest3D™ VR programming Application Programming Interface.

Within the simulated Quest3D environment, learners are able to experience the outcomes of all decisions made. They are also challenged by unexpected random events, designed specifically to invite them to make decisions for dealing with these issues. The simulation runs in a fully immersive 3D environment, where learners are able to navigate the whole interior and exterior spaces of the project site. At various points in the scenario, they are also able to interact with the different elements of the simulator in order to retrieve further information, for example, technical specifications, videos on selected OSP construction systems/details, project data, etc. In order to keep them on track, the simulator also provides them with monitoring tools, which reveals issues such as the project time, latest assembled module, accumulative costs of the project, team communication, etc. The monitoring and communication tools are embedded in different parts of the main interface as well as the facilitated standard embedded virtual IPod-type interface, which appears when required. The simulator ultimately records and tracks learners in the database and navigates to the conclusion page to reveal project outcomes.

On commencement of the simulator, learners are required to read the instructions and enter their login information (or create a new user account). This feature keeps a record of accomplishment to date, and allows learners to save or review their previous activities in order learn from previous decisions. After the logon process, learners are required to select the location of the project, for example, rural, suburban or urban. The location of a project has implications on access, equipment, storage, etc – thereby affecting the triggered scenarios. Afterwards, learners are then requested to select the type of system/structure to explore, from a repository of stored systems as different systems have different requirements – some suitable for some locations and not for others. Based on the selected site, learners are then requested to select the type of scenario for the current 'Game'. In the next step, they are provided with brief descriptions of each scenario, details of the selected site, details of the selected OSP system and associated contract. In terms of learning outcomes, different scenarios have different learning outcomes based on the events embedded in them; this allows learners to experience different types of constructional project issues within each scenario. When the scenario has been selected, users are required to select the manufacturer as the provider of building blocks and select the site set-up arrangements with respect to the equipment required. There are various issues associated with manufacturer selection, from product differences, through to the costs of materials, completion and delivery dates, labour costs, fluctuating assembly costs, transportation costs, etc. Logistics solutions are also affected by the type of equipment and site set-up chosen, in addition to equipment requirements and constraints. At the next stage, learners are provided with a summary of all decisions made at this stage, and some figures about the anticipated project costs, duration, etc are presented for review. Learners can then compare the effects of all selected type of project site, OSP type, manufacturer and equipment selection, and scenarios with those from previous attempts. After these initial selections, users are then able to run the VR simulation to experience how the project progresses based on their selections. Different scenarios are then triggered at various times in order to exert 'pressure' on learners, particularly to make them think about options and consequences of their choices, as these will affect the overall project cost, time, resources, etc. A sample screenshot of the simulator in progress can be seen in Figure 18.8. A formal report is then generated at the end of the simulation exercise. This report is used in conjunction with a debriefing session with a nominated instructor to discuss these findings and the thought processes behind them.

Figure 18.8   Simulator screenshot.

18.5   Conclusion

This chapter reviewed the seminal literature on learning and the development of VR tools used in the AEC sector. It also presented a series of challenges facing construction personnel regarding the need for up-skilling in light of changes in the operating environment. New approaches for learning using game-like immersive educational interfaces were then discussed in order to highlight the potential benefits these can offer, especially being able to help learners experience real-world problems in a risk free virtual environment. The implementation of such approaches are now able to leverage significant benefits (ACS, 2009; Thai *et al.*, 2009; Wellings and Levine, 2010), not least improve users' engagement in the process. Whilst several systems are now being promoted in the marketplace (Autodesk, 2011; Bentley, 2011; Cisco, 2011), the use and propensity of these have not yet reached maturity. However, further development (with an AEC focus) could lead to the emergence of truly immersive environments. Moreover, the concatenation of a Game-like VR interface could offer the AEC sector further enhanced opportunities. An example was then presented to explain the theoretical and technical challenges involved. This case study identified the formal development rubrics adopted for a VR Construction Site Simulator. It also showed how learning can take place in a risk free environment, where learners (AEC professionals) are able to evaluate their decisions, and see how these affect project and business outcomes.

Construction projects are increasingly becoming more complex, often engaging new business processes and technological solutions in line with the clients' requirements. This is likely to continue. Moreover, it is advocated that the AEC sector in particular is likely to require a myriad of increasingly new skilled professionals, operatives, and interdisciplinary teams in order to meet these new challenges. It is therefore important that the industry as a whole engages the right type (and level) of skill sets and competence to meet these project requirements and business imperatives. Acknowledging this, the causal drivers and influences associated with successful decision-making in non-collocated design teams need to be fully understood, as this is an important stepping-stone for developing new insight and understanding into collaborative environments, particularly though new social interactions (and decision-making criteria) generated through Game Theory techniques. It is also important to be able to 'measure' and gauge actor involvement and positioning (cognisant of learner domain expertise), so that project teams can anticipate any issues that are likely to cause problems beforehand. Advanced VR training and simulation tools are likely to be pivotal drivers in this respect, as they are increasingly well poised to deliver focused learning outcomes. Future research in this area is also likely to acknowledge the importance of learner styles, thereby delivering bespoke training material to individual learner types.

18.6   Acknowledgements

Part of the work presented in this chapter was funded by the EU IP Framework 6 Programme: ManuBuild – 'Open Building Manufacturing. Special thanks are extended to Vinci (David Leonard and Jeff Stephens); Corus (David Shaw and Samir Boudjabeur); the Technical Research Centre of Finland (Kalle Kähkönen); Technische Universität München (Ron Unser), CIRIA (Mark Sharp), the University of Salford, UK (Mustafa Alshawi, Jack S. Goulding, Wafaa Nadim and Panos Petridis) and all ManuBuild partners for their feedback and support throughout.

References

ACS (2009) *3D Learning and Virtual Worlds*: An ACS: Expertise in Action™ White Paper. Available from http://www.trainingindustry.com/media/2043910/acs%20 3d%20worlds%20and%20virtual%20learning_whitepaper%20april%202009. pdf (accessed August 2011).

Alshawi, M., Goulding, J.S. & Faraj, I. (2006) Knowledge-based learning environments: a conceptual model. *International Journal of Education in the Built Environment*, 1(1), 51–72.

Alshawi, M., Goulding, J.S. & Nadim, W. (2007) Training and education for open building manufacturing: Closing the skills gap. In: ed. Kazi, A.S., Hannus, M., Boudjabeur, S. & Malone, A., *Open Building Manufacturing: Core Concepts and Industrial Requirements*. Helsinki. ManuBuild in collaboration with VTT – Technical Research Centre of Finland.

Alshawi, M, Ingirige., B, (2003) Web-enabled Project Management: An Emerging Paradigm in Construction, Automation in Construction, Vol. 12, Iss 5, pp. 349–364.

Argyris, C. & Schon, D.A. (1978) *Organizational Learning: A Theory of Action Perspective*. Addison-Wesley, MA.

Autodesk (2011) Available from http://usa.autodesk.com/ (accessed December 2011).

Bentley (2011) Available from http://www.bentley.com/en-GB (accessed December 2011).

Bloom, B.S. (Ed.), Engelhart, M.D., Furst, E.J., Hill, W.H., & Krathwohl, D.R. (1956). Taxonomy of Educational Objectives: The Classification of Educational Goals. Handbook 1: Cognitive Domain, Longman, New York.

Bouchlaghem, D., Shang, H., Whyte, J. & Ganah, A. (2005) Visualisation in architecture, engineering and construction (AEC). *Automation in Construction*, **14**(3), 287–295.

Bouzeghoub, A., Defude, B., Duitama, J.F. & Lecocq, C. (2006) A knowledge-based approach to describe and adapt learning objects. *International Journal on E-Learning*, **5**, 95–98.

Briggs, J.C. (1996, 09-01-1996) The promise of virtual reality. *The Futurist Magazine*, **30–31**.

Business Watch (2005) *ICT and Electronic Business in the construction Industry, IT adoption and e-business activity in 2005. The European e-Business Market Watch*. European Commission, Enterprise and Industry Directorate General.

Cassidy, S. (2004) Learning styles: An overview of theories, models, and measures. *Journal of Educational Psychology*, **24**(4), 419–444.

Cera, C.D., Reagali, W.C., Braude, I., Shapirstein, Y. & Foster, C. (2002) *A Collaborative 3D Environment for Authoring Design Semantics*. Graphics in Advanced Computer-Aided Design, 43–55.

Cisco (2011) Available from http://www.cisco.com/en/US/netsol/ns1007/products.html (accessed December 2011).

Clark, R.C. & Meyer, R.E. (2011) *E-Learning and the Science of Instruction: Proven Guidelines for Consumers and Designers of Multimedia Learning*, 3rd edn. John Wiley and Sons, Ltd, Chichester, UK.

Clarke, L. & Wall, C. (1998) UK construction skills in the context of European developments. *Construction Management and Economics* **16**(5), 553–567.

Coffield, F., Moseley, D., Hall, E. & Ecclestone, K. (2004) *Learning Styles And Pedagogy In Post-16 Learning: A Systematic And Critical Review*. Learning and Skills Research Centre, Cromwell Press Ltd, Trowbridge, UK.

Conole, G., Dyke, M., Oliver, M. & Seale, J. (2004) Mapping pedagogy and tools for effective learning design. *Journal of Computers and Education*, **43**(1–2), 17–33.

Davenport, J. (1993) Is there any way out of the andragogy morass? In: ed. Thorpe, M., Edwards, R. & Hanson, A., *Culture and Processes of Adult Learning*. Routledge, London.

Davis, R. & Wong, D. (2007) Conceptualizing and measuring the optimal experience of the elearning environment, decision sciences. *Journal of Innovative Education*, **5**(1), 97–126.

De Bello, T.C. (1990) comparison of eleven major learning styles models: variables, appropriate populations, validity of instrumentation, and the research behind them. *Journal of Reading, Writing, and Learning Disabilities International*, **6**(3), 203–222.

De Freitas, S. & Oliver, M. (2006) how can exploratory learning with games and simulations within the curriculum be most effectively evaluated? *Computers and Education*, **46**(3), 249–264.

Dimitrova, V. (2003) STyLE-OLM: Interactive open learner modelling. *International Journal of Artificial Intelligence in Education*, **13**, 35–78.

Edvinsson, L. & Malone, M.S. (1997) *Intellectual Capital: Realising Your Company's True Value by Finding Its Hidden Roots*. Harper Business, New York.

Egan, J. (1998) *The Egan Report – Rethinking Construction*. Report of the Construction Industry Taskforce to the Deputy Prime Minister. HMSO, London.

Egenfeldt-Nielsen, S. (2007) Third generation educational use of computer games. *Journal of Educational Multimedia and Hypermedia*, **16**(3), 263–281.

EON Reality Inc (2008) *Introduction to Working in EON Studio*. Available from http://www.eonreality.com (accessed August 2011).

Fischer, M. (2000) *4D CAD-3D models incorporated with time schedule*. CIFE Centre for Integrated Facility Engineering in Finland, VTT-TEKES, CIFE Technical Report, Stanford University, USA

Fischer, M. & Kunz, J. (2004) *The Scope and Role of Information Technology in Construction*. Technical Report, No. 156 February 2004), San Francisco. Center for Integrated Facility Engineering, Stanford University, USA.

Forcade, N., Casals, M., Roca, X. & Gangolells, M. (2007) Adoption of web databases for document management in SMEs of the construction sector in Spain. *Automation in Construction*, **16**, 411–424.

Friedman, T.L. (2005) *The World is Flat: A Brief History of the 21st Century*. Farrar, Straus and Giroux, New York.

Fruchter, R. (1998) Internet-based web mediated collaborative design and learning environment. In: *Artificial Intelligence in Structural Engineering Lecture Notes in Artificial Intelligence*. Springer-Verlag, Berlin. 133–145.

Fruchter, R. (2004). Degrees of Engagement in Interactive Workspaces. International Journal of AI & Society, 19(1), 8–21.

Gagne, R.M., (1965), The Conditions of Learning, Holt, Rinehart and Winston, New York, USA.

Garris, R., Ahlers, R. & Driskell, J.E. (2002) Games, motivation, and learning: A research and practice model. *Simulation Gaming*, **33**(4), 441–467.

Gomes, C. & Caldeira, H. (2004) *Virtual Learning Communities in Teacher Training, International Conference on Education, Innovation, Technology and Research in Education*. IADAT, International Association for the Development of Advances in Technology. Paper presented at the Bilbao, Spain.

Goulding, J., Sexton, M., Zhang, X., Kagioglou, M., Aouad, G.F. & Barrett, P. (2007) Technology adoption: Beaking down barriers using a virtual reality design support tool for hybrid concrete. *Construction Management and Economics*, **25**(12), 1239–1250.

Goulding, J., Nadim, W., Petridis, P., Alshawi, M., (2012) Construction Industry Offsite Production: A Virtual Reality Interactive Training Environment Prototype, *Advanced Engineering Informatics*, **26**(1), 103–116.

Govindasamy, T. (2002) Successful Implementation of e-Learning Pedagogical Considerations. *Internet and Higher Education*, **4**, 287–299.

Gross, B. (2007) Digital games in education: the design of games-based learning environments. *Journal of Research on Technology in Education*, **40**(1), 23–38.

Huber, G.P. (1991) Organizational learning: The contributing processes and the literatures. *Journal of Organization Science*, **2**(1), 88–115.

Ibrahim, R. & Pour Rahimian, F. (2010) Comparison of CAD and manual sketching tools for teaching architectural design. *Automation in Construction*, **19**(8), 978–987.

Jacobson, I., Booch, G. & Rumbaugh, J. (1999) *The Unified Software Development Process*, Object Technology Series. Addison-Wesley Professional, MA.

Jarvis, P., Holford, J. & Griffin, C. (2003) *The Theory and Practice of Learning*. Kogan Page Ltd, London.

Juárez-Ramírez, R., Sandoval, G.L., Cabrera Gonzállez, C. & Inzunza-Soberanes, S. (2009) *Educational Strategy Based on IT and the Collaboration Between Academy and Industry for Software Engineering Education and Training.* Paper presented at the m-ICTE 2009, V International Conference on Multimedia and ICT's in Education.

Karagiannidis, C. & Sampson, D. (2004) Adaptation rules relating learning style research and learning objects meta-data. In: ed. Magoulas, G.D. & Chen, S.Y., *Proceedings of International Conference on Adaptive Hypermedia and Adaptive Web-based Systems*, 23–26 August, The Netherlands.

Khanzade, A., Fisher, M. & Reed, D. (2007) *Challenges and Benefits of Implementing Virtual Design and Construction Technologies for Coordination of Mechanical, Electrical, and Plumbing Systems on Large Healthcare Project.* Paper presented at the CIB 24th W78 Conference.

Kim, B. & Chris, S. (2001) Accommodating diverse learning style in the design and delivery of on-line learning experiences. *International Journal of Engineering Education*, **17**, 93–98.

Knowles, M. (1980) *The Modern Practice of Adult Education: From Pedagogy to Androgogy*, 2nd edn. Cambridge Books, New York.

Kolb, D.A. (1984) *Experiential Learning: Experience as the Source of Learning and Development.* Prentice-Hall Inc, NJ.

Laird, D. (2003) *New Perspectives in Organisational Learning, Performance, and Change: Approaches to Training and Development*, 3rd edn. Preseus Books Group, USA.

Lam, P.T.I., Wong, F.W.H. & Tse, K.T.C. (2010) Effectiveness of ICT for construction information exchange among multidisciplinary project teams. *Journal of Computing in Civil Engineering*, **24**(4), 365–376.

Latham, M. (1994) *Constructing the Team, Joint Review of the Procurement and Contractual Arrangements in the UK Construction Industry, Final Report.* HSMO, London.

Leinonen, J., Kähkönen, K., Retik, A R., Flood, R.A., William, I. & O'Brien, J. (2003) New construction management practice based on the virtual reality technology. In ed. Balkema, A.A., *4D CAD And Visualization In Construction: Developments and Applications.* 75–100).

Lou, E.C.W. & Goulding, J.S. (2010) The pervasiveness of e-readiness in the global built environment arena. *Journal of Systems and Information Technology*, **12**(3), 180–195.

Mole, T. (2003) Mind your manners. In: ed. Newton, R., Bowden, A. & Betts, M., *Proceedings of CIB W89 International Conference on Building and Research Bear 2003.* School of Construction and Property Management, University of Salford, Salford, UK.

Nisan, N., Roughgarden, T., Tardos, É. & Vazirani, V.V. (2007) *Algorithmic Game Theory.* Cambridge University Press, Cambridge.

Nonaka, I. & Takeuchi, H. (1995) *The Knowledge Creating Company – How Japanese Companies Create the Dynamics of Innovation.* Oxford University Press, Oxford.

Pahl, C. (2003) Managing evolution and change in web-based teaching and learning environments. *Computer and Educations*, **40**(2), 99–114.

Pashler, H., McDaniel, M., Rohrer, D. & Bjork, R. (2008) Learning styles: concepts and evidence. *Psychological Science in the Public Interest*, **9**(3), 105–119.

Petric, J., Maver, T., Conti, G. & Ucelli, G. (2002) *Virtual Reality in the Service of User Participation in Architecture.* CIB W78 Conference, Aarhus School of Architecture.

Piaget, J, (1952), The Origins of Intelligence in Children, International University Press, New York, USA

Pour Rahimian, F., Ibrahim, R. & Baharudin, M.N. (2008) Using IT/ICT as a new medium toward implementation of interactive architectural communication cultures. *Proceedings of the International Symposium on Information Technology 2008*, ITSim.

Pour Rahimian, F. & Goulding, J.S (2010) game-like virtual reality interfaces in construction management simulation: A new paradigm of opportunities. *9th International Detail Design in Architecture Conference 2010*. The Central University of Lancashire, Preston, UK.

Pour Rahimian, F. & Ibrahim, R. (2011) Impacts of VR 3D sketching on novice designers' spatial cognition in collaborative conceptual architectural design. *Design Studies*, **32**(3), 255–291.

Regenbrecht, H. & Donath, D. (1996) Architectural education and virtual reality aided design (VRAD). InL ed. Bertol. D., *Designing Digital Space – An Architect's Guide to Virtual Reality*. John Wiley and Sons, Inc, New York. 155–176.

Sampaio, A.Z. & Henriques, P.G. (2008) Visual simulation of previous term civil engineering next term activities: Didactic virtual previous term models. *WSCG 2008, 16th International Conference in Central Europe on Computer Graphics, Visualization and Computer Vision*. Plzen, Czech Republic.

Sampaio, A.Z., Ferreira, M.M., Rosário, D.P. & Martins, O.P. (2010) 3D and VR models in civil engineering education: Construction, rehabilitation and maintenance. *Automation in Construction*, **19**(7), 819–828.

Sampson, D., Karagiannidis, C. & Kinshuk. (2002) personalised learning: educational, technological and standardisation perspective. *Interactive Educational Multimedia*, **4**, 24–39.

Senge, P.M. (1990) *The Fifth Discipline: The Art and Practice of the Learning Organization*. Currency Doubleday, New York.

Stonebraker, P.W. & Hazeltine, J.E. (2004) Virtual learning effectiveness: An examination of the process. *The Learning Organization Journal*, **11**(3), 209–225.

Sullivan, P. (1999) Profiting from intellectual capital. *Journal of Knowledge Management*, **3**(2), 132–143.

Syed-Khuzzan, S.M., Goulding, J.S. & Underwood, J. (2008) Personalised learning environments, Part 1: Core development issues for construction. *Industrial and Commercial Training*, **40**(6), 310–319.

Thai, A.M., Lowenstein, D., Ching, D. & Rejeski, D. (2009) *Game Changer: Investing in Digital Play to Advance Children's Learning and Health*: The Joan Ganz Cooney Center. Available from http://www.healthgamesresearch.org/our-publications/research-briefs/Game-Changer (accessed 22 July 2011).

Turocy, T.L., Stengel, B.V. & Hossein, B., (2003) *Game Theory Encyclopedia of Information Systems*. Elsevier, New York. 403–420.

Vincent, A. & Ross, D. (2001) Personalise training: Determine learning style, personality types and multiple intelligence. *The Learning Organisation*, **8**, 36–43.

Wang, F. & Hannafin, M.J. (2005) Design-based research and technology-enhanced learning environments. *Journal of Educational Technology Research and Development*, **53**(4), 5–23.

Ward, J. & Daniel, E. (2005) *Benefits Management: Delivering Value from IS & IT Investments*. John Wiley and Sons, Inc, New York.

Watson, J. & Hardaker, G. (2005) Steps towards personalised Learner Management System (LMS): SCORM implementation. *Campus-Wide Information Systems*, **22**, 56–70.

Weber, G. & Brusilovsky, P. (2001) ELM-ART: An adaptive versatile system for web-based instruction. *International Journal of Artificial Intelligence in Education*, **12**, 351–384.

Wellings, J. & Levine, M.H. (2010) *The Digital Promise: Transforming Learning with Innovative Uses of Technology*: A white paper on literacy and learning in a new media age, Joan Ganz Cooney Center at Sesame Workshop Available from http://www.digitalpromise.org/Files/Apple.pdf (accessed 22 July 2011).

Wen, M.L., Tsai, C., Lin, H. & Chuang, S. (2004) Cognitive-metacognitive and content-technical aspects of constructivist internet-based learning environments: A LISREL analysis. *Journal of Computers and Education*, **43**(3), 237–248.

Whyte, J., Bouchlaghem, N. & Thorpe, A. (1998) *The Promise and Problems of Implementing Virtual Reality In Construction Practice. Paper presented at The Life-cycle of Construction IT Innovations.* Technology Transfer from Research to Practice (CIB W78), Stockholm.

Whyte, J., Bouchlaghem, N., Thorpe, A. & McCaffer, R. (2000) From CAD to Virtual reality: Modelling approaches, data exchange and interactive 3d building design tools. *Automation in Construction*, **10**(1), 43–55.

Xie, H, Shi, W. & Issa R.R.A. (2011) *Using Rfid and Real-Time Virtual Reality Simulation For Optimization in Steel Construction.* ITcon vol. 16, 291–308. Available from http://www.itcon.org

Yoh, M. (2001) *The Reality of Virtual Reality.* Paper presented at the Seventh International Conference on Virtual Systems and Multimedia (VSMM'01), Organized by Center for Design Visualization, University of California Berkley, Berkley, CA.

Zheng, X., Sun, G., & Wang, S. W. (2006). An Approach of Virtual Prototyping Modeling in Collaborative Product Design. Paper presented at the CSCW 2005, LNCS 3865.

Zudilova-Seinstra, E., Adriaansen, T. & van Liere, R. (2009) *Trends in Interactive Visualization: State-of-the-Art Survey.* Springer-Verlag, London.

Index

Construction Innovation and Process Improvement, First Edition.
Edited by Akintola Akintoye, Jack S. Goulding and Girma Zawdie.
© 2012 Blackwell Publishing Ltd. Published 2012 by Blackwell Publishing Ltd.